ORACLE® 11g: SQL

ORACLE® 11*g*: SQL

Joan Casteel

COURSE TECHNOLOGY
CENGAGE Learning™

Australia • Brazil • Japan • Korea • Mexico • Singapore • Spain • United Kingdom • United States

COURSE TECHNOLOGY
CENGAGE Learning™

Oracle 11g: SQL
by Joan Casteel

Vice President, Publisher: Jack Calhoun

Senior Acquisitions Editor: Charles McCormick, Jr.

Product Manager: Kate Mason

Developmental Editor: Lisa M. Lord

Marketing Manager: Bryant Chrzan

Marketing Communications Manager: Libby Shipp

Marketing Coordinator: Suellen Ruttkay

Senior Content Project Manager: Jill Braiewa

Media Editor: Chris Valentine

Senior Art Director: Stacy Jenkins Shirley

Cover Designer: Stuart Kunkler/triARTis

Cover Image: iStock Images

Print Buyer: Julio Esperas

Manuscript Quality Assurance: Nicole Ashton,
   Green Pen Quality Assurance

Proofreader: Harry Johnson

Indexer: Michael Brackney

Production Service: Patrick Franzen, Pre-PressPMG

Compositor: Pre-PressPMG

ISBN-13: 978-1-4390-4128-4
ISBN-10: 1-4390-4128-8

Course Technology
20 Channel Center Street
Boston, MA 02210
USA

Printed in the United States of America
2 3 4 5 6 7 15 14 13 12 11 10

To Scott, a true teacher—one who never stops learning

BRIEF CONTENTS

Preface			xvi
Chapter	**1**	*Overview of Database Concepts*	1
Chapter	**2**	*Basic SQL SELECT Statements*	25
Chapter	**3**	*Table Creation and Management*	57
Chapter	**4**	*Constraints*	99
Chapter	**5**	*Data Manipulation and Transaction Control*	137
Chapter	**6**	*Additional Database Objects*	175
Chapter	**7**	*User Creation and Management*	213
Chapter	**8**	*Restricting Rows and Sorting Data*	243
Chapter	**9**	*Joining Data from Multiple Tables*	283
Chapter	**10**	*Selected Single-Row Functions*	331
Chapter	**11**	*Group Functions*	383
Chapter	**12**	*Subqueries and MERGE Statements*	427
Chapter	**13**	*Views*	471
Appendix A		*Tables for the JustLee Books Database*	511
Appendix B		*SQL*Plus and SQL Developer Overview*	519
Appendix C		*Oracle Resources*	527
Appendix D		*SQL*Loader*	529
Appendix E		*SQL Tuning Topics*	533
Appendix F		*SQL in Various Databases*	551
Glossary			555
Index			563

TABLE OF CONTENTS

Preface   xvi
Chapter 1 *Overview of Database Concepts*   1
Introduction   2
Database Terminology   2
Database Management System   3
Database Design   4
   Entity-Relationship (E-R) Model   5
   Database Normalization   6
   Relating Tables in the Database   10
Structured Query Language (SQL)   12
Databases Used in This Textbook   13
   Basic Assumptions   13
   Tables in the JustLee Books Database   14
Topic Sequence   16
Software Used in This Textbook   16
Chapter Summary   17
Review Questions   17
Multiple Choice   18
Hands-On Assignments   21
Advanced Challenge   22
Case Study: *City Jail*   22

Chapter 2 *Basic SQL SELECT Statements*   25
Introduction   26
Creating the JustLee Books Database   27
SELECT Statement Syntax   30
   Selecting All Data in a Table   31
   Selecting One Column from a Table   33
   Selecting Multiple Columns from a Table   34
Operations in the SELECT Statement   36
   Using Column Aliases   36
   Using Arithmetic Operations   39
   NULL Values   40
   Using DISTINCT and UNIQUE   42
   Using Concatenation   44
Chapter Summary   49
Review Questions   50
Multiple Choice   51
Hands-On Assignments   54
Advanced Challenge   55

Chapter 3 *Table Creation and Management*   57

Introduction   58

Table Design   59

Table Creation   63

    Defining Columns   63

    Viewing a List of Tables: USER_TABLES   65

    Viewing Table Structures: DESCRIBE   66

Table Creation with Subqueries   67

    CREATE TABLE ... AS COMMAND   68

Modifying Existing Tables   70

    ALTER TABLE ... ADD Command   70

    ALTER TABLE ... MODIFY Command   71

    ALTER TABLE ... DROP COLUMN Command   75

    ALTER TABLE ... SET UNUSED/DROP UNUSED COLUMNS Command   76

    Renaming a Table   79

    Truncating a Table   81

Deleting a Table   83

Chapter Summary   88

Review Questions   90

Multiple Choice   90

Hands-On Assignments   94

Advanced Challenge   94

Case Study: *City Jail*   95

Chapter 4 *Constraints*   99

Introduction   100

Creating Constraints   101

    Creating Constraints at the Column Level   102

    Creating Constraints at the Table Level   102

Using the PRIMARY KEY Constraint   103

Using the FOREIGN KEY Constraint   106

Using the UNIQUE Constraint   110

Using the CHECK Constraint   112

Using the NOT NULL Constraint   114

Including Constraints During Table Creation   116

Adding Multiple Constraints on a Single Column   120

Viewing Constraint Information   120

Disabling and Dropping Constraints   122

    Using DISABLE/ENABLE   122

    Dropping Constraints   123

Chapter Summary   126

Review Questions   128

Multiple Choice   129

Hands-On Assignments   132

Advanced Challenge   134

Case Study: *City Jail*   134

Chapter 5 *Data Manipulation and Transaction Control*   137
  Introduction   138
  Inserting New Rows   139
    Using the INSERT Command   139
    Handling Virtual Columns   144
    Handling Single Quotes in an INSERT Value   146
    Inserting Data from an Existing Table   148
  Modifying Existing Rows   150
    Using the UPDATE Command   150
    Using Substitution Variables   152
  Deleting Rows   156
  Using Transaction Control Statements   157
    COMMIT and ROLLBACK Commands   158
    SAVEPOINT Command   159
  Using Table Locks   162
    LOCK TABLE Command   162
    SELECT … FOR UPDATE Command   163
  Chapter Summary   165
  Review Questions   167
  Multiple Choice   167
  Hands-On Assignments   170
  Advanced Challenge   171
  Case Study: *City Jail*   172

Chapter 6 *Additional Database Objects*   175
  Introduction   176
  Sequences   177
    Creating a Sequence   178
    Using Sequence Values   183
    Altering Sequence Definitions   185
    Removing a Sequence   188
  Indexes   188
    B-Tree Indexes   190
    Bitmap Indexes   196
    Function-Based Indexes   197
    Index Organized Tables   198
    Verifying an Index   199
    Altering or Removing an Index   200
  Synonyms   201
    Deleting a Synonym   204
  Chapter Summary   205
  Review Questions   207
  Multiple Choice   208
  Hands-On Assignments   211
  Advanced Challenge   212
  Case Study: *City Jail*   212

Chapter 7 *User Creation and Management*   213
    Introduction   214
    Data Security   215
    Creating a User   216
       Creating Usernames and Passwords   216
    Assigning User Privileges   218
       System Privileges   218
       Granting System Privileges   219
       Object Privileges   220
       Granting Object Privileges   220
    Managing Passwords   224
    Using Roles   225
       Creating and Assigning Roles   226
       Using Predefined Roles   228
       Using Default Roles   229
       Enabling Roles After Login   230
    Viewing Privilege Information   230
    Removing Privileges and Users   232
       Revoking Privileges and Roles   232
       Dropping a Role   234
       Dropping a User   234
    Chapter Summary   235
    Review Questions   237
    Multiple Choice   237
    Hands-On Assignments   241
    Advanced Challenge   241
    Case Study: *City Jail*   242

Chapter 8 *Restricting Rows and Sorting Data*   243
    Introduction   244
    WHERE Clause Syntax   245
       Rules for Character Strings   246
       Rules for Dates   248
    Comparison Operators   248
       BETWEEN ... AND Operator   255
       IN Operator   256
       LIKE Operator   258
    Logical Operators   262
    Treatment of NULL Values   265
    ORDER BY Clause Syntax   267
       Secondary Sort   270
       Sorting by SELECT Order   272
    Chapter Summary   274
    Review Questions   276
    Multiple Choice   277
    Hands-On Assignments   281

Advanced Challenge 281
Case Study: *City Jail* 281

Chapter 9 *Joining Data from Multiple Tables* 283
Introduction 284
Cartesian Joins 285
    Cartesian Join: Traditional Method 286
    Cartesian Join: JOIN Method 288
Equality Joins 289
    Equality Joins: Traditional Method 291
    Equality Joins: JOIN Method 296
Non-Equality Joins 302
    Non-Equality Joins: Traditional Method 303
    Non-Equality Joins: JOIN Method 304
Self-Joins 305
    Self-Joins: Traditional Method 306
    Self-Joins: JOIN Method 307
Outer Joins 308
    Outer Joins: Traditional Method 309
    Outer Joins: JOIN Method 310
Set Operators 312
Chapter Summary 319
Review Questions 322
Multiple Choice 323
Hands-On Assignments 329
Advanced Challenge 329
Case Study: *City Jail* 330

Chapter 10 *Selected Single-Row Functions* 331
Introduction 332
Case Conversion Functions 333
    The LOWER Function 333
    The UPPER Function 334
    The INITCAP Function 335
Character Manipulation Functions 336
    The SUBSTR Function 336
    The INSTR Function 338
    The LENGTH Function 340
    The LPAD and RPAD Functions 340
    The LTRIM and RTRIM Functions 342
    The REPLACE Function 342
    The TRANSLATE Function 343
    The CONCAT Function 344
Number Functions 344
    The ROUND Function 344
    The TRUNC Function 345

The MOD Function   346

The ABS Function   347

The POWER Function   348

Date Functions   348

The MONTHS_BETWEEN Function   349

The ADD_MONTHS Function   350

The NEXT_DAY and LAST_DAY Functions   351

The TO_DATE Function   352

Rounding Date Values   354

Truncating Date Values   355

CURRENT_DATE Versus SYSDATE   355

Regular Expressions   357

Other Functions   359

The NVL Function   359

The NVL2 Function   362

The NULLIF Function   363

The TO_CHAR Function   365

The DECODE Function   367

The CASE Expression   369

The SOUNDEX Function   369

The TO_NUMBER Function   370

The DUAL Table   371

Chapter Summary   372

Review Questions   376

Multiple Choice   377

Hands-On Assignments   381

Advanced Challenge   381

Case Study: *City Jail*   382

Chapter 11 *Group Functions*   383

Introduction   384

Group Functions   386

The SUM Function   386

The AVG Function   388

The COUNT Function   390

The MAX Function   393

The MIN Function   394

Grouping Data   395

Restricting Aggregated Output   398

Nesting Functions   402

Statistical Group Functions   403

The STDDEV Function   403

The VARIANCE Function   404

Enhanced Aggregation for Reporting   405

The GROUPING SETS Expression   407

The CUBE Extension   409

The ROLLUP Extension   411
Chapter Summary   417
Review Questions   419
Multiple Choice   420
Hands-On Assignments   424
Advanced Challenge   425
Case Study: *City Jail*   425

Chapter 12 *Subqueries and MERGE Statements*   427
Introduction   428
Subqueries and Their Uses   429
Single-Row Subqueries   429
  Single-Row Subquery in a WHERE Clause   429
  Single-Row Subquery in a HAVING Clause   434
  Single-Row Subquery in a SELECT Clause   435
Multiple-Row Subqueries   437
  The IN Operator   438
  The ALL and ANY Operators   438
  Multiple-Row Subquery in a HAVING Clause   443
Multiple-Column Subqueries   445
  Multiple-Column Subquery in a FROM Clause   445
  Multiple-Column Subquery in a WHERE Clause   447
NULL Values   448
  NVL in Subqueries   449
  IS NULL in Subqueries   450
Correlated Subqueries   451
Nested Subqueries   453
DML Actions Using Subqueries   455
MERGE Statements   456
Chapter Summary   460
Review Questions   461
Multiple Choice   462
Hands-On Assignments   467
Advanced Challenge   468
Case Study: *City Jail*   469

Chapter 13 *Views*   471
Introduction   472
Creating a View   474
  Creating a Simple View   476
  DML Operations on a Simple View   480
Creating a Complex View   484
  DML Operations on a Complex View with an Arithmetic Expression   484
  DML Operations on a Complex View Containing Data from Multiple Tables   489
  DML Operations on a Complex View Containing Functions or Grouped Data   491

DML Operations on a Complex View Containing DISTINCT or ROWNUM   493
Summary Guidelines for DML Operations on a Complex View   495
Dropping a View   495
Creating an Inline View   496
    TOP-N Analysis   496
Creating a Materialized View   499
Chapter Summary   502
Review Questions   503
Multiple Choice   504
Hands-On Assignments   508
Advanced Challenge   508
Case Study: *City Jail*   509

Appendix A *Tables for the JustLee Books Database*   511
CUSTOMERS Table   511
BOOKS Table   512
ORDERS Table   513
ORDERITEMS Table   514
AUTHOR Table   515
BOOK AUTHOR Table   516
PUBLISHER Table   517
PROMOTION Table   517

Appendix B *SQL*Plus and SQL Developer Overview*   519
Introduction   519
SQL*Plus   519
SQL Developer   523

Appendix C *Oracle Resources*   527
Oracle Academic Initiative (OAI)   527
Oracle Certification Program (OCP)   527
Oracle Technology Network (OTN)   527
International Oracle Users Group (IOUG)   527

Appendix D *SQL*Loader*   529
Introduction   529
Read a Fixed Format File   529
Read a Delimited File   532

Appendix E *SQL Tuning Topics*   533
Introduction   533
Tuning Concepts and Issues   533
    Identifying Problem Areas in Coding   533
    Processing and the Optimizer   535
    The Explain Plan   537
    Timing Feature   542

Selected SQL Tuning Guidelines and Examples 543
  Avoiding Unnecessary Column Selection 544
  Index Suppression 545
  Concatenated Indexes 547
  Subqueries 548
  Optimizer Hints 549

Appendix F *SQL in Various Databases* 551
Introduction 551
  Suppressing Duplicates 551
  Locating a Value in a String 552
  Displaying the Current Date 552
  Specifying a Default Date Format 552
  Replacing NULL Values in Text Data 553
  Adding Time to Dates 553
  Extracting Values from a String 553
  Concatenating 554
  Data Structures 554

  Glossary 555
  Index 563

The past few decades have seen a proliferation of organizations that rely heavily on information technology. These organizations store their data in databases, and many choose Oracle database management systems to access their data. The current Oracle database version, Oracle 11*g*, is a database management system that enables users to create, manipulate, and retrieve data. The purpose of this textbook is to introduce students to basic SQL commands for interacting with Oracle 11*g* databases in a business environment. In addition, concepts relating to objectives of the Oracle 10*g* and Oracle 11*g* SQL certification exams have been incorporated for students wanting to pursue certification.

The Intended Audience

This textbook has been designed for students in technical two-year or four-year programs who need to learn how to interact with databases. Although having an understanding of database design is preferable, an introductory chapter has been included to review the basic concepts of E-R modeling and the normalization process.

Oracle Certification Program (OCP)

This textbook covers the objectives of Exam 1Z0-007, Introduction to Oracle 10*g*: SQL, and Exam 1Z0-051, Oracle Database 11*g*: SQL Fundamentals I. Most objectives for Exam 1Z0-047, Oracle Database SQL Expert, are also covered. Any of these exams serve as the first exam in the Oracle PL/SQL and Oracle Forms Developer or Oracle Database Administrator Oracle Certified Associate level certification tracks. Information about registering for these exams, along with other reference material, is available at *www.oracle.com/education/certification*.

The Approach

The concepts introduced in this textbook are discussed in the context of a hypothetical real-world business: an online book retailer named JustLee Books. The company's business operation and the database structure are introduced and analyzed, and as commands are introduced throughout the textbook, they're modeled with examples using the JustLee Books database. Using consistent examples of a hypothetical company helps you learn the syntax of commands and how to use them in a real-world environment. In addition, a script file that generates the database is available to give you hands-on practice in re-creating examples and practicing variations of SQL commands to enhance your understanding.

To explain what a database is and how it's created, this textbook initially focuses on creating tables and learning how to perform data manipulation operations. After you're familiar with the database structure, the focus then turns to querying a database. In Chapters 8 through 13, you learn how to retrieve data from the database, using the many options of a SELECT statement, including row filtering, joins, functions, and subqueries.

To reinforce the material, each chapter includes a chapter summary and, when appropriate, a syntax guide for the commands covered in the chapter. In addition, each chapter includes review questions and hands-on activities that test your knowledge and challenge you to apply that knowledge to solving business problems. A running case study that builds throughout the textbook provides a second real-world setting—a city jail system—as another opportunity to work with databases.

Overview of This Book

The examples, assignments, and cases in this book help you achieve the following objectives:

- Issue SQL commands that retrieve data based on criteria specified by the user.
- Use SQL commands to join tables and retrieve data from joined tables.
- Perform calculations based on data stored in the database.
- Use functions to manipulate and aggregate data.
- Use subqueries to retrieve data based on unknown conditions.
- Create, modify, and drop database tables.
- Manipulate data stored in database tables.
- Enforce business rules by using table constraints.
- Create users and assign the privileges users need to perform tasks.

The chapters' contents build in complexity while reinforcing previous ideas. **Chapter 1** introduces basic database management concepts, including database design. **Chapter 2** shows how to retrieve data from a table. **Chapter 3** explains how to create new database tables. **Chapter 4** addresses the use of constraints to enforce business rules and ensure the integrity of table data. **Chapter 5** explains adding data to a table, modifying existing data, and deleting data. **Chapter 6** shows how to use a sequence to generate numbers, create indexes to speed up data retrieval, and create synonyms to provide aliases for tables. **Chapter 7** steps you through creating user accounts and roles and shows how to grant (and revoke) privileges to these accounts and roles. **Chapter 8** explains how to restrict rows retrieved from a table, based on a given condition. **Chapter 9** shows how to link tables with common columns by using joins. **Chapter 10** describes the single-row functions supported by Oracle 11g. **Chapter 11** covers using multiple-row functions to derive a single value for a group of rows and explains how to restrict groups of rows. **Chapter 12** covers using subqueries to retrieve rows based on an unknown condition already stored in the database. **Chapter 13** explains using views to restrict access to data and reduce the complexity of certain types of queries.

The appendixes support and reinforce chapter materials. **Appendix A** contains printed versions of the initial table structure and data for the JustLee Books database used throughout this textbook. **Appendix B** introduces the operation of the SQL*Plus and SQL Developer client software tools. **Appendix C** lists Oracle resources for further study. **Appendix D** introduces the SQL*Loader utility for importing data. **Appendix E** introduces basic SQL statement tuning concepts. **Appendix F** identifies SQL differences in some popular databases.

Features

To enhance your learning experience, each chapter in this textbook includes the following elements:

- **Chapter objectives:** Each chapter begins with a list of the concepts to be mastered by the chapter's conclusion. This list gives you a quick overview of chapter contents and serves as a useful study aid.
- **Running case:** A sustained example, the business operation of JustLee Books, is the basis for introducing new commands and practicing the material covered in each chapter.
- **Methodology:** As new commands are introduced in each chapter, the command syntax is shown and then an example, using the JustLee Books database, illustrates using the command in the context of business operations. This methodology shows you not only *how* the command is used, but also *when* and *why* it's used. The script file used to create the database is available so that you can work through the examples in this textbook, engendering a hands-on environment in which you can reinforce your knowledge of chapter material.
- **Tip:** This feature, designated by the Tip icon, provides practical advice and sometimes explains how a concept applies in the workplace.
- **Note:** These explanations, designated by the Note icon, offer more information on performing operations with databases.
- **Database Preparation:** These notes, placed at the end of chapter introductions, tell you which script from the student data files you should run in preparation for chapter examples and activities.
- **Caution:** This warning, designated by the Caution icon, points out database operations that, if misused, could have devastating results.
- **Chapter summaries:** Each chapter's text is followed by a summary of chapter concepts. These summaries are a helpful recap of chapter contents.
- **Syntax summaries:** Beginning with Chapter 2, a Syntax Guide table is included after each chapter summary to recap the command syntax covered in the chapter.
- **Review questions:** End-of-chapter assessment begins with review questions that reinforce the main ideas introduced in each chapter. These questions ensure that you have mastered the concepts and understand the information covered in the chapter.
- **Multiple-choice questions:** Each chapter contains multiple-choice questions covering the material in the chapter. Oracle certification-type questions are included to prepare you for the type of questions you can expect on certification exams and measure your level of understanding.
- **Hands-on assignments:** Along with conceptual explanations and examples, each chapter includes hands-on assignments related to the chapter's contents. The purpose of these assignments is to give you practical experience. In most cases, the assignments are based on the JustLee Books database and build on the examples in the chapter.
- **Advanced challenge:** This section poses another problem about the JustLee Books database for you to solve and is larger in scope than the hands-on assignments.

- **Case studies:** At the end of each chapter is a major case study, designed to help you apply what you have learned to real-world situations. These cases give you the opportunity to synthesize and evaluate information independently, examine potential solutions, and make recommendations, much as you would in an actual business situation. These cases uses a database based on a city jail system.

Supplemental Materials

The following supplemental materials are available when this book is used in a classroom setting. All teaching tools available with this book are provided to instructors on a single Instructor Resources CD as well as on the Cengage Learning Web site at *www.cengage.com/databases/casteel/*.

- **Electronic Instructor's Manual:** The Instructor's Manual accompanying this textbook includes the following items:
  - Additional instructional material to assist in class preparation, including suggestions for lecture topics
  - A sample syllabus
  - When applicable, information about potential problems that can occur in networked environments
- **ExamView®:** This objective-based test generator lets instructors create paper, LAN, or Web-based tests from testbanks designed for this textbook. Instructors can use the QuickTest Wizard to create tests in fewer than five minutes by taking advantage of these question banks or create customized exams.
- **PowerPoint presentations:** Microsoft® PowerPoint slides are included for each chapter. Instructors can use the slides in three ways: as teaching aids during classroom presentations, as printed handouts for classroom distribution, or as network-accessible resources for chapter review. Instructors can add their own slides for additional topics introduced to the class.
- **Data files:** The script files needed to create the JustLee Books and City Jail databases are available on the Cengage Learning Web site at *www.cengage.com/databases/casteel/* and included on the Instructor Resources CD.
- **Solution files:** Solutions to chapter examples, end-of-chapter review questions and multiple-choice questions, hands-on assignments, and case studies are available on the Instructor Resources CD as well as the Cengage Learning Web site at *www.cengage.com/databases/casteel/*. The solutions are password protected.
- **Figure files:** Figure files allow instructors to create their own presentations with figures from the textbook. They are included on the Instructor Resources CD only.

Acknowledgments

I feel fortunate that Cengage Learning pursued my authorship of this textbook and continues to support my efforts. I am one lucky person—I have two angels in heaven, my mother and grandmother, and one angel here on earth, Scott. Without them watching over me, I would not be able to tackle such challenges. I also want to thank my father, who always seems more excited than me every time I finish a book project.

However, this textbook is the result of an incredible effort by many people whom I wish I had to opportunity to thank personally. First, hats off to Lisa Lord, Development Editor, who not only helped me survive the project, but also truly improved this textbook. Second, I would like to thank Kate Mason, Project Manager, for making this project possible and supporting me and the team. I would also like to thank Nicole Ashton of Green Pen Quality Assurance, who kept a sharp eye out for technical accuracy and tested all coding examples to ensure error-free coding. There's no way to express adequate gratitude for this "magic," as these tasks play a critical role in developing an effective textbook.

I would like to express my appreciation to Charles McCormick, Jr., Senior Acquisitions Editor; Jill Braiewa, Senior Content Project Manager; and Patrick Franzen, Senior Project Manager of Pre-Press PMG. I also recognize the critical role of the marketing and art teams in making the book successful. These thanks don't cover all the important people who made this textbook a reality, but I truly appreciate every effort made on its behalf.

In addition, I need to recognize the enormous contribution of colleagues and reviewers, who provided helpful suggestions and insight into the development of this textbook. Reviewers include Dr. Douglas D. Bickerstaff of Eastern Washington University; Dr. Muhammad A. Razi of Haworth College of Business, Western Michigan University, Kalamazoo; and Professor Eli J. Weissman of DeVry Institute of Technology. Last, but not least, thanks to all my students, who put up with my ramblings and, in turn, teach me as well.

READ THIS BEFORE YOU BEGIN

TO THE USER

Data Files

To work through the examples and complete the projects in this book, you need to load the data files created for this book. Your instructor will provide these data files, or you can download them from the Cengage Learning Web site at *www.cengage.com/databases/ casteel/* and then search for this book's title. The data files are designed to supply the same data shown in chapter examples, so you can have hands-on practice in re-creating the queries and their output. The tables in the database can be reset if you encounter problems, such as accidentally deleting data. Working through all examples is highly recommended to reinforce your learning.

Starting with Chapter 2, database script instructions are given at the beginning of the chapter, if applicable. These database script files are in the folder corresponding to the chapter (Chapter05, Chapter10, and so forth) on *www.cengage.com/databases/casteel/* and have filenames such as JLDB_Build_#.sql (substituting the chapter number for the # symbol). If the computer in your school lab—or your own computer—has Oracle 11*g* installed, you can work through the chapter examples and complete the hands-on assignments and case projects. Many of the coding examples in this textbook can be completed successfully with previous versions of Oracle (Oracle 9*i* or Oracle 10*g*).

Using Your Own Computer

To use your own computer to work through the chapter examples and complete the hands-on assignments and case projects, you need the following:

- *Hardware*—A computer with at least 1 GB RAM and 4.75 GB hard disk space available.
- *Software*—You can use the Oracle 11*g* software with Microsoft Windows 2000 Server (all editions with Service Pack 1 installed), Windows Server 2003 (all editions), Windows XP Professional, and Windows Vista (Business, Enterprise or Ultimate Edition). Detailed installation, configuration, and logon information for Oracle 11*g* is provided at *www.cengage.com/ databases/casteel/*.

The Oracle Database 11*g* (Version 11.1.0) DVD, which is included with this book, enables you to install this software on your own computer at home. You can then connect to an Oracle 11*g* Enterprise Edition, Standard Edition, or Personal Edition database. You can use the Oracle Database 11*g* software with the operating systems in the preceding list. Make sure you verify the hardware requirements; more details are included in the installation information at *www.cengage.com/database/casteel/*. Look for this book's title and front cover, and click the link to access information specific to this book.

Before using the software, you *must* register the software and agree to the Oracle Technology Network Developer License Terms to receive the key code for unlocking the software. To do this, go to *http://otn.oracle.com/books/*. After registering the software, you agree that Oracle can contact you for marketing purposes, and any information you provide to Oracle can be used for marketing purposes.

- When you install the Oracle software, you're prompted to change the password for certain default administrative user accounts. Make sure you record the accounts' names and passwords because you might need to log in to the database with one of these administrative accounts in later chapters. After you install Oracle, you're required to enter a username and password to access the software. One default username created during installation is "scott." The default password for this username is "tiger." If you have installed the Personal Edition of Oracle, you don't need to enter a connect string during the login process.
- Keep in mind that you might not need to install the entire Oracle system on your computer. Many academic institutions provide connections to an Oracle server from on or off campus. In this case, you might need to install only one of the client tools mentioned in the textbook to do the coding examples. Ask your instructor to determine the best option for your needs.
- You can't use your own computer to work through chapter examples and complete projects unless you have the data files. You can get these files from your instructor, or you can download them from the Cengage Learning Web site at *www.cengage.com/databases/casteel/* and then search for this book's title.
- When you download the data files, they should be stored in a directory separate from any other files on your hard drive. You need to remember the path or folder containing these files because each script filename must be prefixed with its location before you run the script.

Visit Our Web Site

A supplemental chapter on formatting report output has been included in the book's online materials. Additional materials designed especially for this textbook might be available on the Cengage Learning Web site. Go to *www.cengage.com/databases/casteel/* periodically and search this site for more details.

TO INSTRUCTORS

To complete examples and activities in this textbook, your students must have access to the data files included on the Instructor Resources CD (or downloaded from *www.cengage.com/databases/casteel/*).

The data files consist of the JustLee Database folder and a folder for each chapter. Many chapters require running a script; if so, these instructions are given in a note at the beginning of the chapter. These scripts are in folders corresponding to the chapter (Chapter05, Chapter10, and so forth) and have filenames such as JLDB_Build_#.sql (substituting the chapter number for the # symbol). The initial database creation is done at the beginning of Chapter 2 to create the JustLee Books database. Students should run the scripts as instructed to have a copy of the tables stored in their schemas. You should instruct your students on how to access and copy data files to their own computers. The chapters and projects in this book were tested with Microsoft Windows Server 2003 and Oracle 11.1.0.2 Enterprise Edition.

Note that Oracle9i and Oracle 10g include a browser-based version of SQL*Plus called iSQL*Plus. This interface is no longer supported by Oracle and isn't included with Oracle 11g.

Cengage Learning Data Files

You are granted a license to copy data files to any computer or computer network used by people who have purchased this book.

CHAPTER 1

OVERVIEW OF DATABASE CONCEPTS

LEARNING OBJECTIVES

After completing this chapter, you will be able to do the following:

- Define database terms
- Identify the purpose of a database management system (DBMS)
- Explain database design by using entity-relationship models and normalization
- Explain the purpose of a Structured Query Language (SQL)
- Understand how this textbook's topics are sequenced and how the two sample databases are used
- Identify the software used in this textbook

INTRODUCTION

Imagine you're starting up an online book retail company. How will customer orders be recorded? Will customers be able to search for products by name or keywords? Will you be able to analyze sales information to track profits, determine product success, and target marketing efforts to customers? Analyzing thousands of orders could take days without using a database. A **database** simplifies these tasks because it's a storage structure that provides mechanisms for recording, manipulating, and retrieving data.

The database used throughout this textbook is based on the activities of a hypothetical business, an online bookseller named JustLee Books. The company sells books via the Internet to customers throughout the United States. When a new customer places an order, he or she provides data such as name, billing and shipping addresses, and items ordered. The company also uses a database for all books in inventory.

To access the data required for operating JustLee Books, management relies on a DBMS. A **database management system (DBMS)** is used to create and maintain the structure of a database, and then to enter, manipulate, and retrieve the data it stores. Creating an efficient database design is the key to using a database effectively to support an organization's business operations.

This chapter introduces basic database terminology and discusses the process of designing a database for JustLee Books.

DATABASE TERMINOLOGY

Whenever a customer opens an account with a company, certain data must be collected. In many cases, the customer completes an online form that asks for the customer's name, address, and so on, as shown in Figure 1-1.

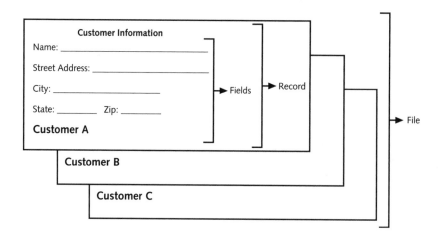

FIGURE 1-1   Collecting customer information

While collecting customer information, a series of characters is identified for each item. A **character** is the basic unit of data, and it can be a letter, number, or special symbol.

A group of related characters (for example, the characters that make up a customer's name) is called a **field**. A field represents one attribute or characteristic (the name, for instance) of the customer. A collection of fields about one customer (for example, name, address, city, state, and zip code) is called a **record**. A group of records about the same type of entity (such as customers or inventory items) is stored in a **file**. A collection of interrelated files—such as those relating to customers, their purchases, and their payments—is stored in a **database**.

These terms relate to the logical database design, but they are often used interchangeably with the terminology for the physical database design. When creating the physical database, a field is commonly referred to as a **column**, a record is called a **row**, and a file is known as a **table**. A table is quite similar to a spreadsheet, in that it contains columns and rows. Figure 1-2 shows a representation of these terms.

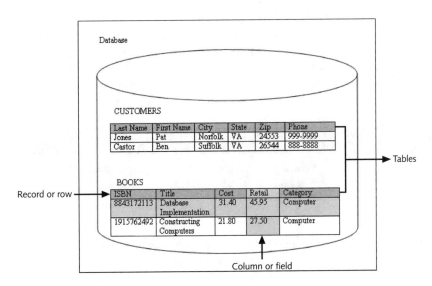

FIGURE 1-2   Database terminology

DATABASE MANAGEMENT SYSTEM

As mentioned earlier, a database is housed in a DBMS, which provides the functionality to create and work with a database. This functionality includes the following:

- *Data storage*: Manage the physical structure of the database.
- *Security*: Control user access and privileges.
- *Multiuser access*: Manage concurrent data access.
- *Backup*: Enable recovery options for database failures.
- *Data access language*: Provide a language that allows database access.
- *Data integrity*: Enable constraints or checks on data.
- *Data dictionary*: Maintain information about database structure.

DATABASE DESIGN

To determine the most appropriate structure of fields, records, and files in a database, developers go through a design process. The design and development of a system is accomplished through a process that's formally called the **Systems Development Life Cycle (SDLC)** and consists of the following steps:

1. *Systems investigation*: Understanding the problem
2. *Systems analysis*: Understanding the solution to the previously identified problem
3. *Systems design*: Defining the logical and physical components
4. *Systems implementation*: Creating the system
5. *Systems integration and testing*: Placing the system into operation for testing
6. *Systems deployment*: Placing the system into production
7. *Systems maintenance and review*: Evaluating the implemented system

Although the SDLC is a methodology designed for any type of system an organization needs, this chapter specifically addresses developing a DBMS. For the purposes of this discussion, assume the problem identified is the need to collect and maintain data about customers and their orders. The identified solution is to use a database to store all needed data. The discussion that follows presents the steps for designing the database.

NOTE

A variety of SDLC models have been developed to address different development environments. The steps presented here represent a traditional waterfall model. Other models, such as fountain and rapid prototyping, involve a different series of steps.

To design a database, the requirements of the database—inputs, processes, and outputs—must be identified first. Usually, the first question asked is, "What information, or output, must come from this database?" or "What questions should this database be able to answer?" By understanding the necessary output, the designer can then determine what information should be stored in the database. For example, if the organization wants to send birthday cards to its customers, the database must store each customer's birth date.

After the requirements of a database have been identified, an entity-relationship (E-R) model is usually drafted to better understand the data to be stored in the database. In an E-R model, an **entity** is any person, place, or thing with characteristics or attributes that will be included in the system. An **E-R model** is a diagram that identifies the entities (customers, books, orders, and such) in the database, and it shows how the entities are related to one another. It serves as the logical representation of the physical system to be built.

The next two sections explain the construction of an E-R model and the normalization process used to determine appropriate entities for a database.

NOTE

An E-R model is also called an entity-relationship diagram (ERD).

Entity-Relationship (E-R) Model

In an E-R model, an entity is usually represented as a square or rectangle. As shown in Figure 1-3, a line depicts how an entity's data relates to another entity. If the line connecting two entities is solid, the relationship between the entities is mandatory. However, if the relationship between two entities is optional, a dashed line is used.

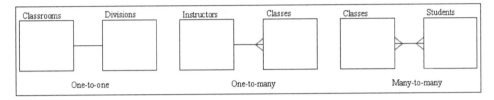

FIGURE 1-3   E-R model notation examples

As shown in Figure 1-3, the following types of relationships can exist between two entities:

- *One-to-one*: In a one-to-one relationship, each occurrence of data in one entity is represented by only one occurrence of data in the other entity. For example, if each classroom is assigned to only one academic division, a one-to-one relationship is created between the classroom and division entities. This type of relationship is depicted in an E-R model as a simple straight line.
- *One-to-many*: In a one-to-many relationship, each occurrence of data in one entity can be represented by many occurrences of the data in the other entity. For example, a class has only one instructor, but an instructor might teach many classes. A one-to-many relationship is represented by a straight line with a crow's foot at the "many" end.
- *Many-to-many*: In a many-to-many relationship, data can have multiple occurrences in both entities. For example, a class can consist of more than one student, and a student can take more than one class. A straight line with a crow's foot at each end indicates a many-to-many relationship.

Figure 1-4 shows a simplified E-R model for the JustLee Books database used throughout this textbook. A more thorough E-R model would include a list of attributes for each entity.

NOTE

The notations in the sample E-R models in this chapter reflect only one way of diagramming entity relationships. If you're using a modeling software tool, you might encounter different notations to represent relationships. For example, Microsoft products typically represent the many side of a relationship with the infinity symbol (∞). In addition, some modeling tools automatically add the common fields or foreign key columns needed as relationships are defined.

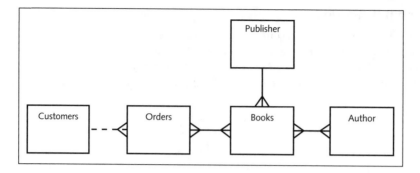

FIGURE 1-4   An E-R model for JustLee Books

The following relationships are defined in the E-R model in Figure 1-4:

- Customers can place multiple orders, but each order can be placed by only one customer (one-to-many). The dashed line between Customers and Orders means a customer can exist in the database without having a current order stored in the ORDERS table. Therefore, this relationship is considered optional.
- An order can consist of more than one book, and a book can appear on more than one order (many-to-many).
- A book can have more than one author, and an author can write more than one book (many-to-many).
- A book can have only one publisher, but a publisher can publish more than one book (one-to-many).

Although some E-R modeling approaches are more complex, the simplified notations used in this chapter do point out the important relationships among entities, and using them helps designers identify potential problems in table layouts. After examining the E-R model in Figure 1-4, you should have noticed the two many-to-many relationships. Before creating the database, all many-to-many relationships must be reduced to a set of one-to-many relationships, as you learn in "Relating Tables in the Database" later in this chapter.

Identifying entities and relationships in the database design process is important because entities are usually represented as a table, and relationships can reveal whether additional tables are needed in the database. If the problem arising from the many-to-many relationship in the E-R model isn't apparent to a designer at this point, it will become clear during the normalization process.

Database Normalization

Many people unfamiliar with database design principles often ask, "Why not just put all the data in one big table?" This single-table approach leads to problems of data redundancy (duplication) and data anomalies (data inconsistencies). For example, review the order data recorded in Table 1-1. The customer information is repeated for each order a customer places (redundancy). Also, the city data in the last row is different from the first two rows. Under these circumstances, it isn't clear whether the last row actually represents a different customer, whether the previous customer had an address change, or whether the city information is incorrect (data anomaly).

TABLE 1-1   Single-Table Approach Example

Last Name	First Name	City	State	Zip	Order Date	Order #
Jones	Pat	Norfolk	VA	24553	3/22/2009	45720
Jones	Pat	Norfolk	VA	24553	5/28/2009	48243
Jones	Pat	Suffolk	VA	26544	9/05/2009	51932

To avoid these data issues, database **normalization** is used to create a design that reduces or eliminates data redundancy and, therefore, avoids data anomalies. In general, normalization helps database designers determine which attributes, or fields, belong to each entity. In turn, this information helps determine which fields belong in each table. Normalization is a multistage process that enables designers to take the raw data to be collected about an entity and develop the data into a structured, normalized form that reduces the risks associated with data redundancy. Data redundancy poses a special problem in databases because storing the same data in different places can cause problems when updates or changes to data are required.

Most novices have difficulty understanding the impact of storing **unnormalized** data—data that hasn't been designed by using a normalization process. Here's an example. Suppose you work for a large company and submit a change-of-address form to the Human Resources (HR) Department. If all the data HR stores is normalized, a data entry clerk needs to update only the EMPLOYEES master table with your new address.

However, if the data is *not* stored in a normalized format, the data entry clerk likely needs to enter the change in each table containing your address—the EMPLOYEE RECORD table, the HEALTH INSURANCE table, the SICK LEAVE table, the ANNUAL TAX INFORMATION table, and so on—even though all this data is stored in the same database. As a result, if your mailing address is stored in several tables (or even duplicated in the same table) and the data entry clerk fails to make the change in one table, you might get a paycheck showing one address and, at the end of the year, have your W-2 form mailed to a different address! Storing data in a normalized format means only one update is required to reflect the new address, and it should always be the one that appears whenever your mailing address is needed.

A portion of the database for JustLee Books is used in this section to step through the normalization process—specifically, the books sold to customers. For each book, you need to store its International Standard Book Number (ISBN), title, publication date, wholesale cost, retail price, category (literature, self-help, and so forth), publisher name, contact person at the publisher for reordering the book (and telephone number), and author or authors' names.

Table 1-2 shows a sample of the data that must be maintained. For ease of illustration, the publishers' telephone numbers are eliminated, and the authors' names use just the first initial and last name. The first step in determining which data should be stored in each table is identifying a **primary key**, which is a field that identifies each record uniquely. You might select the ISBN to identify each book because no two books ever have the same ISBN.

TABLE 1-2   ISBN as the Primary Key

ISBN	Title	Publication Date	Cost	Retail	Category	Publisher	Contact	Author
8843172113	Database Implementation	04-JUN-03	31.40	55.95	Computer	American Publishing	Davidson	T. Peterson, J. Austin, J. Adams
1915762492	Handcranked Computers	21-JAN-05	21.80	25.00	Computer	American Publishing	Davidson	W. White, L. White

NOTE

When data that already exists, such as a book ISBN, is used as a primary key, it's often referred to as an intelligent key or a natural key. At times, data serving as a primary key doesn't exist, so a system-generated unique value is used as a primary key. For example, JustLee Books doesn't have an ID to associate with book authors, so an ID number is generated. This type of data is referred to as a surrogate key or an artificial key.

However, note that in Table 1-2, if a book has more than one author, the Author field contains more than one data value. When a record contains repeating groups (that is, multiple entries for a single column), it's considered unnormalized. **First-normal form (1NF)** indicates that all values of the columns are atomic—meaning they contain no repeating values. To convert records to 1NF, remove the repeating values by making each author entry a separate record, as shown in Table 1-3.

TABLE 1-3   The BOOKS Table in 1NF

ISBN	Title	Publication Date	Cost	Retail	Category	Publisher	Contact	Author
8843172113	Database Implementation	04-JUN-03	31.40	55.95	Computer	American Publishing	Davidson	T. Peterson
8843172113	Database Implementation	04-JUN-03	31.40	55.95	Computer	American Publishing	Davidson	J. Austin
8843172113	Database Implementation	04-JUN-03	31.40	55.95	Computer	American Publishing	Davidson	J. Adams
1915762492	Handcranked Computers	21-JAN-05	21.80	25.00	Computer	American Publishing	Davidson	W. White
1915762492	Handcranked Computers	21-JAN-05	21.80	25.00	Computer	American Publishing	Davidson	L. White

In Table 1-3, the repeating values of authors' names are eliminated—each record now contains no more than one data value for the Author field. Notice that you can no longer use the book's ISBN as the primary key because more than one record now has the same value in the ISBN field. *The only combination of fields that identifies each record uniquely is the ISBN and Author fields together*. When more than one field is used as the primary key for a table, the combination of fields is usually referred to as a **composite primary key**. Now that the repeating values have been eliminated and the records can be identified uniquely, the data is in 1NF, but a few design problems remain.

A problem known as partial dependency can occur when the primary key consists of more than one field. **Partial dependency** means the fields contained in a record (row) depend on only one portion of the primary key. For example, a book's title, publication date, publisher name, and so on, all depend on the book itself, not on who wrote the book (the author). *The simplest way to resolve a partial dependency is to break the composite primary key into two parts—each representing a separate table*. In this case, you can create a table for books and a table for authors. By removing the partial dependency, you have converted the BOOKS table to **second-normal form (2NF)**, as shown in Table 1-4.

TABLE 1-4   The BOOKS Table in 2NF

ISBN	Title	Publication Date	Cost	Retail	Category	Publisher	Contact
8843172113	Database Implementation	04-JUN-03	31.40	55.95	Computer	American Publishing	Davidson
1915762492	Handcranked Computers	21-JAN-05	21.80	25.00	Computer	American Publishing	Davidson

Now that the BOOKS records are in 2NF, you must look for any transitive dependencies. A **transitive dependency** means at least one value in the record isn't dependent on the primary key but on another field in the record. In this case, the contact person from the publisher's office is actually dependent on the publisher, not on the book. To remove the transitive dependency from the BOOKS table, remove the contact information and place it in a separate table. Because the table was in 2NF and had all transitive dependencies removed, the BOOKS table is now in **third-normal form (3NF)**, as shown in Table 1-5.

TABLE 1-5   The BOOKS Table in 3NF

ISBN	Title	Publication Date	Cost	Retail	Category	Publisher
8843172113	Database Implementation	04-JUN-03	31.40	55.95	Computer	American Publishing
1915762492	Handcranked Computers	21-JAN-05	21.80	25.00	Computer	American Publishing

There are several levels of normalization beyond 3NF; however, in practice, tables are typically normalized only to 3NF. The following list summarizes the normalization steps explained in this section:

1. *1NF*: Eliminate all repeating values and identify a primary key or primary composite key.
2. *2NF*: Make certain the table is in 1NF and eliminate any partial dependencies.
3. *3NF*: Make certain the table is in 2NF and remove any transitive dependencies.

Relating Tables in the Database

After the BOOKS table is in 3NF, you can then normalize each remaining table of the database. After each table has been normalized, make certain all relationships among tables have been established. For example, you need a way to determine the author(s) for each book in the BOOKS table. Because authors' names are stored in a separate table, there must be some way to join data. Usually, a connection between two tables is established through a **common field**—one existing in both tables. In many cases, the common field is a primary key for one of the tables. In the second table, it's referred to as a foreign key. The purpose of a **foreign key** is to establish a relationship with another table or tables. The foreign key appears in the "many" side of a one-to-many relationship.

An accepted industry standard is to use an ID code (numbers and/or letters) to represent an entity; this code reduces the chances of data entry errors. For example, instead of entering each publisher's entire name in the BOOKS table, you assign each publisher an ID code in the PUBLISHER table, and then list that ID code in the BOOKS table as a foreign key to retrieve the publisher's name for each book. In this case, the publisher ID code could be the primary key in the PUBLISHER table and a foreign key in the BOOKS table.

During the normalization of JustLee Books' database, the many-to-many relationships prevent normalizing author and order data to 3NF. The unnormalized version of the data has repeating groups for authors in the BOOKS table and for books in the ORDERS table. As part of converting the data into 3NF, two additional tables must be created: ORDERITEMS and BOOKAUTHOR.

A many-to-many relationship can't exist in a relational database. The most common approach to eliminating a many-to-many relationship is to create two one-to-many relationships by adding a bridging entity. A **bridging entity** is placed between the original entities and serves as a "filter" for the data. The ORDERITEMS table, a bridging entity, creates one-to-many relationships with the ORDERS and BOOKS tables. The BOOKAUTHOR table, another bridging entity, creates one-to-many relationships with the BOOKS and AUTHOR tables.

After normalization, the final table structures are as shown in Figure 1-5. Notice the following about the table structures:

- The underlined fields in each table indicate the primary key for that table. As mentioned, a primary key is the field that uniquely identifies each record in the table.
- For the bridging entities that were added, note that composite primary keys uniquely identify each record. The composite primary key for the BOOKAUTHOR table was created by using the primary key from each table it joins (BOOKS and AUTHOR).

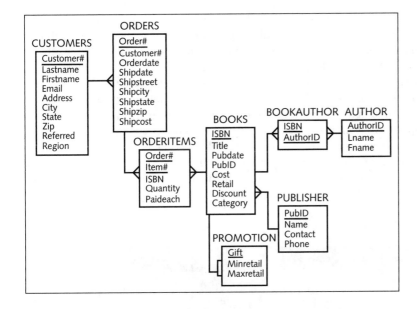

FIGURE 1-5 JustLee Books's table structures after normalization

Figure 1-6 shows a portion of the BOOKS table and the fields it contains after normalization. As mentioned, each field represents a characteristic, or attribute, that's being collected for an entity. The group of attributes for a specific occurrence (for example, a customer or a book) is called a record. In Oracle 11*g*, a list of a table's contents uses *columns to represent fields and rows to represent records*. These terms are used interchangeably throughout this textbook.

ISBN	TITLE	PUBDATE	...	CATEGORY
1059831198	BODYBUILD IN 10 MINUTES A DAY	21-JAN-01	...	FITNESS
0401140733	REVENGE OF MICKEY	14-DEC-01	...	FAMILY LIFE
4981341710	BUILDING A CAR WITH TOOTHPICKS	18-MAR-02	...	CHILDREN
8843172113	DATABASE IMPLEMENTATION	04-JUN-99	...	COMPUTER	← Record
...	...	...	...	...
0299282519	THE WOK WAY TO COOK	11-SEP-00	...	COOKING
8117949391	BIG BEAR AND LITTLE DOVE	08-NOV-01	...	CHILDREN
0132149871	HOW TO GET FASTER PIZZA	11-NOV-02	...	SELF HELP
9247381001	HOW TO MANAGE THE MANAGER	09-MAY-99	...	BUSINESS
2147428890	SHORTEST POEMS	01-MAY-01	...	LITERATURE

Field

FIGURE 1-6 A portion of the BOOKS table after normalization

The term **lookup table** is a common description for the table referenced in a foreign key relationship. For example, a Customer# field is included in the ORDERS table to link each order with a specific customer. You "look up" the Customer# value assigned to an order in the CUSTOMERS table to determine customer information, such as last name, state, and zip code. Many lookup tables simply provide descriptive values to minimize disk use and help ensure data consistency. For example, the BOOKS table contains a Category column including values such as computer, cooking, and business. Instead of storing these kinds of text descriptions in each book record, another table could be created to contain a category code and category description, as shown in Table 1-6. The category code is assigned for each book record, and a relationship to this new table allows linking the full category name with the book. Lookup tables are often used in application development to provide values in a selection list for a user to choose one for data input.

TABLE 1-6   Possible Category Lookup Table

Category Code	Category Description
10	Computer
20	Cooking
30	Business
40	Family Literature

STRUCTURED QUERY LANGUAGE (SQL)

The industry standard for interacting with a relational database is SQL—officially pronounced as "S-Q-L," but many still use the pronunciation "sequel." **Structured Query Language (SQL)** is not considered a programming language, such as VB.NET, COBOL, or Java. It's a data sublanguage with commands focused on creating database objects and manipulating data stored in a database. With SQL statements, users can instruct the DBMS to create and modify tables, enter and maintain data, and retrieve data for a variety of situations. You'll be issuing SQL commands throughout this textbook, using an Oracle 11g database. Five types of SQL commands, listed in Table 1-7, are addressed in this textbook.

TABLE 1-7   SQL Command Types

Type	Example	Description
Query	SELECT	Retrieve data values
Data manipulation language (DML)	INSERT, UPDATE, DELETE	Create or modify data values
Data definition language (DDL)	CREATE, ALTER, DROP	Define data structures

TABLE 1-7   SQL Command Types (continued)

13

Type	Example	Description
Transaction control (TC)	COMMIT, ROLLBACK	Save or undo data value modifications
Data control language (DCL)	GRANT, REVOKE	Set permissions to access database structures

Two industry-accepted committees set the industry standards for SQL: the **American National Standards Institute (ANSI)** and the **International Organization for Standardization (ISO)**. Using industry-established standards allows people to use the same skills to work with different relational DBMSs and enables various programs to communicate with different databases without major redevelopment efforts. The benefit for users (and students) is that the SQL statements learned with Oracle 11g can be transferred to another DBMS program, such as MySQL or Microsoft SQL Server. To work in another environment, you might need to modify statements slightly, but the basic structure of statements and keywords is usually the same. A few key differences between database SQL implementations are covered in Appendix F.

DATABASES USED IN THIS TEXTBOOK

Two main databases are referenced throughout this textbook. The first, JustLee Books, is described in detail in the following paragraphs. This database is used primarily to learn data retrieval statements. The second, City Jail, is used in case studies, and you design and create this database as you work through the chapter case studies. The Chapter 1 case study presents the city jail's information needs and challenges you to design the database. The case studies in Chapters 3 through 6 challenge you to construct objects in the database.

The initial organization of the database structure for JustLee Books was shown in Figure 1-5. The database is used first to record customers' orders. Customers and JustLee employees can identify a book by its ISBN, title, or author name. Employees can also determine when an order was placed and when, or if, the order was shipped. The database also stores the publisher contact information so that the bookseller can reorder a book.

Basic Assumptions

Three assumptions were made when designing the JustLee Books database:

- An order isn't shipped until all items for the order are available. (In other words, there are no back orders or partial order shipments.)
- All addresses are in the United States; otherwise, the structure of the Address/ Zip Code fields would need to be altered because many countries use different address information, such as province names.
- Only orders placed in the current month or orders placed in previous months that didn't ship are stored in the ORDERS table. At the end of each month, all completed orders are transferred to an annual SALES table. This transfer

allows faster processing of data in the ORDERS table; when necessary, users can still access information pertaining to previous orders through the annual SALES table.

In addition to recording data, management wants to be able to track the type of books that customers purchase. Although databases were originally developed to record an organization's data transactions, many have realized the importance of having data to support other business functions. Data collected for a database can be used for other purposes. For example, organizations that deal with thousands or millions of sales transactions each month usually store copies of transactions in a separate database for various types of research. Analyzing historical sales data and other information stored in an organization's database is generally referred to as **data mining**.

For this reason, the bookseller's database also includes data the Marketing Department can use to determine which categories of books customers purchase most often. By knowing buyers' purchasing habits, JustLee Books can promote new items in inventory to customers who purchase that type of book frequently. For example, if a customer has placed several orders for children's books, he or she might purchase similar books in the future. The Marketing Department can then target promotions for other children's books to that customer, knowing there's an increased likelihood of a purchase.

NOTE

Keep in mind the JustLee Books database has been kept small in the extent of tables, variety of data columns, and number of data rows in an effort to ease learning SQL fundamentals. A production retail database usually addresses many other data elements.

Tables in the JustLee Books Database

Next, take a closer look at each table in the JustLee Books database, referring to the table structures in Figure 1-5.

CUSTOMERS table: Notice that the CUSTOMERS table is the first table in Figure 1-5. It serves as a master table for storing basic data related to any customer who has placed an order with JustLee Books. It stores the customer's name, e-mail address, and mailing address, plus the Customer# of the person who referred that customer to the company. As a promotion to attract new customers, the bookstore sends a 10% discount coupon to any customer referring a friend who makes a purchase. The region data allows JustLee to track and analyze sales by geographic service areas.

Why is a Customer# field included in the CUSTOMERS table? Because you might have two customers with the same name, and by assigning each customer a number, you can uniquely identify each person. Using account numbers or codes can also decrease the likelihood of data entry errors caused by incorrect spelling or abbreviations. Keep in mind the Customer# column serves as the primary key column for the CUSTOMERS table.

BOOKS table: The BOOKS table stores each book's ISBN, title, publication date, publisher ID, wholesale cost, and retail price. The table also stores a category name for each book (for example, Fitness, Children, Cooking) to track customers' purchasing patterns, as mentioned. Currently, the category's actual name is entered in the database. The Discount field indicates the current price reduction offered. Therefore, a book's current price is the retail amount less the discount amount, if applicable.

AUTHOR and BOOKAUTHOR tables: As shown in Figure 1-5, the AUTHOR table maintains a list of authors' names. Because a many-to-many relationship originally existed between the books entity and the authors entity, the BOOKAUTHOR table was created as a bridging table between these two entities. The BOOKAUTHOR table stores each book's ISBN and author ID. If you need to know who wrote a particular book, you have the DBMS look up the book's ISBN in the BOOKS table, then look up each entry of the ISBN in the BOOKAUTHOR table, and finally trace the author's name back to the AUTHORS table through the AuthorID field.

ORDERS and ORDERITEMS tables: Data about a customer's order is divided into two tables: ORDERS and ORDERITEMS. The ORDERS table identifies which customer placed each order, the date the order was placed, the date it was shipped, and the shipping cost charged. Because the shipping address might be different from a customer's billing address, the shipping address is also stored in the ORDERS table. If a customer's order includes two or more books, the ORDERS table could contain a repeating group. Therefore, the items purchased for each order are stored separately in the ORDERITEMS table.

The ORDERITEMS table records the order number, the ISBN of the book being purchased, and the quantity for each book. To uniquely identify each item in an order when multiple items are purchased, the table includes an Item# field that corresponds to the item's position in the sequence of products ordered. For example, if a customer places an order for three different books, the first book listed in the order is assigned Item# 1, the second book listed is Item# 2, and so on. A variation of this table could use the combination of the Order# and the book's ISBN to identify each product for an order. However, the concept of item# or line# is widely used in the industry to identify line items on an invoice or in a transaction, so it has been included in this table to familiarize you with the concept.

The Paideach field in the ORDERITEMS table records the price the customer actually paid per copy for a specific book. This price is recorded because the Retail field in the BOOKS table is modified as book prices change, and the current database doesn't maintain a historical book price list.

PUBLISHER table: The PUBLISHER table contains the publisher's ID code, name, contact person, and telephone number. The PUBLISHER table can be joined to the BOOKS table through the PubID field, which is the common field. This linked data from the PUBLISHER and BOOKS table enables you to determine which publisher to contact when you need to reorder books by identifying which books you obtained from each publisher.

PROMOTION table: The last table in Figure 1-5 is the PROMOTION table. JustLee Books has an annual promotion that includes a gift with each book purchased. The gift is based on the book's retail price. Customers ordering books that cost less than $12 receive a certain gift, and customers buying books costing between $12.01 and $25 receive a different gift. The PROMOTION table identifies the gift and the minimum and maximum retail values of the range. There's no exact value that matches the Retail field in the BOOKS table; therefore, to determine the correct gift, you need to determine whether a retail price falls within a particular range.

An actual online bookseller's database would contain thousands of customers and books and, naturally, be more complex than the database shown in this textbook. For example, this database doesn't track data such as the quantity on hand for each book, discounted prices, and sales tax. Furthermore, to simplify the display of data on the screen and in reports, each table contains only a few records.

NOTE

You can find a complete list of the JustLee Books tables in Appendix A. The Appendix A listing represents the initial database columns and data values. Keep in mind that you modify this database as you progress through subsequent chapters.

TOPIC SEQUENCE

The remaining chapters of this textbook introduce the SQL statements and concepts you need to know for the first DBA or Developer Certification exam for Oracle9*i*, Oracle 10*g*, or Oracle 11*g*. They also prepare you to use Oracle in the workplace. The early chapters are organized in the same sequence you would use to create a database. After you learn how to create the database, the focus moves to data retrieval, which covers a vast array of options. The appendixes cover a variety of topics to complement the textbook, such as software tool use and SQL differences in various database products. However, before you can build a database, you need to understand how to perform basic data queries, which are covered in Chapter 2, "Basic SQL SELECT Statements."

Working through the examples in each chapter and completing the assignments help enhance your learning process.

SOFTWARE USED IN THIS TEXTBOOK

The Oracle 11*g* database system is used for this textbook. Oracle 11*g* is offered in a variety of editions, including Personal, Standard, and Enterprise. Any of these editions is suitable for performing all examples and assignments in this textbook. Two client software tools are included with Oracle 11*g*: SQL*Plus and SQL Developer. These tools are introduced briefly in the next chapter. All figures in this textbook are shown in the SQL Developer interface; however, either client tool can be used. Previous versions of Oracle (Oracle9*i* or Oracle 10*g*) can be used for much of the work in this textbook, with the exception of several new features requiring the current version.

Previous versions of Oracle offered a Windows-based and an Internet interface installation of SQL*Plus (iSQL*Plus). Both these SQL*Plus editions were deprecated with Oracle 11*g*. You can still use the previous version of the Windows-based SQL*Plus to connect to Oracle 11*g*, if you like. The Oracle 11*g* SQL*Plus client uses a command-line interface. The SQL Developer tool provides a graphical user interface (GUI) to view database objects, enter SQL statements with syntax color coding, and view output.

- A DBMS is used to create and maintain a database.
- A database is composed of a group of interrelated tables.
- A file is a group of related records. A file is also called a table in the physical database.
- A record, also called a row, is a group of related fields about one specific entity.
- Before building a database, designers must look at the system's input, processing, and output requirements. Tables to be included in the database can be identified with the E-R model. An entity in the E-R model usually represents a table in the physical system.
- Through the normalization process, designers can determine whether additional tables are needed and which attributes or fields belong in each table.
- A record is considered unnormalized if it contains repeating groups.
- A record is in first-normal form (1NF) if no repeating groups exist and it has a primary key.
- Second-normal form (2NF) is achieved if the record is in 1NF and has no partial dependencies.
- After a record is in 2NF and all transitive dependencies have been removed, then it's in third-normal form (3NF), which is generally enough for most databases.
- Normalization results in one-to-many relationships between tables.
- A primary key is used to uniquely identify each record.
- A common field is used to join data contained in different tables.
- A foreign key is a common field that exists between two tables but is also a primary key in one of the tables.
- A lookup table is a common term for the table referenced in a foreign key relationship, which provides a more descriptive value for the data.
- A Structured Query Language (SQL) is a data sublanguage for navigating data stored in a database's tables. With SQL statements, users can instruct the DBMS to create and modify tables, enter and maintain data, and retrieve data for a variety of situations.

Review Questions

1. What is the purpose of an E-R model?
2. What is an entity?
3. Give an example of three entities that might exist in a database for a medical office, and list some attributes that would be stored in a table for each entity.
4. Define a one-to-many relationship.
5. Discuss the problems that can be caused by data redundancy.
6. Explain the role of a primary key.
7. Describe how a foreign key is different from a primary key.

8. List the steps of the normalization process.

9. What type of relationship can't be stored in a relational database? Why?

10. Identify at least three reasons an organization might analyze historical sales data stored in its database.

Multiple Choice

1. Which of the following represents a row in a table?

    a. an attribute

    b. a characteristic

    c. a field

    d. a record

2. Which of the following defines a relationship in which each occurrence of data in one entity is represented by multiple occurrences of the data in the other entity?

    a. one-to-one

    b. one-to-many

    c. many-to-many

    d. none of the above

3. An entity is represented in an E-R model as a(n):

    a. arrow

    b. crow's foot

    c. dashed line

    d. none of the above

4. Which of the following is not an E-R model relationship?

    a. some-to-many

    b. one-to-one

    c. one-to-many

    d. many-to-many

5. Which of the following symbols represents a many-to-many relationship in an E-R model?

    a. a straight line

    b. a dashed line

    c. a straight line with a crow's foot at both ends

    d. a straight line with a crow's foot at one end

6. Which of the following can contain repeating groups?

    a. unnormalized data

    b. 1NF

    c. 2NF

    d. 3NF

7. Which of the following defines a relationship in which each occurrence of data in one entity is represented by only one occurrence of data in the other entity?

   a. one-to-one

   b. one-to-many

   c. many-to-many

   d. none of the above

8. Which of the following has no partial or transitive dependencies?

   a. unnormalized data

   b. 1NF

   c. 2NF

   d. 3NF

9. Which of the following symbols represents a one-to-many relationship in an E-R model?

   a. a straight line

   b. a dashed line

   c. a straight line with a crow's foot at both ends

   d. a straight line with a crow's foot at one end

10. Which of the following has no partial dependencies but can contain transitive dependencies?

    a. unnormalized data

    b. 1NF

    c. 2NF

    d. 3NF

11. Which of the following has no repeating groups but can contain partial or transitive dependencies?

    a. unnormalized data

    b. 1NF

    c. 2NF

    d. 3NF

12. The unique identifier for a record is called the:

    a. foreign key

    b. primary key

    c. turn key

    d. common field

13. Which of the following fields also serves as a primary key in another table when two tables are joined together on that value?

    a. foreign key

    b. primary key

c. turn key

d. repeating group key

14. A unique identifier for a data row that consists of more than one field is commonly called a:

    a. primary plus key

    b. composite primary key

    c. foreign key

    d. none of the above

15. Which of the following symbols represents an optional relationship in an E-R model?

    a. a straight line

    b. a dashed line

    c. a straight line with a crow's foot at both ends

    d. a straight line with a crow's foot at one end

16. Which of the following, when used in an E-R model, indicates the need for an additional table?

    a. sometimes-to-always relationship

    b. one-to-one relationship

    c. one-to-many relationship

    d. many-to-many relationship

17. Which of the following represents a field in a table?

    a. a record

    b. a row

    c. a column

    d. an entity

18. Which of the following defines a relationship in which data can have multiple occurrences in each entity?

    a. one-to-one

    b. one-to-many

    c. many-to-many

    d. none of the above

19. When part of the data in a table depends on a field in the table that isn't the table's primary key, it's known as:

    a. transitive dependency

    b. partial dependency

    c. psychological dependency

    d. a foreign key

20. Which of the following is used to join data contained in two or more tables?

    a. primary key

    b. unique identifier

    c. common field

    d. foreign key

Hands-On Assignments

To perform assignments 1 to 5, refer to the table structures in Figure 1-5 and the table listings in Appendix A.

1. Which tables and fields would you access to determine which book titles have been purchased by a customer and when the order shipped?

2. How would you determine which orders have not yet been shipped to the customer?

3. If management needed to determine which book category generated the most sales in April 2009, which tables and fields would they consult to derive this information?

4. Explain how you would determine how much profit was generated from orders placed in April 2009.

5. If a customer inquired about a book written in 2003 by an author named Thompson, which access path (tables and fields) would you need to follow to find the list of books meeting the customer's request?

  In assignments 6 to 10, create a simple E-R model depicting entities and relationship lines for each data scenario.

6. A college needs to track placement test scores for incoming students. Each student can take a variety of tests, including English and math. Some students are required to take placement tests because of previous coursework.

7. Every employee in a company is assigned to one department. Every department can contain many employees.

8. A movie megaplex needs to collect movie attendance data. The company maintains 16 theaters in a single location. Each movie offered can be shown in one or more of the available theaters and is typically scheduled for three to six showings in a day. The movies are rotated through the theaters to ensure that each is shown in one of the stadium-seating theaters at least once.

9. An online retailer of coffee beans maintains a long list of unique coffee flavors. The company purchases beans from a number of suppliers; however, each specific flavor of coffee is purchased from only a single supplier. Many of the customers are repeat purchasers and typically order at least five flavors of beans in each order.

10. Data for an information technology conference needs to be collected. The conference has a variety of sessions scheduled over a two-day period. All attendees must register for the sessions they plan to attend. Some speakers are presenting only one session, whereas others are handling multiple sessions. Each session has only one speaker.

Advanced Challenge

To perform this activity, refer to the table structures in Figure 1-5 and the table listings in Appendix A.

In this chapter, the normalization process was shown for just the BOOKS table. The other tables in the JustLee Books database are shown after normalization. Because the database needs to contain data for each customer's order, perform the steps for normalizing the following data elements to 3NF:

- Customer's name and billing address
- Quantity and retail price of each item ordered
- Shipping address for each order
- Date each order was placed and the date it was shipped

Assume the unnormalized data in the list is stored in one table. Provide your instructor with a list of the tables you have identified at each step of the normalization process (that is, 1NF, 2NF, 3NF) and the attributes, or fields, in each table. Remember that each customer can place more than one order, each order can contain more than one item, and an item can appear on more than one order.

Case Study: *City Jail*

Your company receives the following memo. First, based on the memo, create an initial database design (E-R model) for the City Jail that indicates entities, attributes (columns), primary keys, and relationships. In developing your design, consider the columns needed to build relationships between the entities. Use only the entities identified in the memo to develop the E-R model.

Second, create a list of additional entities or attributes not identified in the memo that might be applicable to a crime-tracking database.

TIP

Keep in mind that the memo is written from an end-user perspective—not by a database developer!

MEMO
To: Database Consultant
From: City Jail Information Director
Subject: Establishing a Crime-Tracking Database System

It was a pleasure meeting with you last week. I look forward to working with your company to create a much-needed crime-tracking system. As you requested, our project group has outlined the crime-tracking data needs we anticipate. Our goal is to simplify the process of tracking criminal activity and provide a more efficient mechanism for data analysis and reporting. Please review the data needs outlined below and contact me with any questions.

Criminals: name, address, phone number, violent offender status (yes/no), probation status (yes/no), and aliases

Crimes: classification (felony, misdemeanor, other), date charged, appeal status (closed, can appeal, in appeal), hearing date, appeal cutoff date (always 60 days after the hearing date), arresting officers (can be more than one officer), crime codes (such as burglary, forgery, assault;

hundreds of codes exist), amount of fine, court fee, amount paid, payment due date, and charge status (pending, guilty, not guilty)

Sentencing: start date, end date, number of violations (such as not reporting to probation officer), and type of sentence (jail period, house arrest, probation)

Appeals: appeal filing date, appeal hearing date, status (pending, approved, and disapproved)

Note: Each crime case can be appealed up to three times.

Police officers: name, precinct, badge number, phone contact, status (active/inactive)

Additional notes:

- A single crime can involve multiple crime charges, such as burglary and assault.
- Criminals can be assigned multiple sentences. For example, a criminal might be required to serve a jail sentence followed by a period of probation.

BASIC SQL SELECT STATEMENTS

LEARNING OBJECTIVES

After completing this chapter, you will be able to do the following:

- Create the initial database
- Identify keywords, mandatory clauses, and optional clauses in a SELECT statement
- Select and view all columns of a table
- Select and view one column of a table
- Display multiple columns of a table
- Use a column alias to clarify the contents of a particular column
- Perform basic arithmetic operations in the SELECT clause
- Remove duplicate lists by using the DISTINCT or UNIQUE keyword
- Use concatenation to combine fields, literals, and other data

INTRODUCTION

In Chapter 1, fundamental relational database design concepts were introduced along with the role of SQL statements in interacting with a database. Before jumping into creating database tables, learning to navigate in a database by using SQL SELECT statements is advantageous. Querying the database enables you to verify existing database tables, the structure of tables, and data values stored in tables.

In this chapter, you begin learning how to retrieve data from a database by using some basic SELECT statements. In later chapters, you explore more options for querying the database. Table 2-1 summarizes the commands covered in this chapter.

TABLE 2-1   Summary of Commands in This Chapter

Command Description	Basic Syntax Structure	Example
Command to view all columns of a table	SELECT * FROM *tablename*;	SELECT* FROM books;
Command to view one column of a table	SELECT *columnname* FROM *tablename*;	SELECT title FROM books;
Command to view multiple columns of a table	SELECT *columnname,* *columnname, ...* FROM *tablename*;	SELECT title, pubdate FROM books;
Command to assign an alias to a column during display	SELECT *columnname* [AS] *alias* FROM *tablename*;	SELECT title AS titles FROM books; *or* SELECT title titles FROM books;
Command to perform arithmetic operations during retrieval	SELECT *arithmetic* *expression* FROM *tablename*;	SELECT retail-cost FROM books;
Command to eliminate duplication in output	SELECT DISTINCT *columnname* FROM *tablename*; *or* SELECT UNIQUE *columnname* FROM *tablename*;	SELECT DISTINCT state FROM customers; *or* SELECT UNIQUE state FROM customers;
Command to perform concatenation of column contents during display	SELECT *columnname* || *columnname* FROM *tablename*;	SELECT firstname || lastname FROM customers;
Command to view the structure of a table	DESCRIBE *tablename*	DESCRIBE books

In the next section, you create the JustLee Books database. The tables in this database are used in the chapter examples and Hands-On Assignments.

CREATING THE JUSTLEE BOOKS DATABASE

First, identify which software tool you're using to connect to Oracle 11g and your login information. Figure 2-1 shows the Oracle 11g SQL*Plus tool interface, and Figure 2-2 shows the Oracle 11g SQL Developer tool interface.

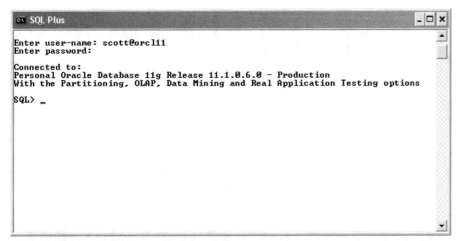

FIGURE 2-1   SQL*Plus interface

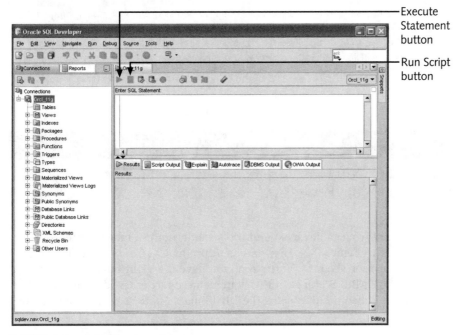

FIGURE 2-2   SQL Developer interface

Review Appendix B, "SQL*Plus and SQL Developer Overview," to become familiar with the interface you're using. The appendix is separated into two sections: The first introduces the use of the SQL*Plus interface, and the second addresses the SQL Developer interface.

Next, create the JustLee Books database by executing the script provided in the data files for this textbook. Locate the data files and verify that the JLDB_Build.sql file is available in the Chapter2 folder. Work through the following steps for the SQL*Plus or SQL Developer interface to create the database.

SQL*Plus:

1. Start SQL*Plus and log in.
2. To execute the script file, enter **start C:*data**chapter2*\\JLDB_Build.sql** at the SQL> prompt. (For **C:*data**chapter2***, substitute the drive letter and path-name for your system. Make sure there are no spaces in the pathname.)
3. Press **Enter.**

SQL Developer:

1. From the menu, click **File, Open** and navigate to the **JLDB_Build.sql** file.
2. Click the file, and then click the **Open** button. You should see the script statements in the work area.
3. Click the **Run Script** button above the work area (refer to Figure 2-2) to execute the statements. You might be prompted to select a connection.

NOTE

If you want to see the script's contents, you can open it with any text editor or word-processing program. A script is simply a file containing SQL statements to be processed as a set. The database script for this chapter is a series of CREATE TABLE and INSERT statements to construct the tables. You learn these statements in later chapters. If you're already familiar with constraints, note that some needed constraints aren't included in this initial script, as you're adding more in Chapter 4.

TIP

In the SQL*Plus client tool, an @ symbol can be used in the place of the START keyword to execute a script.

Now you can verify what tables exist and the structure of the tables. Enter and execute the statement in Figure 2-3. This statement produces a list of table names existing in the current user account. Verify that you have the eight tables listed in Figure 2-3. The USER_TABLES object in this statement is part of the database's **data dictionary**, which is a collection of objects the DBMS manages to maintain information about the database.

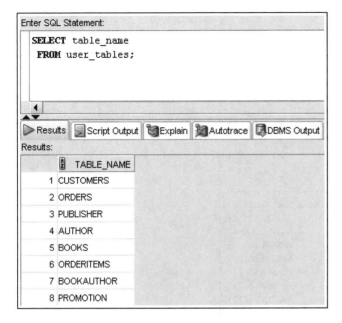

FIGURE 2-3 List of existing tables

All code execution figures in this textbook are shown in the SQL Developer tool. Any SELECT statement should be executed with the Execute Statement button. Any other types of statements are executed with the Run Script button.

After you have a list of existing tables, you can use the DESCRIBE command to view the structure of a table. Figure 2-4 shows using the DESCRIBE command to list the structure of the BOOKS table. Notice that the abbreviation DESC can be used instead of the full word DESCRIBE. The listing shows the names and datatypes for all columns in the table.

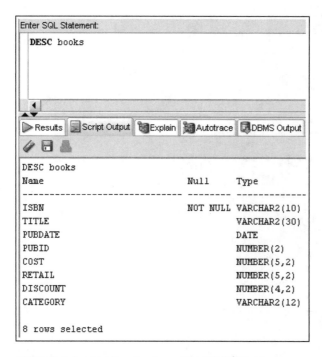

```
Enter SQL Statement:

  DESC books




  ◄

▲▼
► Results  ▣ Script Output  ▣Explain  ▣Autotrace  ▣DBMS Output

 ✎ 🖫 🖨

DESC books
Name                      Null       Type
------------------------- --------   --------------

ISBN                      NOT NULL   VARCHAR2(10)
TITLE                                VARCHAR2(30)
PUBDATE                              DATE
PUBID                                NUMBER(2)
COST                                 NUMBER(5,2)
RETAIL                               NUMBER(5,2)
DISCOUNT                             NUMBER(4,2)
CATEGORY                             VARCHAR2(12)

8 rows selected
```

FIGURE 2-4   List the structure of the BOOKS table

TIP

The DESCRIBE command is an SQL*Plus command, not an SQL command, so a semicolon isn't necessary to end this statement.

The Chapter2 data files folder contains two files: JLDB_Build.sql and JLDB_Drop.sql. If needed, you can rebuild the JustLee database at any time as you work through this text-book. You might need to do this if you remove a row or table accidentally as you experiment with SQL statements. To reconstruct the database, run the JLDB_Drop.sql script first to remove any tables existing from the initial database creation done earlier in this chapter. Then execute the JLDB_Build.sql script again to rebuild all the tables. Don't be concerned if you get any "object does not exist" errors from the JLDB_Drop.sql script. These messages just indicate that the script attempted to remove a table that no longer exists in your account.

SELECT STATEMENT SYNTAX

Most of the SQL operations performed on a database in a typical organization are **SELECT statements,** which enable users to retrieve data from tables. A user can view all the fields and records in a table or specify displaying only certain fields and records. In essence, the SELECT statement asks the database a question, which is why it's also known as a **query**.

After querying a database, the results that are displayed can be based on certain conditions specified in the SELECT statement. In other words, what's displayed is basically the answer to the question the user asked. For example, in this chapter, you learn the basic structure of a SELECT statement and how to display only certain fields from a table. In Chapter 8, you learn how to modify the SELECT statement to display only certain rows.

The **syntax** for an SQL statement is the basic structure, or rules, required to execute the statement. Figure 2-5 shows the syntax for the SELECT statement.

```
SELECT    [DISTINCT | UNIQUE] (*, columnname [ AS alias], …)
          FROM      tablename
          [WHERE    condition]
          [GROUP BY group_by_expression]
          [HAVING   group_condition]
          [ORDER BY columnname];
```

FIGURE 2-5   Syntax for the SELECT statement

The capitalized words (SELECT, FROM, WHERE, and so forth) in Figure 2-5 are **keywords** (words with a predefined meaning in Oracle 11*g*). Each section in the figure that begins with a keyword is referred to as a **clause** (SELECT clause, FROM clause, WHERE clause, and so on). Note these important points about SELECT statements:

- The only clauses required for the SELECT statement are SELECT and FROM, so they are the only clauses in Figure 2-5 discussed in this chapter.
- Square brackets indicate optional portions of the statement. (Optional clauses are discussed in subsequent chapters.)
- SQL statements can be entered over several lines (as shown in Figure 2-5) or on one line. Most SQL statements are entered with each clause on a separate line to improve readability and make editing easier. As various SELECT commands are explained in this chapter, you'll see variations on spacing, number of lines, and capitalization. These variations are pointed out as you encounter them in this textbook.
- An SQL statement ends with a semicolon.

Selecting All Data in a Table

To have the SELECT statement return *all* data from a specific table, type an asterisk (*) after SELECT, as shown in Figure 2-6. This statement retrieves all data stored in the CUSTOMERS table.

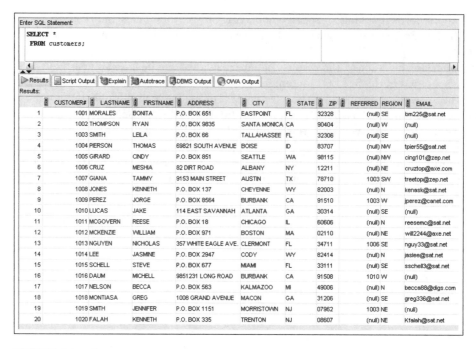

FIGURE 2-6 Command to select all data in a table

The asterisk (*) is a symbol that instructs Oracle 11*g* to include all columns in the table. This symbol can be used only in the SELECT clause of a SELECT statement. If you need to view or display all columns in a table, typing an asterisk is much simpler than typing the name of each column.

> **N O T E**
>
> As mentioned, all execution output in this textbook is shown in the SQL Developer interface, which is displayed in a table format by default for SELECT statements. Your output format will be different if you're using SQL*Plus because this client tool displays output in a text format. Appendix B includes sample output from both interfaces.

> **N O T E**
>
> Did you notice the values of "(null)" in some rows of the Referred column? This value is displayed as a blank if you're using the SQL*Plus client tool. A NULL value indicates the field is empty. NULL values are discussed later in this chapter.

When looking at the results of the SELECT statement, pay attention to the column headings. Depending on the tool being used and the options set, some column headings might be truncated. Keep in mind that when you refer to a column in any SQL statement, you still need to specify the entire column name. Because of the possibility of the name being truncated, be sure you don't depend on the displayed column headings when entering column names in SQL statements.

TIP

You can view the exact name of each column by entering the command DESCRIBE *tablename*.

Selecting One Column from a Table

In the example in Figure 2-6, an asterisk was used to specify displaying all columns in the table. If the table contains a large number of fields, the results might look cluttered, or maybe the table contains sensitive data you don't want other users to see. In these situations, you can instruct Oracle 11g to return only specific columns in the results. Choosing specific columns in a SELECT statement is called **projection**. You can select one column—or as many as all columns—contained in the table.

For example, suppose you want to view the titles of all books in the inventory. The data about books is stored in the BOOKS table, and the name of the column you need is Title. As shown in Figure 2-7, you can list the column name after the SELECT keyword. Type the statement shown in Figure 2-7 to list all the book titles.

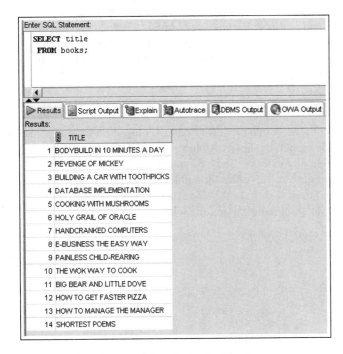

FIGURE 2-7   Command to select a single column

NOTE

If you get an error message rather than the query results, you might have made a typing error. The error message should display the line in which the error occurred. An asterisk beneath the line serves as an indicator of the error; however, it's not always the exact cause. If the error message indicates that the error is in the second line of the statement, you might have entered BOOK instead of BOOKS. Simply retype the statement with the correction, and it should return the correct results.

The results display only the specified field, which is Title. You might want to practice some variations of the same SELECT statement. Try entering the examples shown in Figure 2-8, one at a time, and notice that the results are the same.

```
SELECT TITLE FROM BOOKS;

select title from books;

SELECT title FROM books;

SELECT TITLE
 FROM BOOKS;
```

FIGURE 2-8   The SELECT statement can be entered on one or more lines

As shown in these examples, the statement can be entered on one or more lines. Also, notice that *keywords, table names, and column names are not case sensitive*. To distinguish between keywords and other parts of the SELECT statement, the keywords are capitalized in examples. Keep in mind that this is *not* a requirement of Oracle 11g; it's simply a convention used to improve readability.

Selecting Multiple Columns from a Table

In most cases, displaying only one column from a table isn't enough output on which to base decisions. If you want to know the date each book was published, you could retrieve all the fields from the BOOKS table and manually extract the fields you need. As an alternative, you could issue one SELECT statement to retrieve the Title field, another to retrieve the Publication Date (Pubdate) field, and then match up the two results. However, issuing a query requesting both the title and the publication date for each book is more practical, as shown in Figure 2-9.

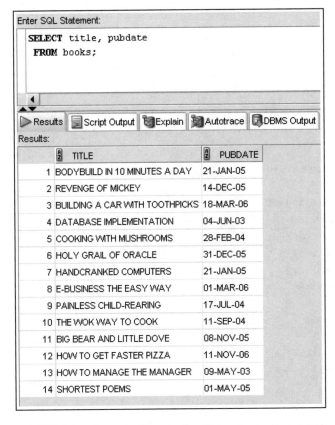

FIGURE 2-9 Command to select multiple columns from a table

When specifying more than one column in the SELECT clause, commas should separate the columns listed. Although a space has been entered after the comma, it's not part of the SELECT statement's required syntax. The space simply improves the statement's readability.

When looking at the results of this query, notice the order in which columns are listed in the output: Title is listed first, followed by Pubdate. Oracle 11*g* sequences columns in the display in the same order the user sequences them in the SELECT statement. To change the order and display the Pubdate column first, simply reverse the order of columns listed in the SELECT clause, as shown in Figure 2-10.

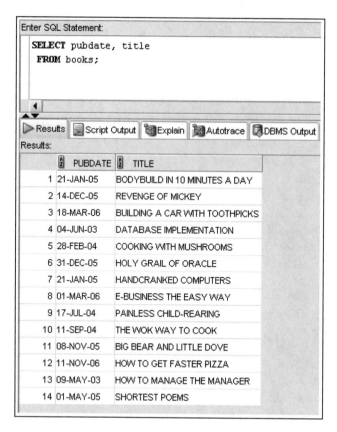

FIGURE 2-10 Reversed column sequence in the SELECT clause

OPERATIONS IN THE SELECT STATEMENT

Now that you've selected columns from tables, take a look at some other operations. In this section, you learn how to use column aliases, use arithmetic operations, and eliminate duplicate output.

Using Column Aliases

Sometimes a column name is a vague indicator of the data that's displayed. To better describe the data displayed in the output, you can substitute a **column alias** for the column name in query results. For example, if you're displaying a list of all books stored in the database, you might want the column heading to read "Title of Book." To instruct the software to use a column alias, simply list the column alias next to the column name in the SELECT clause. Figure 2-11 shows the title and category for each book in the BOOKS table, but it adds a column alias for the title. The **optional keyword** of **AS** has been included to distinguish between the column name and the column alias.

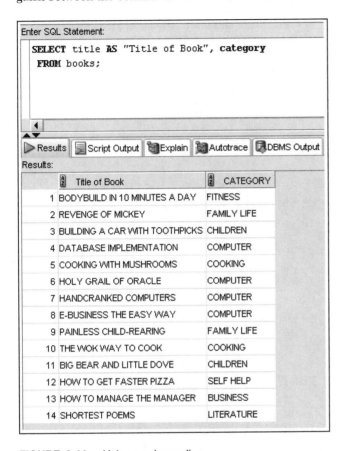

FIGURE 2-11   Using a column alias

You need to keep some guidelines in mind when using a column alias. If the column alias contains spaces or special symbols, or if you don't want it to appear in all uppercase letters, *you must enclose it in quotation marks* (" "). By default, column headings shown in query results are capitalized. Using quotation marks overrides this default setting. However, notice that the letter case of data displayed *in* the column isn't altered.

NOTE

As shown in the SELECT statement, *you must separate the list of field names with commas*. If you forget a comma, Oracle 11g interprets the subsequent field name as a column alias, and you don't get the results you intended.

If the column alias consists of only one word without special symbols, it doesn't need to be enclosed in quotation marks. In Figure 2-12, the Retail field is assigned the column alias of Price. Also, note that the optional keyword AS used in Figure 2-11 isn't included. Because a comma doesn't separate the words *retail* and *price*, Oracle 11*g* assumes that Price is the column alias for the Retail column.

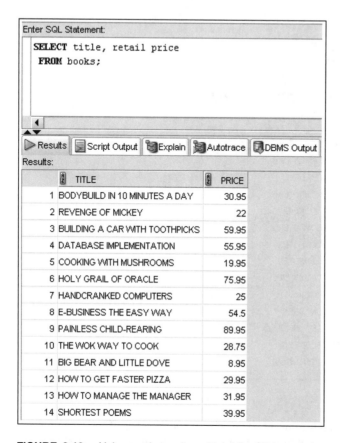

FIGURE 2-12   Using a column alias without the AS keyword

NOTE

Many development shops consistently use quotation marks or the AS keyword with column aliases to improve code readability.

As you look at the results in Figure 2-12, notice the alignment of data values in columns:

- By default, the data for text, or character, fields is left-aligned.
- Data for numeric fields is right-aligned.

- Oracle 11*g* doesn't display insignificant zeros (zeros that don't affect the value of the number being displayed). The retail price of the book *Hand-cranked Computers* is $25.00. Because the zeros in the two decimal positions are insignificant, Oracle 11*g* doesn't display them. To force Oracle 11*g* to display a specific number of decimal positions or use special symbols (such as dollar signs), formatting codes, discussed in Chapter 10, are required.

Using Arithmetic Operations

Simple arithmetic operations, such as multiplication (*), division (/), addition (+), and subtraction (-), can be used in the SELECT clause of a query. Keep in mind that Oracle 11*g* adheres to the standard order of operations:

1. Moving from left to right in the arithmetic equation, any required multiplication and division operations are solved first.
2. Addition and subtraction operations are solved after multiplication and division, again moving from left to right in the equation.

To override this order of operations, you can use parentheses to enclose the portion of the equation that should be calculated first. For example, you might need to calculate a book's profit margin by subtracting the retail price from the cost and dividing the result by the cost. This operation could be written as retail-cost/cost. However, the division operation is solved first, which is cost/cost, so the result is always subtracting the number one from the retail price. To force the subtraction to occur before the division, add parentheses as shown: (retail-cost)/cost.

N O T E

Oracle 11*g* doesn't support an exponent operator in the SELECT statement. For example, in some programs, you can enter *number*^3 to raise a number to the power of three. (Raising a number to the power of three means multiplying the number by itself three times; for example, 5^3 equals 5 * 5 * 5.) With Oracle 11*g*, if you need to use an exponential operation in the SELECT statement, break it down into its multiplication equivalent or use the POWER function, introduced in Chapter 10.

Next, you want to determine the profit generated by the sale of each book. The BOOKS table contains two fields you can use to derive the profit: Cost and Retail. A book's profit is the difference (subtraction) between the amount the bookstore paid for the book (cost) and the selling price of the book (retail). To clarify the column's contents, assign the column alias "profit" to the calculated field, as shown in Figure 2-13.

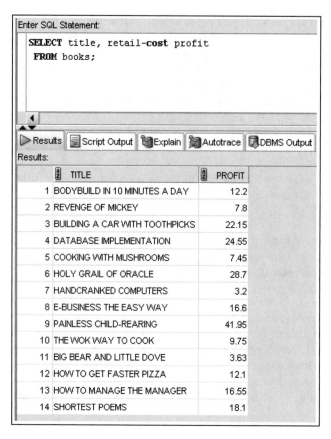

FIGURE 2-13 Using a column alias for an arithmetic expression

NULL Values

If no value is entered for a column in a row of data, the value is considered NULL, indicating an absence of data. Query all the data in the BOOKS table, as shown in Figure 2-14, and notice that some rows have a NULL value for the Discount amount. As mentioned, this value is shown as "(null)" in SQL Developer or simply a blank in SQL*Plus.

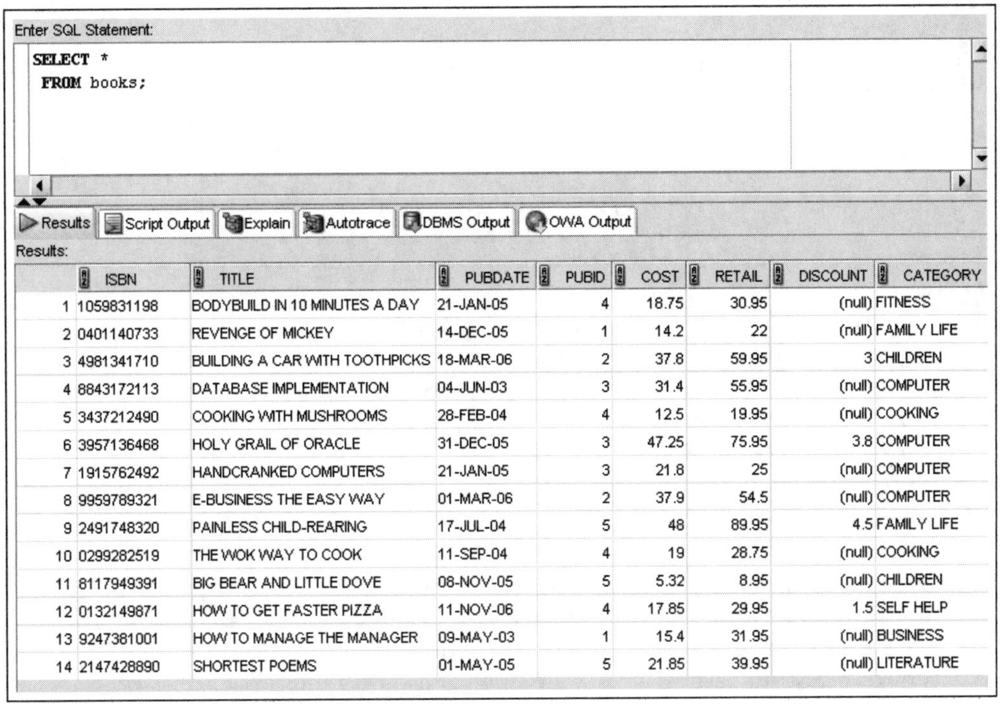

Enter SQL Statement:

```
SELECT *
FROM books;
```

Results | Script Output | Explain | Autotrace | DBMS Output | OWA Output

Results:

	ISBN	TITLE	PUBDATE	PUBID	COST	RETAIL	DISCOUNT	CATEGORY
1	1059831198	BODYBUILD IN 10 MINUTES A DAY	21-JAN-05	4	18.75	30.95	(null)	FITNESS
2	0401140733	REVENGE OF MICKEY	14-DEC-05	1	14.2	22	(null)	FAMILY LIFE
3	4981341710	BUILDING A CAR WITH TOOTHPICKS	18-MAR-06	2	37.8	59.95	3	CHILDREN
4	8843172113	DATABASE IMPLEMENTATION	04-JUN-03	3	31.4	55.95	(null)	COMPUTER
5	3437212490	COOKING WITH MUSHROOMS	28-FEB-04	4	12.5	19.95	(null)	COOKING
6	3957136468	HOLY GRAIL OF ORACLE	31-DEC-05	3	47.25	75.95	3.8	COMPUTER
7	1915762492	HANDCRANKED COMPUTERS	21-JAN-05	3	21.8	25	(null)	COMPUTER
8	9959789321	E-BUSINESS THE EASY WAY	01-MAR-06	2	37.9	54.5	(null)	COMPUTER
9	2491748320	PAINLESS CHILD-REARING	17-JUL-04	5	48	89.95	4.5	FAMILY LIFE
10	0299282519	THE WOK WAY TO COOK	11-SEP-04	4	19	28.75	(null)	COOKING
11	8117949391	BIG BEAR AND LITTLE DOVE	08-NOV-05	5	5.32	8.95	(null)	CHILDREN
12	0132149871	HOW TO GET FASTER PIZZA	11-NOV-06	4	17.85	29.95	1.5	SELF HELP
13	9247381001	HOW TO MANAGE THE MANAGER	09-MAY-03	1	15.4	31.95	(null)	BUSINESS
14	2147428890	SHORTEST POEMS	01-MAY-05	5	21.85	39.95	(null)	LITERATURE

FIGURE 2-14   NULL values in the Discount column

NULL values can lead to undesirable results in operations. For example, what if you need to determine a book's current sale price by subtracting the Discount column amount from the Retail column amount? This operation seems simple enough, using a subtraction operation as shown in Figure 2-15. When you inspect the results, however, you discover that each row containing a NULL value for the Discount column shows a calculated sale price of NULL. If any value in an arithmetic operation is NULL, the result is NULL. To perform the operation successfully, you can use functions to substitute a value, such as 0, for a NULL value. Functions are covered in Chapter 10.

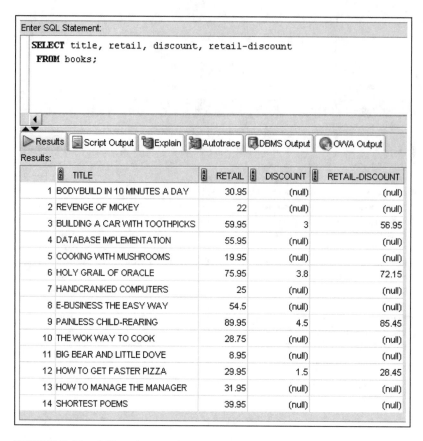

FIGURE 2-15 Arithmetic operations with NULL values

Using DISTINCT and UNIQUE

Suppose you want to know the states in which your customers live so that you can focus a marketing campaign on a particular region of the country. You want a list to identify only the states, not customer names, addresses, and so on. One option is to select the State column from the CUSTOMERS table. You'll notice quickly, however, that some states are listed more than once, if more than one customer lives in a state. If you're working with only 20 records, simply crossing out duplicate states on a printout isn't a problem. However, if you're dealing with thousands of records, this task is cumbersome.

To eliminate duplicate listings, you can use the **DISTINCT** option in your SELECT statement. For example, suppose you have five customers living in Texas (TX). Without the DISTINCT option, TX appears in your results five times. If you include the DISTINCT option, TX appears only once. To use the DISTINCT option, use the keyword DISTINCT between the SELECT keyword and the first column of the column list, as shown in Figure 2-16.

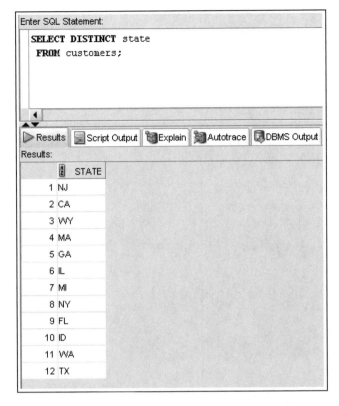

FIGURE 2-16 List of unduplicated states for customers

In Figure 2-16, the database was queried to determine the states in which customers live. Although there are 20 customers in the CUSTOMERS table, they live in only 12 states. You could use this information to determine where you're most likely to attract more customers or to identify geographical areas that aren't responding to a nationwide marketing effort.

The DISTINCT keyword is applied to all columns listed in the SELECT clause, even though it's stated directly after the SELECT keyword. In the example in Figure 2-16, if you had also included CITY in the SELECT clause, each different combination of city and state would have been listed only once in the output. If no two customers in the database live in the same city and state, you would still have 20 rows of output—one row for each customer. However, two customers live in Burbank, CA. Notice that the output in Figure 2-17 shows only 19 rows because Burbank, CA is listed only once as a result of the DISTINCT operation.

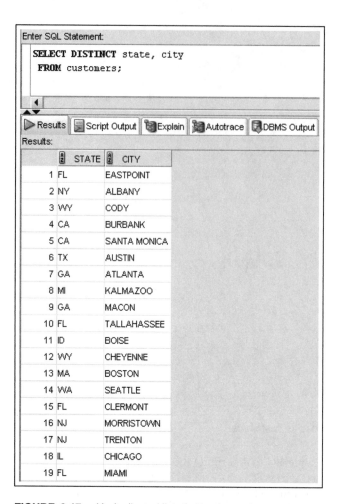

Enter SQL Statement:

```
SELECT DISTINCT state, city
  FROM customers;
```

Results | Script Output | Explain | Autotrace | DBMS Output

Results:

	STATE	CITY
1	FL	EASTPOINT
2	NY	ALBANY
3	WY	CODY
4	CA	BURBANK
5	CA	SANTA MONICA
6	TX	AUSTIN
7	GA	ATLANTA
8	MI	KALMAZOO
9	GA	MACON
10	FL	TALLAHASSEE
11	ID	BOISE
12	WY	CHEYENNE
13	MA	BOSTON
14	WA	SEATTLE
15	FL	CLERMONT
16	NJ	MORRISTOWN
17	NJ	TRENTON
18	IL	CHICAGO
19	FL	MIAMI

FIGURE 2-17 Unduplicated list of cities for customers

NOTE

You can also use the **UNIQUE** keyword to eliminate duplicates. It works the same way as the DISTINCT keyword. The following returns the same results as the example in Figure 2-16:

SELECT UNIQUE state

FROM customers;

Using Concatenation

In previous examples, if an output list contained more than one field, each field was placed in a separate column. In some situations, however, you might want the contents of each field to be displayed next to each other, without much blank space. For example, for a list of customer names, you might prefer to have them combined so that they appear as a single column rather than separate first name and last name columns.

Combining the contents of two or more columns is known as **concatenation**. To instruct Oracle 11g to concatenate the output of a query, use two vertical bars, or pipes (||), between the fields you're combining. On a keyboard, this symbol is located above the backslash (\). In the example in Figure 2-18, the goal is to have the customer's last name listed immediately after the first name instead of in a separate column. As you look at the results in this figure, the first thing you should notice is that each customer's first name and last name run together, and it's difficult to tell where one name ends and the other begins. To make the results more readable, you need to include a blank space between the First-name and Lastname fields.

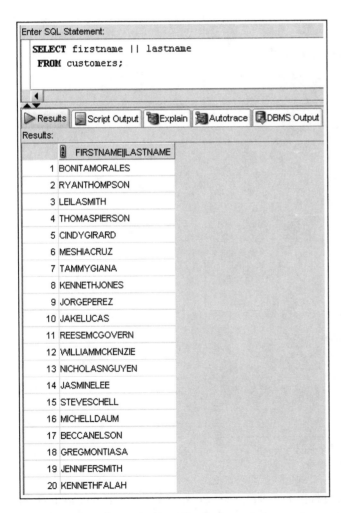

FIGURE 2-18 Concatenation of two columns

To have Oracle 11*g* insert a blank space, you must concatenate the Firstname and Lastname fields with a **string literal**. A string literal instructs Oracle 11*g* to interpret the characters you have entered "literally," not to consider them a keyword or command. *A string literal must be enclosed in single quotation marks* (' '). When you use a string literal, the character or characters you type inside the single quotation marks should appear in the output exactly as you have typed them. In this instance, the string literal is a blank space. Figure 2-19 shows what the customer's list looks like after including a blank space in the output.

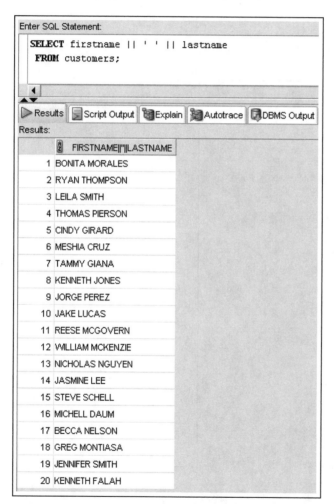

FIGURE 2-19   Using a string literal in concatenation

Although you now have a readable list of all customer names, the display has an unusual column heading. The column heading shown in the results is exactly what you entered for the field list—including the concatenation symbols and the string literal. If this list is for management or a person who isn't familiar with Oracle 11*g*, you might want to give your output a more professional appearance. To do this, you can use a column alias, as you did previously. The query in Figure 2-20 substitutes Customer Name as the column heading in the results.

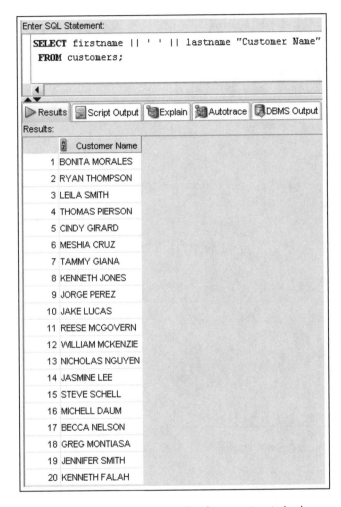

FIGURE 2-20   Using a column alias for concatenated values

TIP

If you get an error message, make sure the blank space is in single quotation marks and the column alias is in *double* quotation marks.

You can also use literals to include any characters needed to produce output in a certain format. For example, you need a customer listing in the following format: Customer# : Lastname, Firstname. Notice that the semicolon and comma are provided by using literals in the query shown in Figure 2-21.

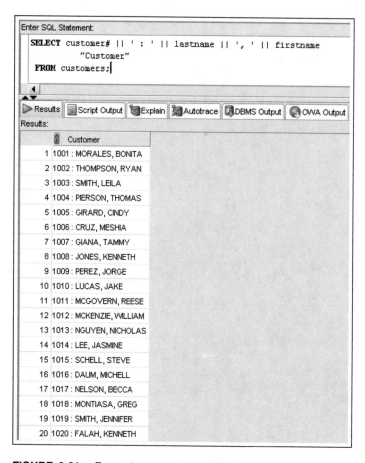

FIGURE 2-21 Formatting concatenated strings with literals

Chapter Summary

- A basic query in Oracle 11*g* SQL includes the SELECT and FROM clauses, the only mandatory clauses in a SELECT statement.
- To view all columns in a table, use an asterisk (*) or list all the column names in the SELECT clause.
- To display a specific column or set of columns, list the column names in the SELECT clause (in the order in which you want them to appear).
- When listing column names in the SELECT clause, a comma must separate column names.
- A column alias can be used to clarify the contents of a particular column. If the alias contains spaces or special symbols, or if you want to display the column with any lowercase letters, you must enclose the column alias in quotation marks (" ").
- To specify which table contains the columns you want, you must list the table name after the keyword FROM.
- Basic arithmetic operations can be performed in the SELECT clause.
- NULL values indicate an absence of a value and might have an undesirable effect on arithmetic operations.
- To remove duplicate listings, include the DISTINCT or UNIQUE keyword.
- A string literal is a set of characters enclosed in single quotation marks (' '); it's interpreted as is, not treated as a keyword or command.
- Use vertical bars (| |) to combine, or concatenate, fields, literals, and other data.

Chapter 2 Syntax Summary

The following table summarizes the syntax you have learned in this chapter. You can use the table as a study guide and reference.

Syntax Guide

Element	Description	Example
SELECT clause	Identify the columns for retrieval in a SELECT command	`SELECT title`
FROM clause	Identify the table containing selected columns	`FROM books`
SELECT statement	View columns in a table	`SELECT title` `FROM books;`
,	Separate column names in a list when retrieving multiple columns from a table	`SELECT title, pubdate` `FROM books;`
*	Return all data in a table when used in a SELECT clause	`SELECT *` `FROM books;`

Element	Description	Example

AS	Indicate a column alias to change a column heading in output	`SELECT title AS`
		`  titles, pubdate`
		`FROM books;`

	Create a column alias to change a column heading in output *without* using AS	`SELECT title titles,`
		`  pubdate`
		`FROM books;`

" "	Preserve spaces, symbols, or letter case in an output column alias	`SELECT title AS "Book`
		`  Name"`
		`FROM books;`

* multiplication
/ division
+ addition
– subtraction
	Solve arithmetic operations (Oracle 11g first solves * and /, then solves + and –, unless parentheses are used)	`SELECT title,`
		`  retail-cost profit`
		`FROM books;`

DISTINCT	Eliminate duplicates in a list	`SELECT DISTINCT state` `FROM customers;`
UNIQUE	Eliminate duplicates in a list	`SELECT UNIQUE state` `FROM customers;`
|| (concatenation)	Combine display of content from multiple columns into a single column	`SELECT city || state` `FROM customers;`
' ' (string literal)	Indicate the exact set of characters, including spaces, to be displayed	`SELECT city ||' ' || state` `FROM customers;`
DESCRIBE	Display the structure of a table	`DESCRIBE books`
USER_TABLES	List the name of all tables in the current account	`SELECT table_name` `FROM user_tables;`

Review Questions

1. What is a data dictionary?
2. What are the two required clauses for a SELECT statement?
3. What is the purpose of the SELECT statement?
4. What does an asterisk (*) in the SELECT clause of a SELECT statement represent?
5. What is the purpose of a column alias?

6. How do you indicate that a column alias should be used?

7. When is it appropriate to use a column alias?

8. What are the guidelines to keep in mind when using a column alias?

9. How can you concatenate columns in a query?

10. What is a NULL value?

Multiple Choice

To determine the exact name of the fields used in tables for these questions, refer to the tables in the JustLee Books database, or use the **DESCRIBE** *tablename* command to view the table's structure.

1. Which of the following SELECT statements displays a list of customer names from the CUSTOMERS table?
   a. SELECT customer names FROM customers;
   b. SELECT "Names" FROM customers;
   c. SELECT firstname, lastname FROM customers;
   d. SELECT firstname, lastname, FROM customers;
   e. SELECT firstname, lastname, "Customer Names" FROM customers;

2. Which clause is required in a SELECT statement?
   a. WHERE
   b. ORDER BY
   c. GROUP BY
   d. FROM
   e. all of the above

3. Which of the following is *not* a valid SELECT statement?
   a. SELECT lastname, firstname FROM customers;
   b. SELECT * FROM orders;
   c. Select FirstName NAME from CUSTOMERS;
   d. SELECT lastname Last Name FROM customers;

4. Which of the following symbols represents concatenation?
   a. *
   b. ||
   c. []
   d. ' '

5. Which of the following SELECT statements returns all fields in the ORDERS table?
   a. SELECT customer#, order#, orderdate, shipped, address FROM orders;
   b. SELECT * FROM orders;
   c. SELECT ? FROM orders;
   d. SELECT ALL FROM orders;

6. Which of the following symbols is used for a column alias containing spaces?
   a. ' '
   b. | |
   c. " "
   d. / /

7. Which of the following is a valid SELECT statement?
   a. SELECT TITLES * TITLE! FROM BOOKS;
   b. SELECT "customer#" FROM books;
   c. SELECT title AS "Book Title" from books;
   d. all of the above

8. Which of the following symbols is used in a SELECT clause to display all columns from a table?
   a. /
   b. &
   c. *
   d. "

9. Which of the following is *not* a valid SELECT statement?
   a. SELECT cost-retail FROM books;
   b. SELECT retail+cost FROM books;
   c. SELECT retail * retail * retail FROM books;
   d. SELECT retail^3 from books;

10. When must a comma be used in the SELECT clause of a query?
    a. when a field name is followed by a column alias
    b. to separate the SELECT clause and the FROM clause when only one field is selected
    c. It's never used in the SELECT clause.
    d. when listing more than one field name and the fields aren't concatenated
    e. when an arithmetic expression is included in the SELECT clause

11. Which of the following commands displays a listing of the category for each book in the BOOKS table?
    a. SELECT title books, category FROM books;
    b. SELECT title, books, and category FROM books;
    c. SELECT title, cat FROM books;
    d. SELECT books, | | category "Categories" FROM books;

12. Which clause is *not* required in a SELECT statement?
    a. SELECT
    b. FROM
    c. WHERE
    d. All of the above clauses are required.

13. Which of the following lines of the SELECT statement contains an error?

```
1    SELECT title, isbn,
2    Pubdate "Date of Publication"
3    FROM books;
```
   a. line 1
   b. line 2
   c. line 3
   d. There are no errors.

14. Which of the following lines of the SELECT statement contains an error?

```
1    SELECT ISBN,
2    retail-cost
3    FROM books;
```
   a. line 1
   b. line 2
   c. line 3
   d. There are no errors.

15. Which of the following lines of the SELECT statement contains an error?

```
1    SELECT title, cost,
2    cost*2
3    'With 200% MarkUp'
4    FROM books;
```
   a. line 1
   b. line 2
   c. line 3
   d. line 4
   e. There are no errors.

16. Which of the following lines of the SELECT statement contains an error?

```
1    SELECT name, contact,
2    "Person to Call", phone
3    FROM publisher;
```
   a. line 1
   b. line 2
   c. line 3
   d. There are no errors.

17. Which of the following lines of the SELECT statement contains an error?

```
1   SELECT ISBN, || ' is the ISBN for the book named ' ||
2   title
3   FROM books;
```

    a.   line 1

    b.   line 2

    c.   line 3

    d.   There are no errors.

18. Which of the following lines of the SELECT statement contains an error?

```
1   SELECT title, category
2   FORM books;
```

    a.   line 1

    b.   line 2

    c.   There are no errors.

19. Which of the following lines of the SELECT statement contains an error?

```
1   SELECT name, contact
2   "Person to Call", phone
3   FROM publisher;
```

    a.   line 1

    b.   line 2

    c.   line 3

    d.   There are no errors.

20. Which of the following lines of the SELECT statement contains an error?

```
1   SELECT *
2   FROM publishers;
```

    a.   line 1

    b.   line 2

    c.   There are no errors.

Hands-On Assignments

To determine the exact name of fields used in the tables for these exercises, refer to the tables in the JustLee Books database, or use the **DESCRIBE *tablename*** command to view the table's structure.

1. Display a list of all data contained in the BOOKS table.

2. List the title only of all books available in inventory, using the BOOKS table.

3. List the title and publication date for each book in the BOOKS table. Use the column heading "Publication Date" for the Pubdate field.

4. List the customer number for each customer in the CUSTOMERS table, along with the city and state in which the customer lives.

5. Create a list containing the publisher's name, the person usually contacted, and the publisher's telephone number. Rename the contact column "Contact Person" in the displayed results. (*Hint*: Use the PUBLISHER table.)

6. Determine which categories are represented in the current book inventory. List each category only once.

7. List the customer number from the ORDERS table for each customer who has placed an order with the bookstore. List each customer number only once.

8. Create a list of each book title stored in the BOOKS table and the category in which each book belongs. Reverse the sequence of the columns so that the category of each book is listed first.

9. Create a list of authors that displays the last name followed by the first name for each author. The last names and first names should be separated by a comma and a blank space.

10. List all information for each order item. Include an item total, which can be calculated by multiplying the Quantity and Paideach columns. Use a column alias for the calculated value to show the heading "Item Total" in the output.

Advanced Challenge

The management of JustLee Books has submitted two requests. The first is for a mailing list of all customers stored in the CUSTOMERS table. The second is for a list of the percentage of profit generated by each book in the BOOKS table. The requests are as follows:

1. Create a mailing list from the CUSTOMERS table. The mailing list should display the name, address, city, state, and zip code for each customer. Each customer's name should be listed in order of last name followed by first name, separated with a comma, and have the column header "Name." The city and state should be listed as one column of output, with the values separated by a comma and the column header "Location."

2. To determine the percentage of profit for a particular item, subtract the item's cost from the retail price to calculate the dollar amount of profit, and then divide the profit by the item's cost. The solution is then multiplied by 100 to determine the profit percentage for each book. Use a SELECT statement to display each book's title and percentage of profit. For the column displaying the percentage markup, use "Profit %" as the column heading.

Required: Determine the SQL statements needed to perform the two required tasks. Each statement should be tested to ensure its validity. Submit documentation of the commands and their results, using the format specified by your instructor.

Case Study: *City Jail*

The case study resumes in Chapter 3.

TABLE CREATION AND MANAGEMENT

LEARNING OBJECTIVES

After completing this chapter, you should be able to do the following:

- Identify the table name and structure
- Create a new table with the CREATE TABLE command
- Use a subquery to create a new table
- Add a column to an existing table
- Modify the definition of a column in an existing table
- Delete a column from an existing table
- Mark a column as unused and then delete it later
- Rename a table
- Truncate a table
- Drop a table

INTRODUCTION

Since joining JustLee Books as a database specialist, you're able to query the existing database; however, now you need to address some requested database modifications. First, the management of JustLee Books is implementing a sales commission program for all account managers who have been employed by the company for more than six months. Account managers will receive a commission for each order from customers in the geographical region they supervise. This program requires adding a new table for account manager data. Second, data extracts are needed to enable the Marketing Department to perform customer analyses. Third, a number of modifications are needed to address current data needs. For example, a new column is needed in the PUBLISHER table to store a service rating indicator.

Chapters 3 through 5 explain the SQL commands used to create and modify tables, assign constraints on columns, add data to tables, and edit existing data. This chapter addresses methods for creating tables and modifying existing tables. Commands used to create or modify database tables are called **data definition language (DDL)** commands. These commands are basically SQL commands used specifically to create or modify database objects. A **database object** is a defined, self-contained structure in Oracle 11g. In this chapter, you create database tables, which are considered database objects. Later, in Chapter 6, you learn how to create and modify other types of database objects. Table 3-1 provides an overview of this chapter's topics.

TABLE 3-1 Overview of Chapter Contents

Commands and Clauses	Description
Creating Tables	
CREATE TABLE	Creates a new table in the database. The user names the columns and identifies the type of data to be stored. To view a table, use the SQL*PLUS command DESCRIBE.
CREATE TABLE . . . AS	Creates a table from existing database tables, using the AS clause and subqueries.
Modifying Tables	
ALTER TABLE . . . ADD	Adds a column to a table.
ALTER TABLE . . . MODIFY	Changes a column size, datatype, or default value.
ALTER TABLE . . . DROP COLUMN	Deletes one column from a table.
ALTER TABLE . . . SET UNUSED or SET UNUSED COLUMN	Marks a column for deletion at a later time.

Commands and Clauses	Description
Modifying Tables	
DROP UNUSED COLUMNS	Completes the deletion of a column previously marked with SET UNUSED.
RENAME . . . TO	Changes a table name.
TRUNCATE TABLE	Deletes all table rows, but the table name and column structure remain.
Deleting Tables	
DROP TABLE	Removes an entire table from the Oracle 11g database.
PURGE TABLE	Permanently deletes a table in the recycle bin.
Recovering Tables	
FLASHBACK TABLE . . . TO BEFORE DROP	Recovers a dropped table if PURGE option not used when table dropped.

DATABASE PREPARATION

This chapter assumes you have created the initial JustLee Books database as instructed in Chapter 2.

TABLE DESIGN

Before issuing an SQL command to create a table, you must complete the entity design, as discussed in Chapter 1. For each entity, you must choose the table's name and determine its structure—that is, what columns to include in the table. In addition, you need to determine the width of any character or numeric columns.

Take a look at these requirements in more depth. Oracle 11g has the following rules for naming both tables and columns:

- The names of tables and columns can be up to 30 characters and must begin with a letter. These limitations apply only to a table or column name, not to data in a column.
- The names of tables and columns can't contain any blank spaces.
- Numbers, the underscore symbol (_), and the number sign (#) are allowed in table and column names.
- Each table owned by a user should have a unique table name, and the column names in each table should be unique.
- Oracle 11g "reserved words," such as SELECT, DISTINCT, CHAR, and NUMBER, can't be used for table or column names.

Because the new table is to contain data about account managers, the table name is ACCTMANAGER. The ACCTMANAGER table needs to contain each account manager's name, employment date, and assigned region as well as an ID code to act as the table's primary key and uniquely identify each account manager. Although having two account managers with the same name is unlikely, the ID code can reduce data entry errors because users need to type only a short code instead of a manager's entire name in SQL commands.

> **NOTE**
>
> The ID column is included in the ACCTMANAGER table to serve as the table's primary key. However, the PRIMARY KEY constraint isn't defined in this chapter. Chapter 4 expands on table creation capabilities by introducing the constraints that can be defined for a table.

Now that the table's contents have been determined, the columns can be designed. When you create a table in Oracle 11g, you must define each column. Before you can create the columns, however, you must do the following:

- Choose a name for each column.
- Determine the type of data each column stores.
- Determine (in some cases) the column's maximum width.

You need to identify the type of data to be stored in each column so that you can assign an appropriate datatype for each column. Table 3-2 lists the datatypes used in this chapter.

TABLE 3-2   Oracle 11g Datatypes

Datatype	Description
VARCHAR2(n)	*Var*iable-length *char*acter data, and the n represents the column's maximum length. The maximum size is 4000 characters. There's no default size for this datatype; a minimum value must be specified. Example: VARCHAR2(9) can contain up to nine letters, numbers, or symbols.
CHAR(n)	Fixed-length character column, and the n represents the column's length. The default size is 1, and the maximum size is 2000. Example: CHAR(9) can contain nine letters, numbers, or symbols. However, if fewer than nine are entered, spaces are added to the right to force the data to reach a length of nine.
NUMBER(p, s)	Numeric column. The p indicates **precision**, the total number of digits to the left and right of the decimal position, to a maximum of 38 digits; the s, or **scale**, indicates the number of positions to the right of the decimal. Example: NUMBER(7, 2) can store a numeric value up to 99999.99. If precision or scale isn't specified, the column defaults to a precision of 38 digits.

TABLE 3-2   Oracle 11*g* Datatypes (continued)

Datatype	Description
DATE	Stores date and time between January 1, 4712 BC and December 31, 9999 AD. Seven bytes are allocated to the column to store the century, year, month, day, hour, minute, and second of a date. Oracle 11*g* displays the date in the format DD-MON-YY. Other aspects of a date can be displayed by using the TO_CHAR format. Oracle 11*g* defines the field width as seven bytes.

NOTE

Oracle currently has two variable-length character datatypes: VARCHAR and VARCHAR2. However, Oracle recommends using VARCHAR2 rather than VARCHAR, so VARCHAR2 is used throughout this textbook. Oracle's SQL reference states "The VARCHAR datatype is currently synonymous with the VARCHAR2 datatype. Oracle recommends that you use VARCHAR2 rather than VARCHAR. In the future, VARCHAR might be defined as a separate datatype used for variable-length character strings compared with different comparison semantics."

TIP

Many additional datatypes are available, including BINARY_FLOAT, BINARY_DOUBLE, INTEGER, LONG, CLOB, RAW(*n*), LONG RAW, BLOB, BFILE, TIMESTAMP, and INTERVAL. You can explore details on all datatypes in the SQL Language Reference available at the Oracle Technology Network (OTN) Web site.

A **datatype** identifies the type of data Oracle 11*g* is expected to store in a column. Identifying the type of data helps you verify that you input the correct data and allows you to manipulate data in ways specific to that datatype. For example, you need to calculate the difference in number of days between two date values, such as the Orderdate and Shipdate columns from the ORDERS table. To accomplish this task, the system needs to be able to associate the date values to a calendar. If the columns have a DATE datatype, Oracle 11*g* associates the values to calendar days automatically.

Now return to creating the new ACCTMANAGER table. You need to include eight columns. The first column, which uniquely identifies each account manager, is named AmID. The ID code assigned to each account manager consists of the first letter of a manager's last name, followed by a three-digit number. Because the column's data consists of both letters and numbers, it must be defined to store the datatype CHAR and have a width of four. The CHAR datatype should be used only when the length of values stored in the column is consistent. In this case, every AmID has a length of four. Oracle 11*g* pads a CHAR column to the specified length if an entry doesn't fill the column's entire width. This padding can result in wasted storage space if most values Oracle 11*g* stores must include blank spaces to force a column's contents to a specified length. If character data stored in a column won't be a consistent length, the VARCHAR2 datatype should be used. The VARCHAR2 datatype uses only the physical space required to hold the entered value, and storage space is conserved because no padding takes place.

Why not just always use the VARCHAR2 datatype and avoid using CHAR? There's a slight processing overhead in managing a variable-length field. Therefore, using CHAR when it applies has the advantage of reducing processing overhead. Many people in the industry use both datatypes; however, it's still a topic of debate.

62

TIP

JustLee Books decided to use the AmID code consisting of characters and digits because it's already being used in the payroll system, and the company wants to integrate all the systems. When you're trying to determine an appropriate primary key value, considering codes already in use is important. If JustLee Books didn't have any existing values to identify account managers, a system-generated value would most likely be used. Typically, these columns are numeric to allow using sequences to populate the column. Sequences are covered in Chapter 6.

The second and third columns of the ACCTMANAGER table are used to store each account manager's first and last name. If you store first names and last names in separate columns, you can perform simple searches on each part of a manager's name. Because each account manager's name consists of characters, these columns are assigned the datatype VARCHAR2. Name values vary greatly in length. Therefore, the variable-length datatype is more appropriate than CHAR.

Generally, you define the width so that it can hold the largest value that could ever be entered in that column. However, an account manager might be hired whose first or last name is extremely long. Increasing a column's width at a later time is easy. Therefore, the assumption is that a column width of 12 characters is enough to store the names of current account managers. The columns are named Amfirst and Amlast.

NOTE

The actual names of account managers are provided in Chapter 5.

The fourth column is used to store each account manager's employment date. Because the datatype is DATE, you don't have to worry about the column length—it's predetermined by Oracle 11g. All that remains is choosing a name for the column. In this case, Amedate, for "account manager employment date," is appropriate.

The next three columns address account manager earnings: salary (Amsal), commission (Amcomm), and total earnings (Amearn). All these columns contain monetary values, so a NUMBER datatype with two decimal places is suitable. The Amearn column is calculated by adding the Amsal and Amcomm column values. It's a **virtual column** that generates a value automatically based on other column values in the table. The AS keyword is followed by the expression to define the virtual column in the table creation statement. The virtual column's value is derived when it's queried; it's not physically stored in the row data.

The eighth and final column for the ACCTMANAGER table is the region to which the account manager is assigned. The United States is divided into eight geographical regions, each identified by a two-letter code (such as NE for the Northeast region). The column name is Region, and it's defined as a CHAR datatype with a width of two characters. The VARCHAR2 datatype could also be used; however, because the values stored in the Region column always consist of two characters, the CHAR datatype is appropriate.

TABLE CREATION

Figure 3-1 shows the basic syntax of the SQL command to create a table in Oracle 11g.

```
CREATE TABLE [ schema. ] tablename
( columnname datatype [ DEFAULT value ]
[ , columnname datatype [ DEFAULT value ]] );
```

FIGURE 3-1   CREATE TABLE syntax

The keywords **CREATE TABLE** instruct Oracle 11g to create a table. Optionally, a **schema** can be included to indicate who "owns" the table. For example, if the person creating the table is also the person who owns the table, the schema can be omitted, and the current username is assumed by default. On the other hand, if you're creating the ACCTMANAGER table for someone with the username DRAKE, the schema and table name are entered as DRAKE.ACCTMANAGER to inform Oracle 11g that the ACCTMANAGER table belongs to DRAKE's schema, not yours. A database object's owner has the right to perform certain operations on that object. With a table, the only way another database user can query or manipulate data in the table is to be given permission from the table's owner or the database administrator. The table name, of course, is the name used to identify the table being created.

> **NOTE**
>
> Many database objects, such as tables, supporting an application need to be shared by many users. In this case, users must be granted permission to access objects in the schema in which they reside. Also, to create a table in a schema other than your own, you must be granted permission to use the CREATE TABLE command for that schema. Chapter 7 explains the different privileges in Oracle 11g.

Defining Columns

After entering the table name, you define the columns to be included in the table. A table can contain a maximum of 1000 columns. The CREATE TABLE syntax requires *enclosing the column list in parentheses*. If the table contains more than one column, the name, datatype, and width (if applicable) are listed for the first column before the next column is defined. Commas separate columns in the list.

The CREATE TABLE command also allows assigning a default value to a column. The default value is the one Oracle 11g stores automatically if the user makes no entry in the column. For example, imagine users entering order data. Rather than have the user spend time entering the current date as the order date, have the system do it automatically by setting the DEFAULT value to the system's current date.

Using the syntax in Figure 3-1, the SQL command in Figure 3-2 shows creating the ACCTMANAGER table.

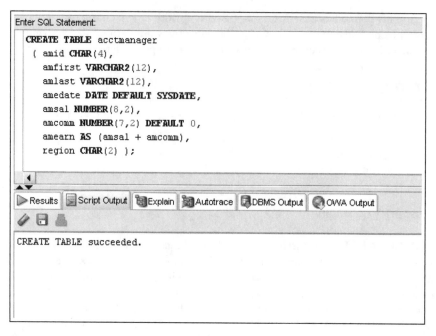

```
Enter SQL Statement:
    CREATE TABLE acctmanager
    ( amid CHAR(4),
      amfirst VARCHAR2(12),
      amlast VARCHAR2(12),
      amedate DATE DEFAULT SYSDATE,
      amsal NUMBER(8,2),
      amcomm NUMBER(7,2) DEFAULT 0,
      amearn AS (amsal + amcomm),
      region CHAR(2) );
```

▶ Results | 📄 Script Output | 🗒 Explain | 🗒 Autotrace | 🗇 DBMS Output | 🌐 OWA Output

```
CREATE TABLE succeeded.
```

FIGURE 3-2   The creation of the ACCTMANAGER table

N O T E

In the SQL Developer tool, you can use the Execute Statement or Run Script button to execute the DDL statements in this chapter. Using the Execute Statement button generates a command completion message at the bottom left of the SQL Developer window. If an error occurs, a popup window is displayed. Using the Run Script button displays all messages in the Script Output tab below the statement area. The figures in this textbook show SELECT statements using the Execute Statement button and all other types of statements using the Run Script button.

In the command shown in Figure 3-2, the table name is ACCTMANAGER. It's entered in lowercase letters to distinguish it from the CREATE TABLE keywords. Oracle 11*g* SQL commands aren't case sensitive; however, for clarity, in this textbook, all keywords are in uppercase and all user-supplied values are in lowercase. Because the user who creates the table is also the table owner, the schema has been omitted.

The eight columns to be created are listed next in parentheses. Each column is defined on a separate line to improve readability, but doing so isn't an Oracle 11*g* requirement. Notice the definition for the Amedate column; it has been assigned a default value of SYSDATE. This value instructs Oracle 11*g* to insert the current date (according to the Oracle 11*g* server) if the user enters a new account manager without including the person's date of employment. Of course, this default value is beneficial only if the account manager's record is created on the same date the person is hired; otherwise, the date must be entered, and the DEFAULT setting is ignored.

NOTE

A user can't have two tables with the same name. If you attempt to create a table with the same name as another table in your schema, Oracle 11*g* returns an ORA-00955 error message. Similarly, if you create a table and then want to create another table with the same name, you must first delete the existing table with the command DROP TABLE *tablename*;. (The DROP TABLE command is shown later in this chapter.)

Notice that after the command has been executed, Oracle 11*g* returns a message indicating the table was created successfully. The message doesn't contain any reference to rows being created. At this point, the table doesn't contain any data; only the table structure has been created. (In other words, the table has been defined in terms of a table name and the type of data it will contain.) The data, or rows, must be entered in the table as a separate step with the INSERT command. You enter all the data for account managers in Chapter 5.

TIP

If you get an error message (such as an ORA-00922 error message) when executing the CREATE TABLE command, it could be a result of 1) not including a closing parenthesis to end the column list, or 2) not separating each column definition with a comma. If an error message is displayed stating that you don't have sufficient privileges, ask the database administrator to grant you the CREATE TABLE privilege.

Viewing a List of Tables: USER_TABLES

Recall that you can query the data dictionary to verify all existing tables in your schema. The USER_TABLES data dictionary object maintains information on all your tables. Figure 3-3 shows querying the data dictionary to generate a list of table names. Imagine how important this query can be when you start a new job and need to explore an existing database!

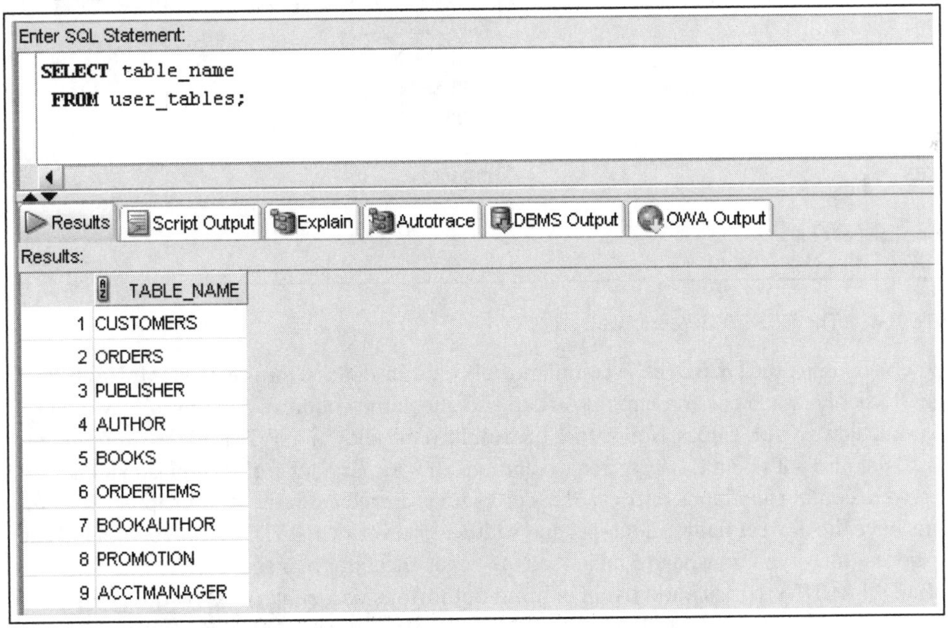

FIGURE 3-3 Listing names of all tables

Viewing Table Structures: DESCRIBE

To determine whether the table structure was created correctly, you can use the SQL*Plus command DESCRIBE *tablename* to display the table's structure, as shown in Figure 3-4. Because the DESCRIBE command is an SQL*Plus command rather than an SQL command, it can be abbreviated as DESC, and an ending semicolon isn't required.

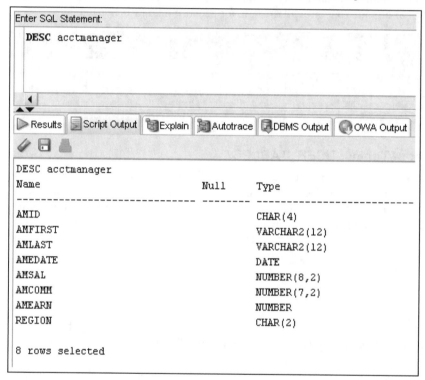

FIGURE 3-4 The DESCRIBE command

When you issue the **DESCRIBE** command, all columns defined for the ACCTMANAGER table are listed. For each column name, you can also check the column's datatype and whether the column allows NULL values. Notice that the results don't show a "NOT NULL" requirement for the AmID column—it's blank. Because this column is the primary key for the table, it shouldn't be allowed to contain any blank entries. (This problem is corrected in the next chapter.) If all the columns have the correct name, datatype, and width—*and* your CREATE TABLE command executed successfully—you now have a table ready to accept account managers' data.

The DEFAULT settings and virtual column definitions can be verified by querying the data dictionary object USER_TAB_COLUMNS. Use the query shown in Figure 3-5 to verify the DEFAULT settings on the Amedate and Amcomm columns as well as the calculation assigned to the Amearn virtual column. Notice that the Amearn column displays the datatype NUMBER. The DBMS assigns the datatype to accommodate the data derived from the expression in the virtual column definition.

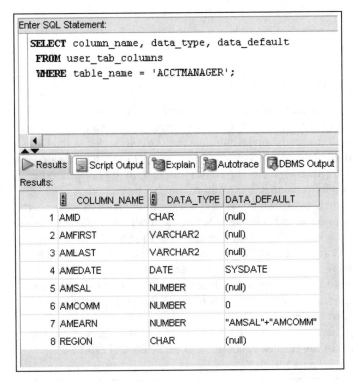

FIGURE 3-5 Verify DEFAULT and virtual column settings

TABLE CREATION WITH SUBQUERIES

In the previous section, a table was created "from scratch." However, sometimes you need to create a table based on data in existing tables. The JustLee Books Marketing Department needs to analyze customer data, and management doesn't want the analysis queries to slow down the production server on which orders are entered. The name and street address columns aren't needed for the analysis, so only a portion of the CUSTOMERS table is required. The Marketing Department requested that the table be named CUST_MKT.

A nested query, or subquery, is required to create this new table based on the existing CUSTOMERS table. A **subquery** is a SELECT statement used in another SQL command. Any type of action you can perform with a SELECT statement (such as filtering rows, filtering columns, and calculating aggregate amounts) can be performed when creating a table with a subquery. At this point, this textbook has covered only basic queries, so the example is limited to the SELECT statement features that are already familiar to you. After you have become familiar with all the features of the SELECT statement, you'll understand the expanded explanation of this topic in Chapter 12.

CREATE TABLE ... AS COMMAND

To create a table containing data from existing tables, you can use the CREATE TABLE command with an AS clause containing a subquery. The syntax, shown in Figure 3-6, uses the CREATE TABLE keywords to instruct Oracle 11*g* to create a table. The new table's name is then provided.

```
CREATE TABLE tablename [(columnname, …)]
AS (subquery);
```

FIGURE 3-6 CREATE TABLE ... AS command syntax

If you need to give columns in the new table different names from those in the existing table, list the new column names in parentheses after the table name, or you can use column aliases. However, if you don't want to change any column names, the column list in the CREATE TABLE clause can be omitted. If you do provide a column list, it must contain a name for *every* column to be returned by the query—including names that remain the same. In other words, if the subquery is to return five columns, five columns must be listed in the CREATE TABLE clause, or Oracle 11*g* returns an error message and the statement fails. In addition, the column list must be in the *same order* as the columns listed in the subquery's SELECT clause, as Oracle 11*g* uses positional order to match the new column list to the SELECT column list.

The AS keyword instructs Oracle 11*g* that the columns in the new table are based on the columns the subquery returns. The AS keyword must precede the subquery. The columns in the new table are created based on the same datatype and width as columns in the existing table. To distinguish the subquery from the rest of the CREATE TABLE command, the subquery *must be enclosed in parentheses*.

Figure 3-7 shows the creation of the CUST_MKT table based on a subquery. To verify the table structure and contents in the results, execute the DESCRIBE and SELECT commands, as shown in Figures 3-8 and 3-9.

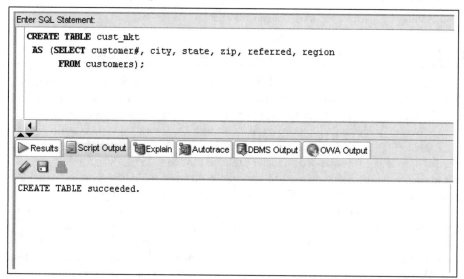

FIGURE 3-7 Creating a table based on a subquery

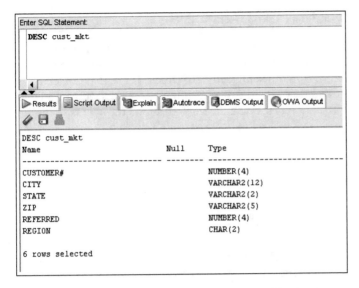

FIGURE 3-8 Using DESCRIBE to verify the new table structure

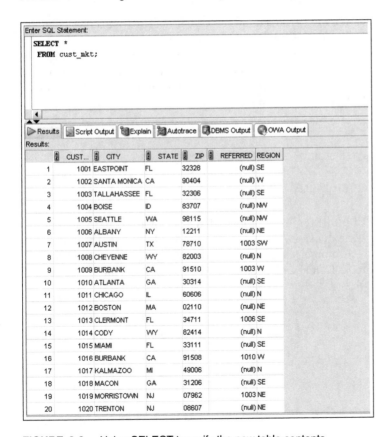

FIGURE 3-9 Using SELECT to verify the new table contents

MODIFYING EXISTING TABLES

At times, you need to make structural changes to a table. For example, you might need to add a column, delete a column, or simply change a column's size. Each of these changes is made with the **ALTER TABLE** command. A useful feature of Oracle 11g is that you can modify a table without having to shut down the database. Even if users are accessing a table, it can still be modified without disruption of service. Figure 3-10 shows the syntax of the ALTER TABLE command.

```
ALTER TABLE tablename
ADD|MODIFY|DROP COLUMN| columnname [definition];
```

FIGURE 3-10   Syntax of the ALTER TABLE command

Whether you should use an ADD, MODIFY, or DROP COLUMN clause depends on the type of change being made. First, take a look at the ADD clause.

ALTER TABLE ... ADD Command

Using an **ADD** clause with the ALTER TABLE command allows a user to add a new column to a table. The same rules for creating a column in a new table apply to adding a column to an existing table. The new column must be defined by a column name and datatype (and width, if applicable). A default value can also be assigned. The difference is that the new column is added at the end of the existing table—it will be the last column. Figure 3-11 shows the syntax of the ALTER TABLE command with the ADD clause.

```
ALTER TABLE tablename
ADD (columnname datatype, [DEFAULT] …);
```

FIGURE 3-11   Syntax of the ALTER TABLE ... ADD command

Suppose that after the PUBLISHER table was created, management requests adding a column for a telephone number extension. The column name should be Ext. The column can consist of a maximum of four numeric digits, so the column is defined as a NUMBER datatype with a precision of four. To make this change to the PUBLISHER table, issue the command shown in Figure 3-12.

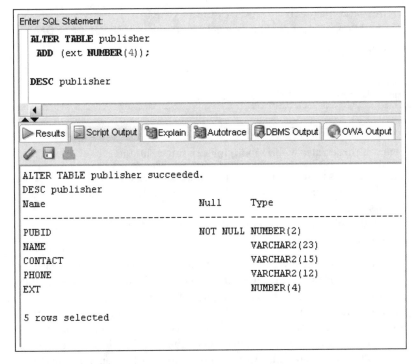

```
Enter SQL Statement:

 ALTER TABLE publisher
  ADD (ext NUMBER(4));

 DESC publisher


Results  Script Output  Explain  Autotrace  DBMS Output  OWA Output

ALTER TABLE publisher succeeded.
DESC publisher
Name                            Null     Type
------------------------------- -------- ---------------------------
PUBID                           NOT NULL NUMBER(2)
NAME                                     VARCHAR2(23)
CONTACT                                  VARCHAR2(15)
PHONE                                    VARCHAR2(12)
EXT                                      NUMBER(4)

5 rows selected
```

FIGURE 3-12   The ALTER TABLE ... ADD command

Oracle 11g returns a message indicating that the table was altered successfully. To double-check that the column was added with the correct datatype, you can also issue the DESCRIBE command shown in Figure 3-12.

TIP

When executing multiple commands in SQL Developer, use the Run Script button rather than the Execute Statement button. The Execute Statement button processes only one statement per execution.

NOTE

When you add a column to a table containing rows of data, the new column is empty for all existing rows. Issue a SELECT command on the PUBLISHER table to confirm that the new Ext column is empty.

If you need to add more than one column to the ACCTMANAGER table, list the additional columns in a list, and separate each column and its datatype with a comma, using the same format as the CREATE TABLE command.

ALTER TABLE ... MODIFY Command

To change an existing column's definition, you can use a **MODIFY** clause with the ALTER TABLE command. The changes that can be made to a column include the following:

- Changing the column size (increase or decrease)

- Changing the datatype (such as VARCHAR2 to CHAR)
- Changing or adding the default value of a column (such as DEFAULT SYSDATE)

The syntax of the ALTER TABLE ... MODIFY command is shown in Figure 3-13.

```
ALTER TABLE tablename
MODIFY (columnname datatype [DEFAULT] ,...);
```

FIGURE 3-13   Syntax of the ALTER TABLE ... MODIFY command

You should be aware of three rules when modifying existing columns:

- A column must be as wide as the data fields it already contains.
- If a NUMBER column already contains data, you can't decrease the column's precision or scale.
- Changing the default value of a column doesn't change the values of data already in the table.

Now take a closer look at these rules. The first rule applies when you want to decrease the size of a column already containing data. You can decrease a column's size only to a size that's no less than the largest width of existing data. For example, a column has been defined as a VARCHAR2 datatype with a width of 15 characters. However, the largest entry in that particular column contains only 12 characters. Therefore, you can decrease the column size only to a width of 12. If you try to decrease the size to a width less than 12, Oracle 11g returns an error message, and the SQL statement fails. As shown in Figure 3-14, when a user attempts to decrease the column width to a size that doesn't accommodate existing data, Oracle 11g returns an ORA-01441 error message, and the table isn't altered.

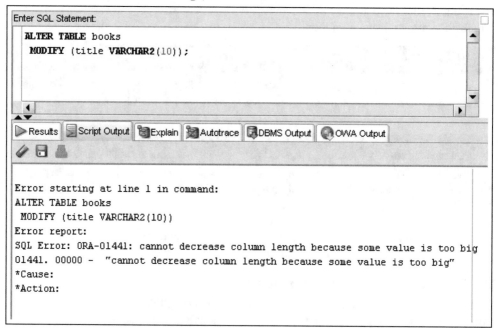

FIGURE 3-14   Modify column size error

Second, Oracle 11*g* doesn't allow you to decrease the precision or scale of a NUMBER column if the column contains any data. Regardless of whether the current values stored in a NUMBER column will be affected, Oracle 11*g* returns an ORA-01440 error message, and the statement fails unless the column is empty. As shown in Figure 3-15, if you attempt to change the size of the Retail column in the BOOKS table, an error message is displayed, and the table isn't altered.

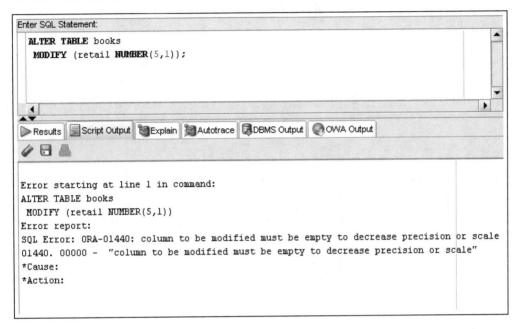

FIGURE 3-15   Modify numeric column error

The third rule applies when you modify existing columns and decide to change the default value assigned to a column. When a column's default value is changed, it changes only the value assigned to *future rows* inserted into the table. The default value assigned to existing rows remains the same. Therefore, if the default value in existing rows must be changed, these changes must be performed manually. (Chapter 5 explains how to change existing values in a row.)

JustLee Books has decided to assign a service rating code to every publisher that reflects the publisher's service promptness. The code should initially be set to N when a new publisher is added. To demonstrate the effect of adding a DEFAULT option to an existing column, the new column is added first and then modified to set the default value. Figure 3-16 shows the ALTER TABLE command to accomplish this task. Notice that the SELECT statement shows that the new Rating column is empty for existing rows.

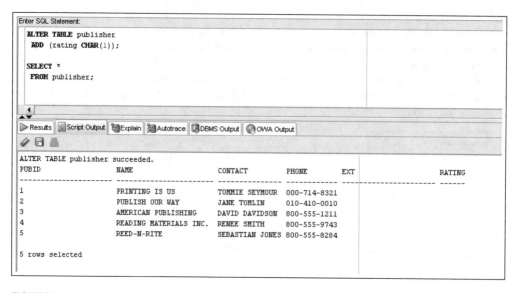

FIGURE 3-16   Adding the Rating column to the PUBLISHER table

Second, the DEFAULT option is added to the new column, as shown in Figure 3-17. Notice that the SELECT statement shows that the new Rating column is still empty for existing rows; the DEFAULT option takes effect only on new rows added. You need to query the data dictionary to verify the DEFAULT setting, as shown earlier in Figure 3-5.

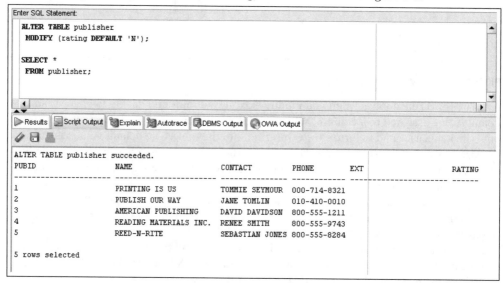

FIGURE 3-17   Adding the DEFAULT option to the new column

N O T E

Column headings might be truncated based on available line space, so your column headers might vary from those shown in figures. Formatting column widths in SQL*Plus output is covered in Chapter 14 (which is part of this book's online materials).

Now return your attention to the ACCTMANAGER table created earlier in this chapter. A common issue with tables is the need to widen columns to accommodate longer data values. After creating the ACCTMANAGER table, you find out that one of the account managers has a long name requiring more than the 12 spaces you assigned to the Amlast column. To accommodate the name, the Amlast column must be increased to a width of 18. What command should you use to do this?

Figure 3-18 shows the ALTER TABLE command to widen the Amlast column. Notice that the command's MODIFY clause doesn't state that the Amlast column should be increased by six characters. Instead, the MODIFY clause states the datatype and the new width (increased to the new total width needed for the column). To make certain the change is made, you might also use the DESCRIBE command to view the new table structure.

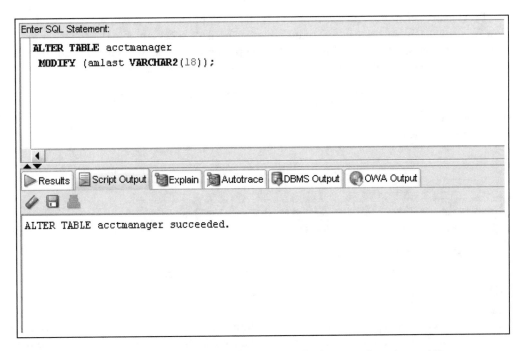

FIGURE 3-18   The ALTER TABLE ... MODIFY command to increase the column width

ALTER TABLE ... DROP COLUMN Command

To delete an existing column from a table, you can use the **DROP COLUMN** clause with the ALTER TABLE command. This clause deletes both the column and its contents, so it should be used with extreme caution. Figure 3-19 shows the syntax of the ALTER TABLE ... DROP COLUMN command.

```
ALTER TABLE tablename
DROP COLUMN columnname;
```

FIGURE 3-19   Syntax of the ALTER TABLE ... DROP COLUMN command

You should keep the following rules in mind when using the DROP COLUMN clause:

- Unlike using ALTER TABLE with the ADD or MODIFY clauses, a DROP COLUMN clause can reference only *one* column.
- If you drop a column from a table, the deletion is permanent. You can't "undo" the damage if you delete the wrong column accidentally. The only option is to add the column back to the table and then manually reenter all the data it contained previously.
- You can't delete the last remaining column in a table. If a table contains only one column and you try to delete it, the command fails, and Oracle 11g returns an error message.
- A primary key column can't be dropped from a table.

Previously, you added the Ext column to store each publisher's telephone extension. However, management has decided that the extension isn't necessary and doesn't want to waste the storage space that column would take up. Therefore, the Ext column needs to be deleted from the PUBLISHER table with the command shown in Figure 3-20.

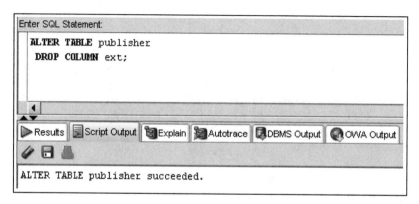

FIGURE 3-20 The ALTER TABLE ... DROP COLUMN command

After the command is processed, the Ext column of the PUBLISHER table is removed. To verify that the column no longer exists, use the DESCRIBE command to list the structure of the PUBLISHER table.

ALTER TABLE ... SET UNUSED/DROP UNUSED COLUMNS Command

When the Oracle 11g server executes database structural changes, such as dropping a column from a large table, processing of other current statements might be delayed. To avoid this problem, you can include a **SET UNUSED** clause in the ALTER TABLE command to mark the column for deletion at a later time. If a column is marked for deletion, it's unavailable and isn't displayed in the table structure. Because the column is unavailable, it doesn't appear in the results of any queries, and no other operation can be performed on the column except the ALTER TABLE ... DROP UNUSED command.

In other words, after a column is set as "unused," the column and all its contents are no longer available and can't be recovered in the future. The command simply postpones physically erasing data from storage until later—usually when the server is processing fewer queries, such as after business hours. A **DROP UNUSED COLUMNS** clause is used with the ALTER TABLE command to complete the deletion process for any column marked as unused. Figure 3-21 shows the syntax of the ALTER TABLE ... SET UNUSED command. As shown, the syntax for this command has two options for the SET UNUSED clause.

```
ALTER TABLE tablename
SET UNUSED (columnname);
     OR
ALTER TABLE tablename
SET UNUSED COLUMN columnname;
```

FIGURE 3-21 Syntax of the ALTER TABLE ... SET UNUSED command

Regardless of the syntax used, only one column per command can be marked for deletion. Figure 3-22 shows the syntax to drop a column previously identified as unused.

```
ALTER TABLE tablename
DROP UNUSED COLUMNS;
```

FIGURE 3-22 Syntax of the ALTER TABLE ... DROP UNUSED COLUMNS command

When the DROP UNUSED COLUMNS clause is used, any column previously set as "unused" is deleted, and any storage previously occupied by data in the column becomes available.

Suppose the JustLee Books Marketing Department has decided it doesn't need to see the customer state data in the CUST_MKT table. To delete this column from the CUST_MKT table, you could use the DROP COLUMN option of the ALTER TABLE command. However, if the table contains thousands of records, deleting a column slows down operations for other Oracle 11g users. In this case, you can mark the State column of the CUST_MKT table as unused with the command shown in Figure 3-23.

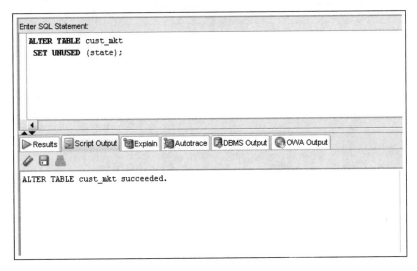

FIGURE 3-23 The ALTER TABLE ... SET UNUSED command

Table Creation and Management

To make certain the State column was marked for deletion correctly, you can use the DESCRIBE command to check that the column is no longer available, as shown in Figure 3-24.

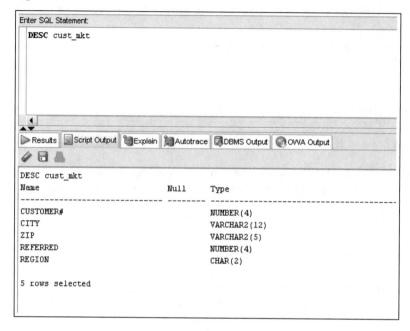

FIGURE 3-24 Verifying that the column is no longer available

You can also reference the data dictionary to determine whether any columns are marked as unused. The data dictionary object USER_UNUSED_COL_TABS contains information on which tables have unused columns and how many columns are set as unused. Figure 3-25 displays a query on this object.

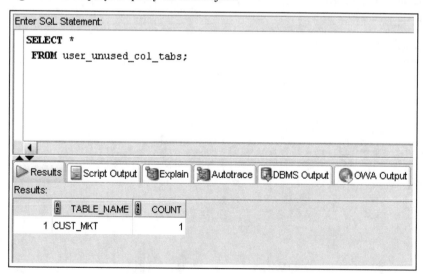

FIGURE 3-25 Listing tables with columns marked as unused

After the State column has been set as unused, the storage space taken up by data in the column can be reclaimed by using the command shown in Figure 3-26.

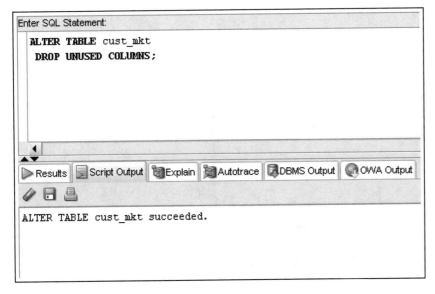

FIGURE 3-26   The ALTER TABLE ... DROP UNUSED COLUMNS command

Renaming a Table

Oracle 11g allows changing the name of any table you own by using the **RENAME ... TO** command. The syntax of this command is shown in Figure 3-27.

```
RENAME oldtablename TO newtablename;
```

FIGURE 3-27   Syntax of the RENAME ... TO command

Previously, the CUST_MKT table was created. However, the Marketing Department wants to maintain a series of snapshots for customer data and, therefore, wants the table name to reflect the month and year the table was created. Assuming the current CUST_MKT table was created in September 2009, the command to make the name change is shown in Figure 3-28.

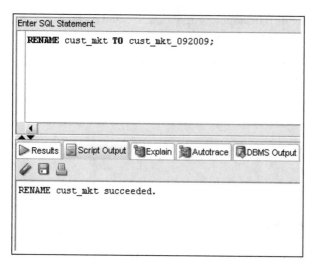

FIGURE 3-28   The RENAME ... TO command

After the RENAME ... TO command is executed, any queries directed to the original table named CUST_MKT result in an error message. The table can now be referenced only as the CUST_MKT_092009 table. The SELECT statements shown in Figure 3-29 and Figure 3-30 demonstrate this point.

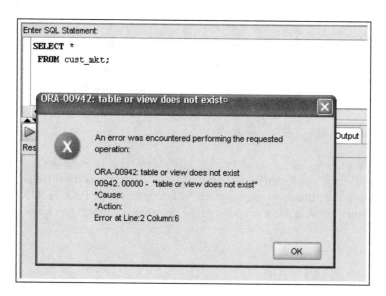

FIGURE 3-29   Verifying that the CUST_MKT table no longer exists

Enter SQL Statement:

```
SELECT *
 FROM cust_mkt_092009;
```

Results | Script Output | Explain | Autotrace | DBMS Output | OWA Output

Results:

	CUSTOMER#	CITY	ZIP	REFERRED	REGION
1	1001	EASTPOINT	32328	(null)	SE
2	1002	SANTA MONICA	90404	(null)	W
3	1003	TALLAHASSEE	32306	(null)	SE
4	1004	BOISE	83707	(null)	NW
5	1005	SEATTLE	98115	(null)	NW
6	1006	ALBANY	12211	(null)	NE
7	1007	AUSTIN	78710	1003	SW
8	1008	CHEYENNE	82003	(null)	N
9	1009	BURBANK	91510	1003	W
10	1010	ATLANTA	30314	(null)	SE
11	1011	CHICAGO	60606	(null)	N
12	1012	BOSTON	02110	(null)	NE
13	1013	CLERMONT	34711	1006	SE
14	1014	CODY	82414	(null)	N
15	1015	MIAMI	33111	(null)	SE
16	1016	BURBANK	91508	1010	W
17	1017	KALMAZOO	49006	(null)	N
18	1018	MACON	31206	(null)	SE
19	1019	MORRISTOWN	07962	1003	NE
20	1020	TRENTON	08607	(null)	NE

FIGURE 3-30 Verifying that the RENAME operation was successful

TIP

When working in an organization, don't change the name of a table accessed by other users unless you first inform them of the new table name. Failure to inform users of the change prevents them from finishing their work and could create havoc until the problem is identified. Of course, this is assuming you didn't change the table's name to stop someone from accessing it in the first place!

Truncating a Table

When a table is truncated, all rows in the table are removed, but the table itself remains. In other words, the columns still exist, even though no values are stored in them. This action is basically the same as deleting all rows in a table. However, if you simply delete all rows in a table, the storage space these rows occupy is still allocated to the table. To delete the rows stored in a table *and* free up the storage space they occupied, use the **TRUNCATE TABLE** command. The syntax of this command is shown in Figure 3-31.

```
TRUNCATE TABLE tablename;
```

FIGURE 3-31   Syntax of the TRUNCATE TABLE command

Assume the CUST_MKT_092009 table was originally created a day early and didn't include customers added on the last day of September. The database administrator is going to repopulate the table; however, all the current rows need to be removed before the repopulation can occur. Keep in mind that the TRUNCATE TABLE command keeps the table structure intact so that it can be reused in the future. The command shown in Figure 3-32 deletes the rows currently in the CUST_MKT_092009 table and releases the storage space they occupy.

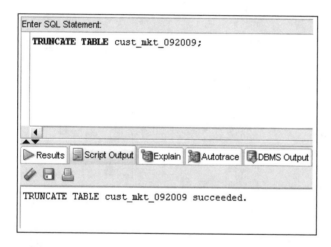

FIGURE 3-32   The TRUNCATE TABLE command

To verify that all rows of the CUST_MKT_092009 table have been removed, perform a query to see all rows in the table. If the table still exists but contains no rows, Oracle 11*g* displays a message indicating that no rows exist.

TIP

The TRUNCATE command is quite useful when creating and using database tables to support testing new applications. As you test an application, typically the tables are populated with sample data. The TRUNCATE command is an easy way to maintain table structures while eliminating the test data to start another round of testing or move the tables into production.

DELETING A TABLE

You can remove a table from an Oracle 11g database by issuing the **DROP TABLE** command. Figure 3-33 shows the syntax of this command.

```
DROP TABLE tablename [PURGE];
```

FIGURE 3-33   Syntax of the DROP TABLE command

For example, after truncating the CUST_MKT_092009 table, you realize you no longer need the table (or so many modifications have to be made to the table structure that it's not worth the trouble to make the changes). The CUST_MKT_092009 table can be deleted by using the DROP TABLE command, shown in Figure 3-34.

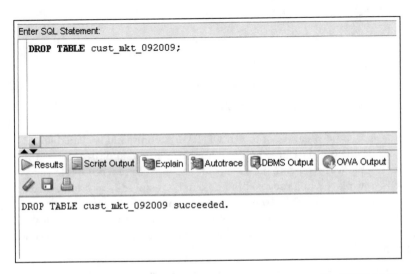

FIGURE 3-34   Using the DROP TABLE command to remove the CUST_MKT_092009 table

After the table has been dropped, the table name is no longer valid, and the table can't be accessed by any commands. To verify that the correct table was deleted, you can use the DESCRIBE command to see the structure of the CUST_MKT_092009 table. If the table no longer exists, Oracle 11g returns an error message, as shown in Figure 3-35.

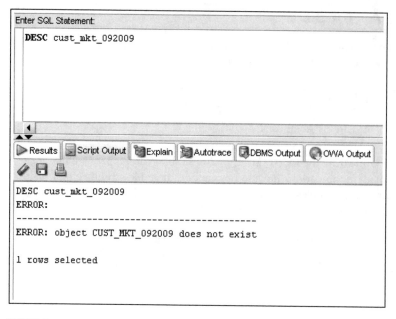

FIGURE 3-35 Using DESCRIBE to verify the dropped table

Starting with Oracle 10g, a new feature of the DROP TABLE command is available. In previous Oracle versions, the DROP TABLE command was permanent, and the only method for recovering a dropped table was restoring the data from backups. In recent Oracle versions, a dropped table is now placed in a recycle bin and can be restored—both table structure and data! Now that the CUST_MKT_092009 table has been dropped, you should check the recycle bin with the command shown in Figure 3-36. Notice that a new name beginning with BIN has been assigned to the table. This new name is assigned by the system, so your results might vary.

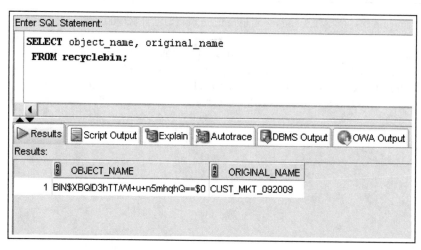

FIGURE 3-36 Checking the recycle bin

Now that you have a table in the recycle bin, how do you recover the table to use it again? You need to use the FLASHBACK TABLE command, shown in Figure 3-37. Keep in mind that the entire table structure and all data in the table at the time of the drop are restored. After performing this command, issue DESCRIBE and SELECT commands on the table to verify that it's restored.

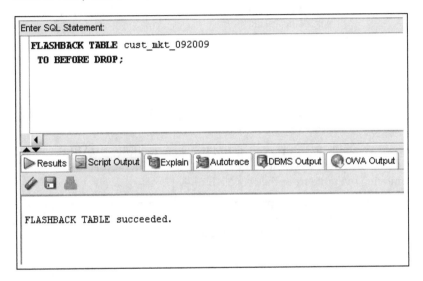

FIGURE 3-37   Using FLASHBACK TABLE to restore a dropped table

Obviously, if the dropped table is actually just moved to a recycle bin, the storage space required for the table is still being used. If you have limited storage space, you might not want this to happen. In addition, sometimes you know that you want to delete a table permanently. If the table has already been dropped and is now in the recycle bin, you need to remove it from the recycle bin to delete the table permanently and clear the storage space being used. Drop the CUST_MKT_092009 table again and verify its existence in the recycle bin with the commands shown in Figure 3-38.

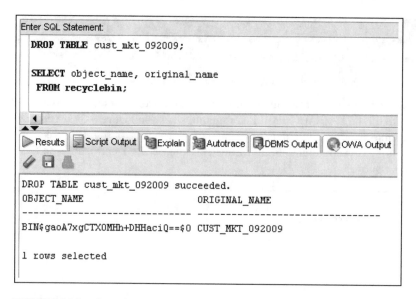

FIGURE 3-38   Dropping a table and checking the recycle bin

Now you can remove the table from the recycle bin by using the PURGE TABLE command and referencing the name the table has been assigned in the recycle bin. Figure 3-39 shows removing the table from the recycle bin.

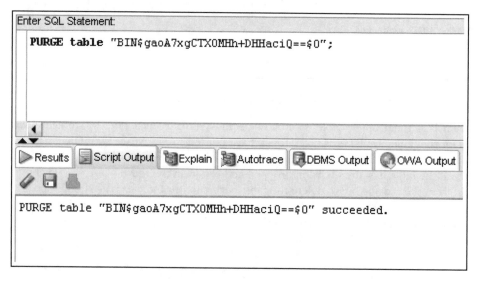

FIGURE 3-39   Removing a table from the recycle bin

N O T E

Keep in mind that the system assigns the table name in the recycle bin, so you'll most likely have a different table name than what's shown in Figure 3-39. Also, if you want to remove all tables in the recycle bin, you can do so by issuing a PURGE RECYCLE BIN command.

If you're sure you want to delete a table permanently, you can bypass moving the table to the recycle bin by using the PURGE option in the DROP TABLE statement. For example, you could have dropped the CUST_MKT_092009 table with the command shown in Figure 3-40, and then it wouldn't have been moved to the recycle bin. Keep in mind that using this option means the table can't be recovered.

```
DROP TABLE cust_mktg_092009 PURGE;
```

FIGURE 3-40   Dropping a table with the PURGE option

CAUTION

Always use caution when deleting with the PURGE option. After a table is deleted with PURGE, the table and all data it contains are gone. In addition, any index that has been created based on this table is dropped. (Indexes are discussed in Chapter 6.)

Chapter Summary

- You create a table with the CREATE TABLE command. Each column to be included in the table must be defined in terms of the column name, datatype, and the width (for certain datatypes).
- A table can contain up to 1000 columns.
- Each column name in a table must be unique.
- Table and column names can contain as many as 30 characters. The names must begin with a letter and can't contain blank spaces.
- A DEFAULT setting assigns a column value if no value is provided for the column when a new row is added.
- A virtual column is a column defined by an expression that generates a value based on other column values in the table when queried.
- A query on the data dictionary object USER_TAB_COLUMNS enables you to verify DEFAULT and virtual column settings.
- To create a table based on existing tables' structure and data, use the CREATE TABLE ... AS command to use a subquery that extracts the necessary data from the existing table.
- You can change a table's structure with the ALTER TABLE command. Columns can be added, resized, and even deleted with the ALTER TABLE command.
- When using the ALTER TABLE command with the DROP COLUMN clause, only one column can be specified for deletion.
- You can use the SET UNUSED clause to mark a column so that its storage space can be freed up later.
- Tables can be renamed with the RENAME ... TO command.
- To delete all the rows in a table, use the TRUNCATE TABLE command.
- To remove both the structure of a table and all its contents, use the DROP TABLE command.
- A dropped table is moved to the recycle bin and can be recovered by using the FLASHBACK TABLE command.
- Using the PURGE option in a DROP TABLE command removes the table permanently, meaning you can't recover it from the recycle bin.

Chapter 3 Syntax Summary

The following table summarizes the syntax you have learned in this chapter. You can use the table as a study guide and reference.

Commands and Clauses	Description	Example
Creating Tables		
CREATE TABLE	Creates a new table in the database—the user names the columns; defaults and datatypes define/limit columns. To view the table structure, use the SQL*PLUS command DESCRIBE.	`CREATE TABLE acctmanager (amid VARCHAR2(4), amname VARCHAR2(20), amedate DATE DEFAULT SYSDATE, region CHAR(2));`
CREATE TABLE ... AS (...)	Creates a table from existing database tables, using the AS clause and subqueries.	`CREATE TABLE customerorder AS (SELECT customer#, orderdate, shipdate FROM orders);`
Modifying Tables		
ALTER TABLE ... ADD	Adds a column to a table.	`ALTER TABLE acctmanager ADD (ext NUMBER(4));`
ALTER TABLE ... MODIFY	Changes a column size, datatype, or default value.	`ALTER TABLE acctmanager MODIFY (amname VARCHAR2(25));`
ALTER TABLE ... DROP COLUMN	Deletes one column from a table.	`ALTER TABLE acctmanager DROP COLUMN ext;`
ALTER TABLE ... SET UNUSED *or* SET UNUSED COLUMN	Marks a column for deletion at a later time.	`ALTER TABLE cust_mkt SET UNUSED (state);`
DROP UNUSED COLUMNS	Completes the deletion of a column marked with SET UNUSED.	`ALTER TABLE cust_mkt DROP UNUSED COLUMNS;`
RENAME ... TO	Changes a table name.	`RENAME cust_mkt TO cust_mkt_092009;`
TRUNCATE TABLE	Deletes table rows, but table name and column structure remain.	`TRUNCATE TABLE cust_mkt_092009;`

Commands and Clauses	Description	Example
Deleting Tables		
DROP TABLE ... PURGE	Removes an entire table permanently from the database with the PURGE option.	**DROP TABLE cust_mkt_092009 PURGE;**
PURGE TABLE	Permanently deletes a table in the recycle bin.	**PURGE TABLE "BIN$IDMdosJceWxgg041==$0";**
Recovering Tables		
FLASHBACK TABLE ... TO BEFORE DROP	Recovers a dropped table if the PURGE option isn't used when the table is dropped.	**FLASHBACK TABLE cust_mkt_092009 TO BEFORE DROP;**

Review Questions

To answer the following questions, refer to the tables in the JustLee Books database.

1. Which command is used to create a table based on data already contained in an existing table?
2. List four datatypes supported by Oracle 11*g*, and provide an example of data that could be stored by each datatype.
3. What guidelines should you follow when naming tables and columns in Oracle 11*g*?
4. What is the difference between dropping a column and setting a column as unused?
5. How many columns can be dropped in one ALTER TABLE command?
6. What happens to the existing rows of a table if the DEFAULT value of a column is changed?
7. Explain the difference between truncating a table and deleting a table.
8. If you add a new column to an existing table, where does the column appear in relation to existing columns?
9. What happens if you try to decrease the scale or precision of a NUMBER column to a value less than the data already stored in the field?
10. Are a table and the data contained in the table erased from the system permanently if a DROP TABLE command is issued on the table?

Mutiple Choice

To answer the following questions, refer to the tables in the JustLee Books database.

1. Which of the following is a correct statement?
    a. You can restore the data deleted with the DROP COLUMN clause, but not the data deleted with the SET UNUSED clause.

b. You can't create empty tables—all tables must contain at least three rows of data.

c. A table can contain a maximum of 1000 columns.

d. The maximum length of a table name is 265 characters.

2. Which of the following is a valid SQL statement?

a. `ALTER TABLE secustomersspent ADD DATE lastorder;`

b. `ALTER TABLE secustomerorders DROP retail;`

c. `CREATE TABLE newtable AS (SELECT * FROM customers);`

d. `ALTER TABLE drop column *;`

3. Which of the following is not a correct statement?

a. A table can be modified only if it doesn't contain any rows of data.

b. The maximum number of characters in a table name is 30.

c. You can add more than one column at a time to a table.

d. You can't recover data contained in a table that has been truncated.

4. Which of the following is not a valid SQL statement?

a. `CREATE TABLE anothernewtable (newtableid VARCHAR2(2));`

b. `CREATE TABLE anothernewtable (date, anotherdate)`
   `AS (SELECT orderdate, shipdate FROM orders);`

c. `CREATE TABLE anothernewtable (firstdate, seconddate)`
   `AS (SELECT orderdate, shipdate FROM orders);`

d. All of the above are valid statements.

5. Which of the following is true?

a. If you truncate a table, you can't add new data to the table.

b. If you change the default value of an existing column, all existing rows containing a NULL value in the same column are set to the new DEFAULT value.

c. If you delete a column from a table, you can't add a column to the table with the same name as the previously deleted column.

d. If you add a column to an existing table, it's always added as the last column of the table.

6. Which of the following commands creates a new table containing a virtual column?

a. `CREATE TABLE newtable AS (SELECT order#, title, quantity, retail FROM orders);`

b. `CREATE TABLE newtable (price NUMBER(3), total NUMBER(8,2));`

c. `CREATE TABLE newtable (calc1 NUMBER(4), calc2 NUMBER(4);`

d. `CREATE TABLE newtable (cola NUMBER(3), colb NUMBER(3), colc AS (cola+colb));`

7. Which of the following commands drops any columns marked as unused from the SECUSTOMERORDERS table?

a. `DROP COLUMN FROM secustomerorders WHERE column_status = UNUSED;`

b. `ALTER TABLE secustomerorders DROP UNUSED COLUMNS;`

Table Creation and Management

c.   `ALTER TABLE secustomerorders DROP (unused);`

d.   `DROP UNUSED COLUMNS;`

8.   Which of the following statements is correct?

   a.   A table can contain a maximum of only one column marked as unused.

   b.   You can delete a table by removing all columns in the table.

   c.   Using the SET UNUSED clause allows you to free up storage space used by a column.

   d.   None of the above statements are correct.

9.   Which of the following commands removes all data from a table but leaves the table's structure intact?

   a.   `ALTER TABLE secustomerorders DROP UNUSED COLUMNS;`

   b.   `TRUNCATE TABLE secustomerorders;`

   c.   `DELETE TABLE secustomerorders;`

   d.   `DROP TABLE secustomerorders;`

10.   Which of the following commands changes a table's name from OLDNAME to NEWNAME?

   a.   `RENAME oldname TO newname;`

   b.   `RENAME table FROM oldname TO newname;`

   c.   `ALTER TABLE oldname MODIFY TO newname;`

   d.   `CREATE TABLE newname (SELECT * FROM oldname);`

11.   The default width of a VARCHAR2 field is:

   a.   1

   b.   30

   c.   255

   d.   None—there's no default width for a VARCHAR2 field.

12.   Which of the following is *not* a valid statement?

   a.   You can change the name of a table only if it doesn't contain any data.

   b.   You can change the length of a column that doesn't contain any data.

   c.   You can delete a column that doesn't contain any data.

   d.   You can add a column to a table.

13.   Which of the following characters can be used in a table name?

   a.   _

   b.   (

   c.   %

   d.   !

14. Which of the following is true?
    a. All data in a table can be recovered if the table is dropped with the PURGE option.
    b. All data in a table can be recovered from the recycle bin if the table is dropped.
    c. All data in a table is lost if the table is dropped.
    d. All of the above statements are true.

15. Which of the following commands is valid?
    a. `RENAME customer# TO customernumber FROM customers;`
    b. `ALTER TABLE customers RENAME customer# TO customernum;`
    c. `DELETE TABLE customers;`
    d. `ALTER TABLE customers DROP UNUSED COLUMNS;`

16. Which of the following commands creates a new table containing two columns?
    a. `CREATE TABLE newname (col1 DATE, col2 VARCHAR2);`
    b. `CREATE TABLE newname AS (SELECT title, retail, cost FROM books);`
    c. `CREATE TABLE newname (col1, col2);`
    d. `CREATE TABLE newname (col1 DATE DEFAULT SYSDATE, col2 VARCHAR2(1));`

17. Which of the following is a valid table name?
    a. 9NEWTABLE
    b. DATE9
    c. NEW"TABLE
    d. None of the above are valid table names.

18. Which of the following is a valid datatype?
    a. CHAR3
    b. VARCHAR4(3)
    c. NUM
    d. NUMBER

19. Which object in the data dictionary enables you to verify DEFAULT column settings?
    a. DEFAULT_COLUMNS
    b. DEF_TAB_COLUMNS
    c. USER_TAB_COLUMNS
    d. None of the above

20. Which of the following SQL statements changes the size of the Title column in the BOOKS table from the current length of 30 characters to the length of 35 characters?
    a. `ALTER TABLE books CHANGE title VARCHAR(35);`
    b. `ALTER TABLE books MODIFY (title VARCHAR2(35));`
    c. `ALTER TABLE books MODIFY title (VARCHAR2(35));`
    d. `ALTER TABLE books MODIFY (title VARCHAR2(+5));`

Table Creation and Management

Hands-On Assignments

1. Create a new table containing the category code and description for the categories of books sold by JustLee Books. The table should be called CATEGORY, and the columns should be CatCode and CatDesc. The CatCode column should store a maximum of 2 characters, and the CatDesc column should store a maximum of 10 characters.

2. Create a new table containing these four columns: Emp#, Lastname, Firstname, and Job_class. The table name should be EMPLOYEES. The Job_class column should be able to store character strings up to a maximum length of four, but the column values shouldn't be padded if the value has less than four characters. The Emp# column contains a numeric ID and should allow a five-digit number. Use column sizes you consider suitable for the Firstname and Lastname columns.

3. Add two columns to the EMPLOYEES table. One column, named EmpDate, contains the date of employment for each employee, and its default value should be the system date. The second column, named EndDate, contains employees' date of termination.

4. Modify the Job_class column of the EMPLOYEES table so that it allows storing a maximum width of two characters.

5. Delete the EndDate column from the EMPLOYEES table.

6. Rename the EMPLOYEES table as JL_EMPS.

7. Create a new table containing these four columns from the existing BOOKS table: ISBN, Cost, Retail, and Category. The name of the ISBN column should be ID, and the other columns should keep their original names. Name the new table BOOK_PRICING.

8. Mark the Category column of the BOOK_PRICING table as unused. Verify that the column is no longer available.

9. Truncate the BOOK_PRICING table, and then verify that the table still exists but no longer contains any data.

10. Delete the BOOK_PRICING table permanently so that it isn't moved to the recycle bin. Delete the JL_EMPS table so that it can be restored. Restore the JL_EMPS table and verify that it's available again.

Advanced Challenge

The management of JustLee Books has approved implementing a new commission policy and benefits plan for the account managers. The following changes need to be made to the existing database:

- Two new columns must be added to the ACCTMANAGER table: one to indicate the commission classification assigned to each employee and another to contain each employee's benefits code. The commission classification column should be able to store integers up to a maximum value of 99 and be named Comm_id. The value of the Comm_id column should be set to a value of 10 automatically if no value is provided when a row is added. The benefits code column should also accommodate integer values up to a maximum of 99 and be named Ben_id.

- A new table, COMMRATE, must be created to store the commission rate schedule and must contain the following columns:
  - Comm_id: A numeric column similar to the one added to the ACCTMANAGER table
  - Comm_rank: A character field that can store a rank name allowing up to 15 characters
  - Rate: A numeric field that can store two decimal digits (such as .01 or .03)
- A new table, BENEFITS, must be created to store the available benefit plan options and must contain the following columns:
  - Ben_id: A numeric column similar to the one added to the ACCTMANAGER table
  - Ben_plan: A character field that can store a single character value
  - Ben_provider: A numeric field that can store a three-digit integer
  - Active: A character field that can hold a value of Y or N

Required: Create the SQL statements to address the changes needed to support the new commission and benefits data.

Case Study: *City Jail*

In the Chapter 1 case study, you designed the new database for City Jail. Now you need to create all the tables for the database. First, create all the tables using the information outlined in Section A. Second, make the modifications outlined in Section B. Save all SQL statements used to accomplish these tasks.

Section A

Table	Column	Data Description	Length	Scale	Default Value
Aliases	Alias_ID	Numeric	6		
	Criminal_ID	Numeric	6	0	
	Alias	Variable character	10		
Criminals	Criminal_ID	Numeric	6	0	
	Last	Variable character	15		
	First	Variable character	10		
	Street	Variable character	30		
	City	Variable character	20		
	State	Fixed character	2		
	Zip	Fixed character	5		
	Phone	Fixed character	10		

Table	Column	Data Description	Length	Scale	Default Value
	V_status	Fixed character	1		N (for No)
	P_status	Fixed character	1		N (for No)
Crimes	Crime_ID	Numeric	9	0	
	Criminal_ID	Numeric	6	0	
	Classification	Fixed character	1		
	Data_charged	Date			
	Status	Fixed character	2		
	Hearing_date	Date			
	Appeal_cut_date	Date			
Sentences	Sentence_ID	Numeric	6		
	Criminal_ID	Numeric	6		
	Type	Fixed character	1		
	Prob_ID	Numeric	5		
	Start_date	Date			
	End_date	Date			
	Violations	Numeric	3		
Prob_officers	Prob_ID	Numeric	5		
	Last	Variable character	15		
	First	Variable character	10		
	Street	Variable character	30		
	City	Variable character	20		
	State	Fixed character	2		
	Zip	Fixed character	5	0	
	Phone	Fixed character	10	0	
	Email	Variable character	30		
	Status	Fixed character	1		A (for Active)

Table	Column	Data Description	Length	Scale	Default Value
Crime_charges	Charge_ID	Numeric	10	0	
	Crime_ID	Numeric	9	0	
	Crime_code	Numeric	3	0	
	Charge_status	Fixed character	2		
	Fine_amount	Numeric	7	2	
	Court_fee	Numeric	7	2	
	Amount_paid	Numeric	7	2	
	Pay_due_date	Date			
Crime_officers	Crime_ID	Numeric	9	0	
	Officer_ID	Numeric	8	0	
Officers	Officer_ID	Numeric	8	0	
	Last	Variable character	15		
	First	Variable character	10		
	Precinct	Fixed character	4		
	Badge	Variable character	14		
	Phone	Fixed character	10	0	
	Status	Fixed character	1		A (for Active)
Appeals	Appeal_ID	Numeric	5		
	Crime_ID	Numeric	9	0	
	Filing_date	Date			
	Hearing_date	Date			
	Status	Fixed character	1		P (for Pending)
Crime_codes	Crime_code	Numeric	3	0	
	Code_description	Variable character	30		

Coding key for selected columns:

Table	Column	Possible Values
Criminals	V_status	Y (Yes), N (No)
Criminals	P_status	Y (Yes), N (No)
Crimes	Classification	F (Felony), M (Misdemeanor), O (Other), U (Undefined)
Crimes	Status	CL (Closed), CA (Can Appeal), IA (In Appeal)
Sentences	Type	J (Jail Period), H (House Arrest), P (Probation)
Prob_officers	Status	A (Active), I (Inactive)
Crime_charges	Charge_status	PD (Pending), GL (Guilty), NG (Not Guilty)
Officers	Status	A (Active), I (Inactive)
Appeals	Status	P (Pending), A (Approved), D (Disapproved)

Section B

- Add a default value of U for the Classification column of the Crimes table.
- Add a column named Date_Recorded to the Crimes table. This column needs to hold date values and should be set to the current date by default.
- Add a column to the Prob_officers table to contain the pager number for each officer. The column needs to accommodate a phone number, including area code. Name the column Pager#.
- Change the Alias column in the Aliases table to accommodate up to 20 characters.

CHAPTER 4

CONSTRAINTS

LEARNING OBJECTIVES

After completing this chapter, you should be able to do the following:

- Explain the purpose of constraints in a table
- Distinguish among PRIMARY KEY, FOREIGN KEY, UNIQUE, CHECK, and NOT NULL constraints and understand the correct use of each constraint
- Understand how to create constraints when creating a table or modifying an existing table
- Distinguish between creating constraints at the column level and the table level
- Create PRIMARY KEY constraints for a single column and a composite primary key
- Create a FOREIGN KEY constraint
- Create a UNIQUE constraint
- Create a CHECK constraint
- Create a NOT NULL constraint with the ALTER TABLE ... MODIFY command
- Include constraints during table creation
- Add multiple constraints on a single column
- View constraint information
- Use the DISABLE and ENABLE commands with constraints
- Use the DROP command with constraints

INTRODUCTION

In Chapter 3, you learned how to create tables by using SQL commands. In this chapter, you learn how to add constraints to existing tables and include constraints during the table creation process. **Constraints** are rules used to enforce business rules, practices, and policies to ensure the accuracy and integrity of data. For example, a customer places an order on April 2, 2009. However, when the order is shipped, the ship date is entered as March 31, 2009. Shipping an order before the order is placed is impossible, and it indicates a problem with data integrity. If these errors exist in the database, management can't rely on it for decision making or even to support day-to-day operations. Constraints help solve these problems by not allowing data to be added to tables if the data violates certain rules.

This chapter examines five constraints, listed in Table 4-1, that can prevent entering erroneous data in a database.

TABLE 4-1 Constraint Types

Constraint	Description
PRIMARY KEY	Determines which column(s) uniquely identifies each record. The primary key can't be NULL, and the data values must be unique.
FOREIGN KEY	In a one-to-many or parent-child relationship, the constraint is added to the "many" table. The constraint ensures that if a value is entered in a specified column, it must already exist in the "one" table, or the record isn't added.
UNIQUE	Ensures that all data values stored in a specified column are unique. The UNIQUE constraint differs from the PRIMARY KEY constraint in that it allows NULL values.
CHECK	Ensures that a specified condition is true before the data value is added to a table. For example, an order's ship date can't be earlier than its order date.
NOT NULL	Ensures that a specified column can't contain a NULL value. The NOT NULL constraint can be created *only* with the column-level approach to table creation.

DATABASE PREPARATION

The examples in this chapter assume you have already created the JustLee Books database as instructed in Chapter 2.

NOTE

As developers build applications, mechanisms are often included to check data input. For example, if a particular field in the database requires providing a value, the application code can verify that a value was provided for this field on the input form before submitting the values to the database. Application data verification methods can serve as the first line of defense to ensure data integrity. Database constraints are the last line of defense to check data before it's added to the database.

CREATING CONSTRAINTS

You can add constraints during table creation as part of the CREATE TABLE command, or you can do so after the table is created by using the ALTER TABLE command. When creating a constraint, you can choose one of the following options:

- Name the constraint following the same rules as for tables and columns.
- Omit the constraint name and allow Oracle 11g to generate the name.

If the Oracle 11g server names the constraint, it follows the format SYS_Cn, where n is an assigned numeric value to make the name unique. Providing a descriptive name for a constraint is a better practice so that you can identify it easily in the future. For example, constraint violation errors reference the constraint name, so an easy-to-understand name indicating the table, column, and type of constraint is quite helpful.

Industry convention is to use the format *tablename_columnname_constrainttype* for the constraint name—for example, customers_customer#_pk. Constraint types are designated by abbreviations, as shown in Table 4-2.

TABLE 4-2 Constraint Type Abbreviations

Constraint	Abbreviation
PRIMARY KEY	_pk
FOREIGN KEY	_fk
UNIQUE	_uk
CHECK	_ck
NOT NULL	_nn

NOTE

Most development groups have a set of coding conventions that include guidelines for naming database objects, including constraints. When joining a new development group or company, you should review its coding conventions.

When creating a table, you can create a constraint in two ways: at the column level or the table level. Creating a constraint at the column level means the constraint's definition is included as part of the column definition, similar to assigning a default value to a column. Creating a constraint at the table level means the constraint's definition is separate from the column definition. These two methods differ only in that they include the constraint code in different parts of the CREATE TABLE statement.

Creating Constraints at the Column Level

When you create constraints at the column level, the constraint applies to the specified column. The optional CONSTRAINT keyword is used *if* you want to give the constraint a specific name instead of letting Oracle 11g generate one for you. The constraint type uses the following keywords to identify the type of constraint you're creating:

- PRIMARY KEY
- FOREIGN KEY
- UNIQUE
- CHECK
- NOT NULL

NOTE

A NULL value means the column contains no value.

You can create any type of constraint at the column level—unless the constraint is being defined for more than one column (for example, a composite primary key). If the constraint applies to more than one column, *you must create the constraint at the table level*. The general syntax for creating a constraint at the column level is shown in Figure 4-1.

```
columnname [CONSTRAINT constraintname] constrainttype,
```

FIGURE 4-1   Syntax for creating a column-level constraint

As you see later in this chapter, a NOT NULL constraint can be created only at the column level.

Creating Constraints at the Table Level

When you create a constraint at the table level, the constraint definition is separate from any column definitions. Figure 4-2 shows the syntax for creating a constraint at the table level.

```
[CONSTRAINT constraintname] constrainttype
(columnname, ...),
```

FIGURE 4-2   Syntax for creating a table-level constraint

If you create the constraint at the same time you're creating a table, you list the constraint *after* all the columns are defined. In fact, the main difference in the syntax of a column-level constraint and a table-level constraint is that you provide column names for the table-level constraint *at the end of the constraint definition inside parentheses, instead of at the beginning of the constraint definition*. You can use the table-level approach to create any type of constraint except a NOT NULL constraint, which can be created only with the column-level approach, as mentioned previously.

> **NOTE**
>
> Although a constraint can be created at the column level or the table level, the constraint is *always* enforced on a row level, which means you can't add or delete the entire row if any column value violates a constraint.

To simplify the examples for different types of constraints, the following sections show how to add constraints to an existing table. The original database creation script to build the JustLee database doesn't address all the needed constraints, so in the next section, you make the necessary additions. After you have learned how to add constraints by using the ALTER TABLE command, you learn how to include constraints at both the column level and table level during initial table creation by using the CREATE TABLE statement.

USING THE PRIMARY KEY CONSTRAINT

A **PRIMARY KEY constraint** is used to enforce the primary key requirements for a table. Although you can create a table in Oracle 11g without specifying a primary key, the constraint makes certain the columns identified as the table's primary key are unique and *do not contain NULL values*. As stated, a NULL value means no entry is made. It's not equivalent to entering a zero or a blank. Figure 4-3 shows the syntax of the ALTER TABLE command to add a PRIMARY KEY constraint to an existing table.

```
ALTER TABLE tablename
ADD [CONSTRAINT constraintname] PRIMARY KEY (columnname);
```

FIGURE 4-3   Syntax of the ALTER TABLE command to add a PRIMARY KEY constraint

Take a look at an example. The CUSTOMERS table stores a row of data for each customer but doesn't currently have a PRIMARY KEY constraint. Without a PRIMARY KEY designated, a customer could be added twice mistakenly. Confusion could also result if multiple customers have the same name—an order could be charged or shipped to the wrong customer! The Customer# column is included in the CUSTOMERS table to assist in uniquely identifying each customer. Adding a PRIMARY KEY constraint to the Customer# column ensures that each row added is assigned a value that's unique. Only *one* PRIMARY KEY constraint can be defined for each table. To designate the Customer# column as the primary key for the CUSTOMERS table, issue the ALTER TABLE command shown in Figure 4-4. When the command executes, you get a success message, as shown.

Note the following elements in Figure 4-4:

- The ADD clause instructs Oracle 11g to add a constraint to the CUSTOMERS table (listed after the ALTER TABLE keywords).
- The user has chosen the constraint name, customers_customer#_pk, instead of having Oracle 11g assign it.
- The PRIMARY KEY keywords designate the constraint type.

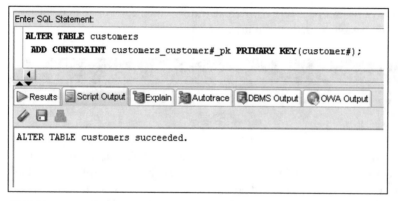

FIGURE 4-4 Adding a PRIMARY KEY constraint

How do you test constraints? Keep in mind that the purpose of constraints is to check whether data values being entered are valid. So to test the constraint, you can attempt a row insert with a duplicate or NULL Customer# value to verify that the primary key rejects the row. You learn more about data manipulation in Chapter 5; however, in this chapter, you execute basic inserts to test constraint enforcement. Figure 4-5 shows an INSERT statement for adding a new customer to the CUSTOMERS table. This statement generates an error because an existing customer row in the table already has the customer# 1020 assigned.

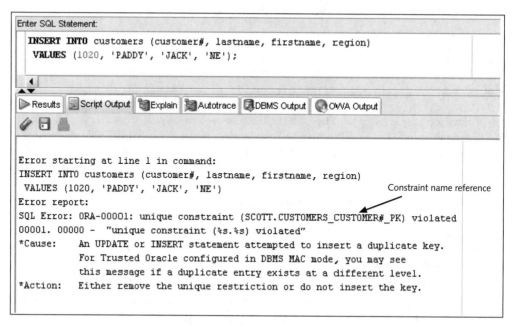

FIGURE 4-5 Insert a row to test the constraint

Constraint violations display an error message referencing the constraint name, as pointed out in Figure 4-5. Using a naming convention is quite helpful in evaluating the cause of the error. Just by viewing the error message, you know what table, column, and type of constraint are the issue. If you don't assign constraint names, the error message displays the system-generated constraint name, which isn't as helpful.

Figure 4-6 shows a successful customer addition, using a customer# not yet in the database.

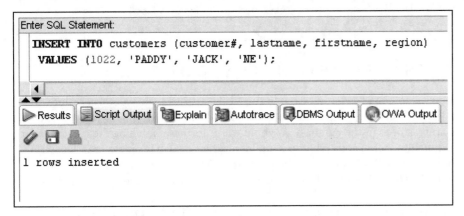

FIGURE 4-6   Insert a new customer record

If the primary key consists of more than one column (a composite primary key), you must create it at the table level in a CREATE TABLE statement. The ORDERITEMS table requires a composite primary key because two columns are used to uniquely identify each item on an order: Order# and Item#. To indicate that the primary key for a table consists of more than one column, simply list the column names, separated by commas, in parentheses after the constraint type. Figure 4-7 shows this format.

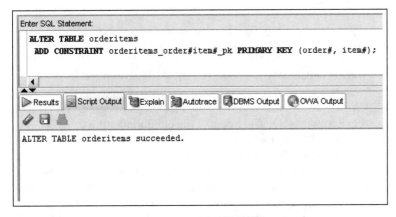

FIGURE 4-7   Adding a composite PRIMARY KEY constraint

After the constraint shown in Figure 4-7 is added to the ORDERITEMS table, all entries in the Order# and Item# columns must create a unique combination in the table, and neither column can contain a NULL value.

> **TIP**
>
> Because a table can have only one PRIMARY KEY constraint, some developers don't include column names in a PRIMARY KEY constraint. In this case, the constraint shown in Figure 4-7 might simply be named orderitems_pk.

USING THE FOREIGN KEY CONSTRAINT

Suppose a new customer who doesn't exist in the CUSTOMERS table places an order with JustLee Books. If the customer information wasn't collected, the customer's name and billing address aren't stored in the database. Without this information, billing the customer for the order is difficult—not exactly what you might consider a good business practice. Or perhaps a book in the BOOKS table has the publisher ID 9—one that doesn't exist in the PUBLISHER table. (It could simply be a typo, or perhaps someone neglected to add this publisher to the PUBLISHER table.)

You can prevent these problems by using a **FOREIGN KEY constraint**. For example, to prevent an order being entered from a customer who doesn't have a record in the CUSTOMERS table, you can create a constraint that compares any entry made in the Customer# column of the ORDERS table with all customer numbers existing in the CUSTOMERS table. If a customer service representative enters a customer number not found in the CUSTOMERS table, the corresponding entry in the ORDERS table is rejected. This constraint requires the customer service representative to collect and enter the customer's information in the CUSTOMERS table, and then enter the order in the ORDERS table.

The syntax to add a FOREIGN KEY constraint to a table is shown in Figure 4-8.

```
ALTER TABLE tablename
ADD [CONSTRAINT constraintname] FOREIGN KEY (columnname)
REFERENCES referencedtablename (referencedcolumnname);
```

FIGURE 4-8   Syntax of the ALTER TABLE command to add a FOREIGN KEY constraint

The keywords FOREIGN KEY are used to identify a column that, if it contains a value, must match data contained in another table. The name of the column identified as the foreign key is placed inside parentheses after the FOREIGN KEY keywords. The REFERENCES keyword refers to **referential integrity**, which means the user is referring to something that exists in another table. For example, the value entered in the Customer# column of the ORDERS table references a value in the Customer# column of the CUSTOMERS table. The REFERENCES keyword is used to identify the table and column that must already contain the data being entered. The column referenced must be a primary key column. In this case, the Customer# column of the CUSTOMERS table must be defined as this table's primary key column.

To create a FOREIGN KEY constraint on the Customer# column of the ORDERS table that makes sure any customer number entered also exists in the CUSTOMERS table before the order is accepted, use the command shown in Figure 4-9.

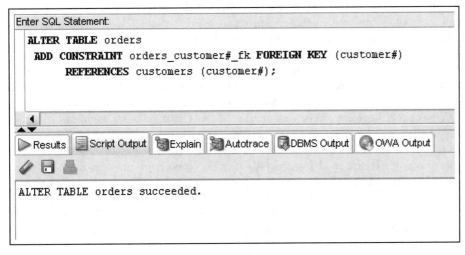

FIGURE 4-9   Adding a FOREIGN KEY constraint

This command instructs Oracle 11g to add a FOREIGN KEY constraint on the Customer# column of the ORDERS table. The name chosen for the constraint is orders_customer#_fk. This constraint makes sure an entry for the Customer# column of the ORDERS table matches a value stored in the Customer# column of the CUSTOMERS table. When the command executes, a message indicates the table was altered successfully, as shown in Figure 4-9. An INSERT statement adding an order assigned to the customer# 2000 tests the FOREIGN KEY constraint, as shown in Figure 4-10. The statement violates the constraint because the customer# 2000 doesn't exist in the referenced CUSTOMERS table.

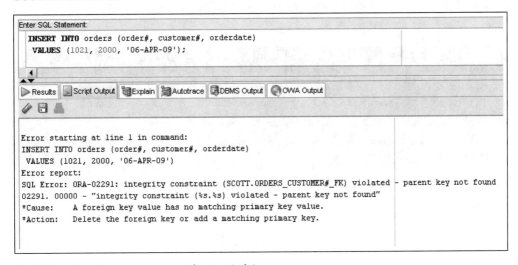

FIGURE 4-10   Insert a row to test the constraint

The syntax for the FOREIGN KEY constraint is more complex than for the PRIMARY KEY constraint because two tables are involved in the constraint. The CUSTOMERS table is the referenced table. It's the "one" side of the one-to-many relationship between the CUSTOMERS and ORDERS table, so each order can be placed by only one customer, but one customer can place many orders. Therefore, the CUSTOMERS table is considered the parent table for the constraint; the ORDERS table is considered the child table.

When a FOREIGN KEY constraint exists between two tables, by default, a record can't be deleted from the parent table if matching entries exist in the child table. This rule means that in the JustLee Books database, *you can't delete a customer from the CUSTOMERS table if there are orders in the ORDERS table for that customer*.

However, suppose you do need to delete a customer from the CUSTOMERS table. Perhaps the customer hasn't paid for previous orders, or perhaps the customer has passed away. Your goal is to remove the customer from the database to make certain no one places an order with that customer's information. The FOREIGN KEY constraint requires first deleting all that customer's orders from the ORDERS table (the child table) and then deleting the customer from the CUSTOMERS table (the parent table).

There's an alternative method, however: You can add the keywords ON DELETE CASCADE to the end of the command issued in Figure 4-9. If these keywords are included in the constraint definition and a record is deleted from the parent table, any corresponding records in the child table are also deleted automatically. Figure 4-11 shows a FOREIGN KEY constraint with the ON DELETE CASCADE option.

```
ALTER TABLE orders
 ADD CONSTRAINT orders_customer#_fk FOREIGN KEY (customer#)
      REFERENCES customers (customer#) ON DELETE CASCADE;
```

FIGURE 4-11   FOREIGN KEY constraint with the ON DELETE CASCADE option

NOTE

If you attempt the command shown in Figure 4-11 and get an error message, it might be because a FOREIGN KEY constraint with the same name already exists. Enter the following command:

```
ALTER TABLE orders DROP CONSTRAINT orders_customer#_fk;
```

After you've removed the previous constraint from the database, you can then reenter the command in Figure 4-11 without an error message.

CAUTION

Using the example in Figure 4-11, if a customer who has placed 20 orders is deleted from the CUSTOMERS table, all orders that customer placed are deleted from the ORDERS table along with the customer record. Clearly, you must be cautious with the ON DELETE CASCADE option. It could create a problem for unsuspecting users who delete outstanding orders unintentionally. Make absolutely certain that any records that might get deleted from the child table with this option won't be needed in the future. If that possibility exists—even remotely—don't include the ON DELETE CASCADE keywords, and force the user to delete the entries in the child table explicitly before removing the parent record.

If a record in a child table has a NULL value for a column that has a FOREIGN KEY constraint, the record is accepted. This constraint ensures only that the customer number is a valid number, *not* that a customer number has been entered for an order. Basically, this means an order could be entered into the ORDERS table without an entry in the Customer# column, and the order is still accepted. To force the user to enter a customer number for an order, you should also add a NOT NULL constraint for the Customer# column in the ORDERS table. (You add this constraint later in "Using the NOT NULL Constraint.")

NOTE

A FOREIGN KEY constraint can't reference a column in a table that has not been designated as the primary key for the referenced table.

Occasionally, a table is created with two columns that are related. For example, the referred column in the CUSTOMERS table records a customer#, which must be a valid customer# from the Customer# column in the same table. A FOREIGN KEY constraint can enforce the relationship between two columns in the same table. Keep in mind that the Customer# column is the PRIMARY KEY constraint column of the CUSTOMERS table, so the FOREIGN KEY constraint references a PRIMARY KEY column. Figure 4-12 shows the statement to add this FOREIGN KEY constraint.

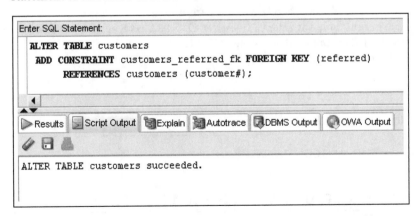

FIGURE 4-12   FOREIGN KEY constraint using columns in the same table

A FOREIGN KEY constraint is unlike the other constraints, in that it typically involves more than one table. Because a FOREIGN KEY can affect more than one table, it can affect a DROP TABLE command on the parent table of the relationship established by the constraint. For example, you just created a FOREIGN KEY constraint relating the CUSTOMERS and ORDERS tables. If you attempt to delete the parent table (CUSTOMERS), an error referencing the FOREIGN KEY is raised, as shown in Figure 4-13.

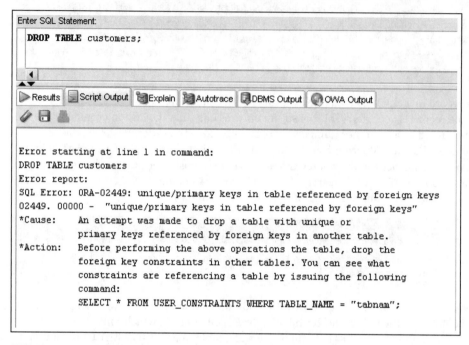

Enter SQL Statement:

```
DROP TABLE customers;
```

Results | Script Output | Explain | Autotrace | DBMS Output | OWA Output

```
Error starting at line 1 in command:
DROP TABLE customers
Error report:
SQL Error: ORA-02449: unique/primary keys in table referenced by foreign keys
02449. 00000 -  "unique/primary keys in table referenced by foreign keys"
*Cause:    An attempt was made to drop a table with unique or
           primary keys referenced by foreign keys in another table.
*Action:   Before performing the above operations the table, drop the
           foreign key constraints in other tables. You can see what
           constraints are referencing a table by issuing the following
           command:
           SELECT * FROM USER_CONSTRAINTS WHERE TABLE_NAME = "tabnam";
```

FIGURE 4-13   DROP TABLE error caused by a FOREIGN KEY

Two options are available to allow deleting the parent table:

- DROP the child table and then DROP the parent table.
- DROP the parent table with the CASCADE CONSTRAINTS option, as shown in Figure 4-14. This option deletes the FOREIGN KEY constraint in the child table and then deletes the parent table.

```
DROP TABLE customers
  CASCADE CONSTRAINTS;
```

FIGURE 4-14   Using the CASCADE CONSTRAINTS option

USING THE UNIQUE CONSTRAINT

The purpose of a **UNIQUE constraint** is to ensure that two records can't have the same value stored in the same column. Although this constraint sounds like a PRIMARY KEY constraint, there's one major difference. *A UNIQUE constraint allows NULL values*, which aren't permitted with a PRIMARY KEY constraint. Therefore, the UNIQUE constraint performs a check on the data only if a value is entered for the column. Figure 4-15 shows the syntax to add a UNIQUE constraint to an existing table.

```
ALTER TABLE tablename
ADD [CONSTRAINT constraintname] UNIQUE (columnname);
```

FIGURE 4-15 Syntax for adding a UNIQUE constraint to a table

As shown in Figure 4-15, the syntax to add a UNIQUE constraint is the same as the syntax for adding a PRIMARY KEY constraint, except the UNIQUE keyword is used to indicate the type of constraint being created.

For example, JustLee Books wants to make certain each book in inventory has a different title entry to help customers and employees differentiate books with titles that are, in fact, the same. The company could add subject or author information at the end of the title so that customers don't accidentally select the wrong book to purchase. To create a UNIQUE constraint on the Title column of the BOOKS table, issue the command shown in Figure 4-16.

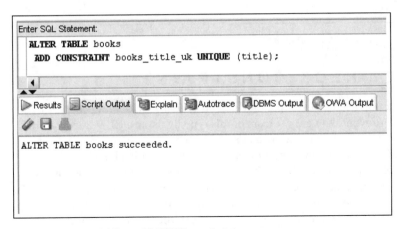

FIGURE 4-16 Adding a UNIQUE constraint

After the command is issued successfully, Oracle 11g doesn't allow any entry in the Title column of the BOOKS table that duplicates an existing entry. If multiple books have the same title, some modification is now required to differentiate the titles of the two books. For example, if a second edition of a book is published with the same title as the first edition, the user must include the edition number in the title to make it different from the record for the previous edition. Figure 4-17 displays an INSERT command attempting to add a book row with the title *Shortest Poems*. A book with this same title already exists in the BOOKS table, so the INSERT command fails because of a UNIQUE KEY constraint violation.

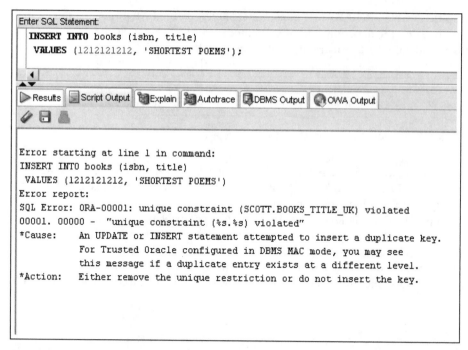

FIGURE 4-17   Insert a row to test the constraint

USING THE CHECK CONSTRAINT

This chapter's introduction used an example in which an order's ship date was earlier than its order date. You can prevent data entry errors of this type by using a CHECK constraint. A **CHECK constraint** requires that a specific condition be met before a record is added to a table. With a CHECK constraint, you can, for example, make certain a book's cost is greater than zero, its retail price is less than $200.00, or a seller's commission rate is less than 50%. The condition included in the constraint can't reference certain built-in functions, such as SYSDATE, or refer to values stored in other rows (although it can be compared to values in the *same* row). For instance, you could use the condition that the order date must be earlier than or equal to the ship date. However, you couldn't add a CHECK constraint that requires an order's ship date to be the same as the current system date because you would have to reference the SYSDATE function. Figure 4-18 shows the syntax for adding a CHECK constraint to an existing table.

```
ALTER TABLE tablename
ADD [CONSTRAINT constraintname] CHECK (condition);
```

FIGURE 4-18   Syntax for adding a CHECK constraint to an existing table

NOTE

The SYSDATE function was introduced in Chapter 3. Other functions are covered in Chapter 10.

The syntax to add a CHECK constraint follows the same format as the syntax to add a PRIMARY KEY or UNIQUE constraint. However, rather than list column names for the constraint, you list the condition that must be satisfied.

To solve the problem of an incorrect ship date being entered in the table, the condition can be stated as (orderdate <= shipdate), which means the ship date can't be earlier than the order date. The command to add the CHECK constraint to the ORDERS table is shown in Figure 4-19.

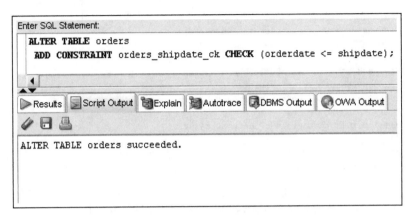

FIGURE 4-19   Adding a CHECK constraint to the ORDERS table

If any records already stored in the ORDERS table violate the orderdate <= shipdate condition, Oracle 11g returns an error message stating that the constraint has been violated, and the ALTER TABLE command fails. This is true for all constraint types. If you attempt to add a CHECK constraint and get an error message indicating a violation, issue a SELECT statement and review the table's data to identify any records preventing the constraint from being added to the table. After you identify and correct those records, you can reissue the ALTER TABLE command, and it should be successful. The INSERT statement in Figure 4-20 attempts to add an order with a shipping date of one day earlier than the order date and fails.

You can use a variety of operators to create the condition needed for your data. The following list shows each operator with an example:

- Less than (<): `retail < 200`
- Greater than (>): `cost > 0`
- Range (BETWEEN): `retail BETWEEN 0 AND 200`
- List of values (IN): `region IN ('NE', 'SE', 'NW', 'SW')`

TIP

The BETWEEN operator is inclusive. The preceding example is interpreted as the retail column value could be 0 or 200 or any number between those two values.

Constraints

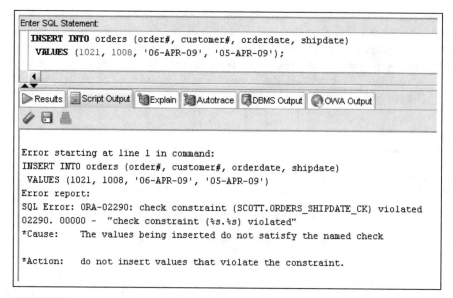

FIGURE 4-20   Insert a row to test the constraint

Another CHECK constraint is needed in the JustLee database to make sure the quantity of an item ordered is at least one. Figure 4-21 shows adding a CHECK constraint to address this requirement. The condition in this constraint could also be written as (quantity >= 1).

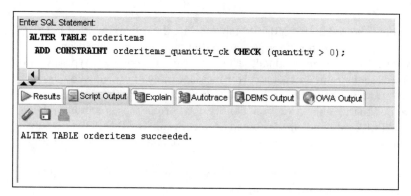

FIGURE 4-21   Adding a CHECK constraint to the ORDERITEMS table

USING THE NOT NULL CONSTRAINT

The **NOT NULL constraint** is a special CHECK constraint with the condition IS NOT NULL. It prevents users from adding a row that contains a NULL value in the specified column. However, a NOT NULL constraint isn't added to a table in the same manner as the other constraints discussed in this chapter. A NOT NULL constraint can be added *only to an existing column by using the ALTER TABLE ... MODIFY command*. The syntax for adding a NOT NULL constraint is shown in Figure 4-22.

```
ALTER TABLE tablename
MODIFY (columnname [CONSTRAINT constraintname]
NOT NULL);
```

FIGURE 4-22 Syntax for adding a NOT NULL constraint to an existing table

The ALTER TABLE ... MODIFY command is the same command used in Chapter 3 to redefine a column. You need to list just the column's name and the keywords NOT NULL. You don't have to list the column's datatype and width or any default value, if one exists. For example, earlier in the chapter, you added a FOREIGN KEY constraint on the Customer# column in the ORDERS table; however, this column still accepts a NULL value. To force the entry of a value for the Customer#, a NOT NULL constraint must be added to the Customer# column of the ORDERS table. Figure 4-23 shows the command and its successful execution.

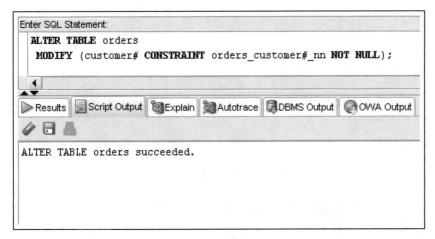

FIGURE 4-23 Adding a NOT NULL constraint

To test the NOT NULL constraint, add a row to the ORDERS table using a NULL value for the Customer# column, as shown in Figure 4-24.

Notice that the error message doesn't reference the NOT NULL constraint name, as with the other constraint types. The error message, however, is clear about the problem, including the identification of the specific table and column. For this reason, many developers don't assign constraint names to this type of constraint (although assigning a name makes referencing the constraint easier if you ever need to delete it in the future). If you don't want to assign a name to a NOT NULL constraint, simply omit the CONSTRAINT keyword and list the constraint type directly after the column name, as shown in Figure 4-25.

NOTE

The DESC command used in Chapter 3 identifies all columns that are NOT NULL or require input, including columns with PRIMARY KEY and NOT NULL constraints. Methods for identifying all the constraints for a table are explained in "Viewing Constraint Information" later in this chapter.

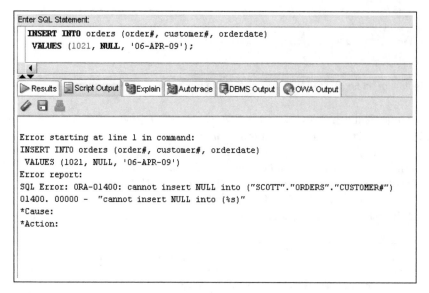

FIGURE 4-24   Insert a row to test the constraint

```
ALTER TABLE orders
MODIFY (customer# NOT NULL);
```

FIGURE 4-25   Adding a NOT NULL constraint without a name

INCLUDING CONSTRAINTS DURING TABLE CREATION

Now that you've examined adding constraints to existing tables, take a look at adding constraints to tables during table creation. When the design process for a database is thorough, you identify all needed constraints before creating a table. In this case, the constraints can be included in the CREATE TABLE command, so they don't need to be added later as a separate step.

JustLee Books would like to create some new tables to store office equipment inventory data. Figure 4-26 shows a basic E-R model for three tables to maintain the office equipment

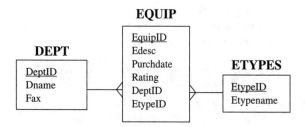

FIGURE 4-26   E-R model for equipment tables

inventory, including the needed columns and relationship lines. The underlined columns uniquely identify each row in the associated table.

After analyzing the data requirements, the following list of requirements, which will be addressed with constraints, was developed:

- Each department name must be unique.
- Each department must be assigned a name.
- Each equipment type name must be unique.
- Each equipment type must be assigned a name.
- Each equipment item must be assigned a valid department.
- If an equipment item is assigned a type, it must be a valid type.
- Valid rating values for equipment are A, B and C.

As mentioned, you can use two approaches to define constraints: at the column level or at the table level. If a constraint is defined at the table level, it's added to the CREATE TABLE statement following all the column definitions. All constraints except the NOT NULL constraint can be defined at the table level.

Next, review the completed CREATE TABLE statements for the three equipment tables. The JustLee technology group uses a convention of assigning names to all constraints except NOT NULL constraints. Figures 4-27 to 4-29 show the complete CREATE TABLE statements for all three tables, including all constraints.

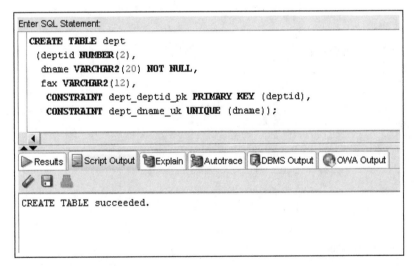

FIGURE 4-27 DEPT table creation

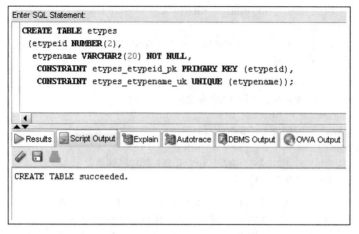

FIGURE 4-28 ETYPES table creation

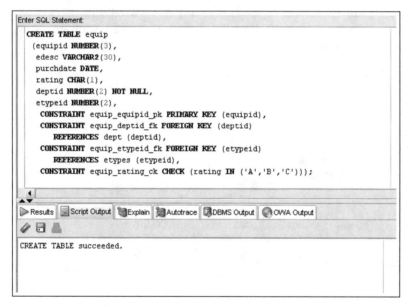

FIGURE 4-29 EQUIP table creation

Notice that all constraints except NOT NULL are listed at the bottom of the CREATE TABLE statement. Also, the EQUIP table includes two FOREIGN KEY constraints, one for each relationship line shown in the E-R model. These constraints meet the requirement that each equipment item is assigned valid DeptID and EtypeID values. If a constraint is created at the column level as part of the CREATE TABLE command, the constraint type is simply listed after the datatype for the column. Figure 4-30 modifies the CREATE TABLE statement for the DEPT table so that all constraints are defined at the column level. Compare this statement to the one in Figure 4-27, which uses the table level style.

Both CREATE TABLE statements build the same table; they differ only in the style of statement.

```
CREATE TABLE dept
 (deptid NUMBER(2) CONSTRAINT dept_deptid_pk PRIMARY KEY,
  dname VARCHAR2(20) NOT NULL
                      CONSTRAINT dept_dname_uk UNIQUE,
  fax VARCHAR2(12));
```

FIGURE 4-30   Constraints defined at the column level

Both the column-level and table-level approaches can be used in the same command. However, the general practice in the industry is to create constraints with the table-level approach. This isn't a requirement; it's simply a preference because a column list can become cluttered if a constraint name is entered in the middle of a list defining all the columns. Therefore, most users define all the columns first, and then include the constraints at the end of the CREATE TABLE command to separate the column definitions from the constraints. This style makes it much easier to go back and identify a problem if you get an error message.

You can create all the constraints without assigning names; however, doing so complicates interpreting error messages, as discussed earlier. The statement shown in Figure 4-31 creates the DEPT table with the three required constraints. Because the statement doesn't assign constraint names, Oracle assigns names to all the constraints.

```
CREATE TABLE dept
 (deptid NUMBER(2) PRIMARY KEY,
  dname VARCHAR2(20) NOT NULL UNIQUE,
  fax VARCHAR2(12));
```

FIGURE 4-31   Constraints with no names assigned

As you apply constraints to tables, keep the following guidelines in mind:

- A NOT NULL constraint shouldn't be assigned to a PRIMARY KEY column. A PRIMARY KEY enforces both NOT NULL and UNIQUE constraints.
- CHECK, FOREIGN KEY, and UNIQUE KEY constraints don't require a value. A NOT NULL constraint must be used along with these constraints to require input for a column.
- If a DEFAULT option is set for a column, a NOT NULL constraint shouldn't be used. If no value is provided for the column, the DEFAULT value is assigned.

TIP

A common error is assigning a NOT NULL constraint to a PRIMARY KEY column. This assignment doesn't generate an error message, but it duplicates processing because a PRIMARY KEY constraint doesn't allow NULL values. Recall that a PRIMARY KEY checks for both uniqueness and no NULL values.

ADDING MULTIPLE CONSTRAINTS ON A SINGLE COLUMN

You can assign a column as many constraints as needed to satisfy all the business rules for that column. For example, you created a CHECK constraint on the Quantity column of the ORDERITEMS table earlier, but this constraint might not be enough. To enter a new order item, a quantity is required, so the column also needs a NOT NULL constraint. You also added a composite PRIMARY KEY constraint to the ORDERITEMS table that included the Order# and Item#. The Order# must also be assigned a FOREIGN KEY constraint that references the ORDERS table to make sure a valid Order# is entered. Figure 4-32 shows the CREATE TABLE statement for the ORDERITEMS table that includes these constraints. Notice that both the Quantity and Order# columns have two constraints assigned.

```
CREATE TABLE ORDERITEMS
  (order# NUMBER(4),
   item# NUMBER(2),
   ISBN VARCHAR2(10),
   Quantity NUMBER(3) NOT NULL,
   PaidEach NUMBER(5,2) NOT NULL,
      CONSTRAINT orderitems_order#item#_pk PRIMARY KEY (order#, item#),
      CONSTRAINT orderitems_order#_fk FOREIGN KEY (order#)
            REFERENCES orders (order#),
      CONSTRAINT orderitems_isbn_fk FOREIGN KEY (isbn)
            REFERENCES books (isbn),
      CONSTRAINT orderitems_quantity_ck CHECK (quantity>0) );
```

FIGURE 4-32 Assigning multiple constraints to a column

NOTE

A NOT NULL constraint isn't required for the Order# column because it's part of the PRIMARY KEY constraint.

VIEWING CONSTRAINT INFORMATION

So far in this chapter, you have learned various ways to create different types of constraints. How do you verify what constraints exist? The USER_CONSTRAINTS view, which is part of the data dictionary, is used to identify existing constraints. For example, you want to display information about all the constraints created earlier for the EQUIP table. To view the portion of the data dictionary that references constraints, use the SELECT statement shown in Figure 4-33 to see information about all constraints on the EQUIP table.

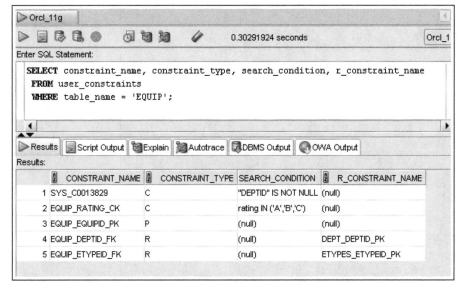

FIGURE 4-33   SELECT statement to view data about existing constraints

In the results, note the columns listed:

- The first column referenced, Constraint_name, lists the name of any constraint in the EQUIP table. Notice that the NOT NULL constraint has a system-generated constraint name.
- The second column, Constraint_type, lists the following letters: P for a PRIMARY KEY constraint, C for a CHECK or NOT NULL constraint, U for a UNIQUE constraint, or R for a FOREIGN KEY constraint.

TIP

The R might seem a little strange for a FOREIGN KEY constraint; however, the purpose of this constraint is to ensure referential integrity—meaning you're referencing something that actually exists. Therefore, the assigned code is the letter R.

- The third column, Search_condition, is used to display the condition in a CHECK or NOT NULL constraint. This column is blank for any other types of constraints.
- The fourth column, R_constraint_name, provides the name of the PRIMARY KEY constraint on the column that a FOREIGN KEY references.

The USER_CONSTRAINTS data dictionary view provides most of the information needed to confirm all the constraint settings for a table. It doesn't, however, include the specific column name the constraint is assigned to. Well-formed constraint names help identify the column, but the actual column name isn't displayed in this view. The USER_CONS_COLUMNS data dictionary view lists column names and assigned constraints, as shown in Figure 4-34.

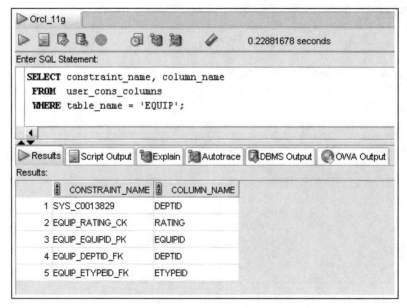

FIGURE 4-34 USER_CONS_COLUMNS data dictionary view

DISABLING AND DROPPING CONSTRAINTS

Sometimes you want to temporarily disable or drop a constraint. In this section, you examine these options.

Using DISABLE/ENABLE

When a constraint exists for a column, each entry made to that column is evaluated to determine whether the value is allowed in that column (that is, it doesn't violate the constraint). If you're adding a large block of data to a table, this validation process can severely slow down the Oracle server's processing speed. If you're certain the data you're adding adheres to the constraints, you can disable the constraints while adding that particular block of data to the table.

To **DISABLE** a constraint, you issue an ALTER TABLE command and change the constraint's status to DISABLE. Later, you can reissue the ALTER TABLE command and change the constraint's status back to **ENABLE**. Figure 4-35 shows the syntax for using the ALTER TABLE command to change the status of a constraint.

```
ALTER TABLE tablename
DISABLE CONSTRAINT constraintname;

ALTER TABLE tablename
ENABLE CONSTRAINT constraintname;
```

FIGURE 4-35 Syntax to disable or enable an existing constraint

For example, you're about to load several hundred rows into the EQUIP table from a file prepared after completing an initial inventory of office equipment. If the file was prepared so that only a rating value of A, B or C could be entered in the file, you already know the rating data already meets the condition in the CHECK constraint on this column. To speed up the data load, the simplest solution is to disable, or turn off, the constraint temporarily, and then enable it when you're finished, as shown in Figure 4-36.

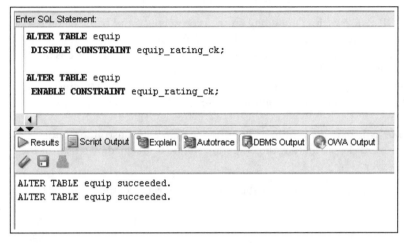

FIGURE 4-36   Disabling and enabling constraints

To instruct Oracle 11g to begin enforcing the constraint again, issue the same statement with the keyword ENABLE rather than DISABLE. After the constraint is enabled, data added or modified is again checked by the constraint.

CAUTION

Keep in mind that when a constraint is enabled, all the data relevant to that constraint is checked for validity. Therefore, if data is added that violates a constraint when the constraint is disabled, an error occurs when the constraint is enabled.

Dropping Constraints

If you create a constraint and then decide it's no longer needed (or you find an error in the constraint), you can delete the constraint from the table with the **DROP** (*constraintname*) command. In addition, if you need to change or modify a constraint, your only option is to delete the constraint and then create a new one. You use the ALTER TABLE command to drop an existing constraint from a table, using the syntax shown in Figure 4-37.

```
ALTER TABLE tablename
DROP PRIMARY KEY | UNIQUE (columnname) |
CONSTRAINT constraintname;
```

FIGURE 4-37   Syntax of the ALTER TABLE command to delete a constraint

Note the following guidelines for the syntax shown in Figure 4-37:

- The DROP clause varies depending on the type of constraint being deleted. If the DROP clause references the PRIMARY KEY constraint for the table, using the keywords PRIMARY KEY is enough because only one such clause is allowed for each table in the database.
- To delete a UNIQUE constraint, only the column name affected by the constraint is required because a column is referenced by only one UNIQUE constraint.
- Any other type of constraint must be referenced by the constraint's actual name—regardless of whether the constraint name is assigned by a user or the Oracle server.

Figure 4-38 shows a statement that drops the CHECK constraint on the Rating column of the EQUIP table.

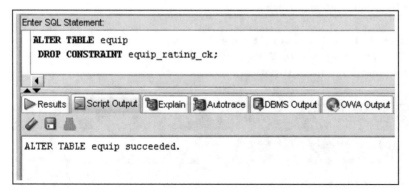

FIGURE 4-38   Dropping a NOT NULL constraint by name

If this ALTER TABLE command is executed successfully, the constraint no longer exists, and any value is accepted as input to the column.

The FOREIGN KEY constraint raises a special concern when attempting to drop constraints because it involves a relationship between two tables. Recall that a FOREIGN KEY column references a PRIMARY KEY column of another table. If you attempt to drop the PRIMARY KEY, an error is raised indicating that a FOREIGN KEY reference exists, as shown in Figure 4-39.

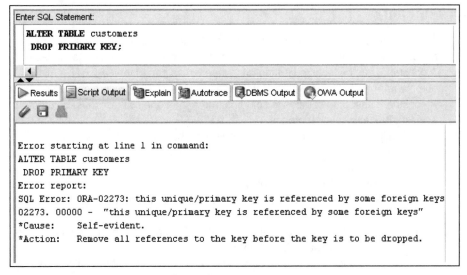

FIGURE 4-39 Error dropping a PRIMARY KEY referenced by a FOREIGN KEY

If needed, the associated FOREIGN KEY can be deleted along with the PRIMARY KEY deletion by using the CASCADE option, as shown in Figure 4-40.

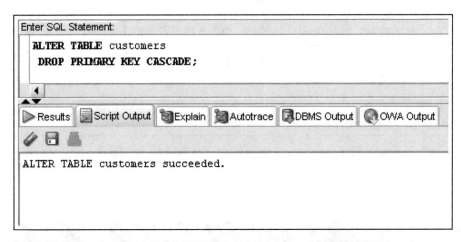

FIGURE 4-40 Dropping a PRIMARY KEY referenced by a FOREIGN KEY by using a CASCADE option

Chapter Summary

- A constraint is a rule applied to data being added to a table. It represents business rules, policies, or procedures. Data violating the constraint isn't added to the table.
- A constraint can be included during table creation as part of the CREATE TABLE command or added to an existing table with the ALTER TABLE command.
- A constraint based on composite columns (more than one column) must be created by using the table-level approach.
- A NOT NULL constraint can be created only with the column-level approach.
- A PRIMARY KEY constraint doesn't allow duplicate or NULL values in the designated column.
- Only one PRIMARY KEY constraint is allowed in a table.
- A FOREIGN KEY constraint requires that the column entry match a referenced column entry in the table or be NULL.
- A UNIQUE constraint is similar to a PRIMARY KEY constraint, except it allows storing NULL values in the specified column.
- A CHECK constraint ensures that data meets a given condition before it's added to the table. The condition can't reference the SYSDATE function or values stored in other rows.
- A NOT NULL constraint is a special type of CHECK constraint. If you're adding to an existing column, the ALTER TABLE ... MODIFY command must be used.
- A column can be assigned multiple constraints.
- The data dictionary views USER_CONSTRAINTS and USER_CONS_COLUMNS enable you to verify existing constraints.
- A constraint can be disabled or enabled with the ALTER TABLE command and the DISABLE and ENABLE keywords.
- A constraint can't be modified. To change a constraint, you must first drop it with the DROP command and then re-create it.

Chapter 4 Syntax Summary

The following table summarizes the syntax you have learned in this chapter. You can use the table as a study guide and reference.

Syntax Guide

Constraint	Description	Example

PRIMARY KEY	Determines which column(s) uniquely identifies each record. The primary key can't be NULL, and the data values must be unique.	*Constraint created during table creation:*
		`CREATE TABLE newtable`
		`  (firstcol NUMBER PRIMARY KEY,`
		`   secondcol VARCHAR2(20));`
		or
		`CREATE TABLE newtable`
		`  (firstcol NUMBER,`
		`   secondcol VARCHAR2(20),`

Constraint	Description	Example

		CONSTRAINT constraint_name_pk
		  PRIMARY KEY (firstcol));
		
		Constraint created after table creation:
		ALTER TABLE newtable
		ADD CONSTRAINT constraint_name_pk
		  PRIMARY KEY (firstcol);

FOREIGN KEY	In a one-to-many relationship, the constraint is added to the "many" table. The constraint ensures that if a value is entered in the specified column, it exists in the table being referred to, or the row isn't added.	*Constraint created during table creation:*
		CREATE TABLE newtable
		  (firstcol NUMBER,
		  secondcol VARCHAR2(20)
		REFERENCES anothertable(col1));
		or
		CREATE TABLE newtable
		  (firstcol NUMBER,
		  secondcol VARCHAR2(20),
		CONSTRAINT constraint_name_fk
		  FOREIGN KEY (secondcol)
		REFERENCES anothertable(col1);
		
		Constraint created after table creation:
		ALTER TABLE newtable
		ADD CONSTRAINT constraint_name_fk
		  FOREIGN KEY (secondcol)

UNIQUE	Ensures that all data values stored in the specified column are unique. The UNIQUE constraint differs from the PRIMARY KEY constraint in that it allows NULL values.	*Constraint created during table creation:*
		CREATE TABLE newtable
		  (firstcol NUMBER,
		  secondcol VARCHAR2(20) UNIQUE);
		or
		CREATE TABLE newtable
		  (firstcol NUMBER,
		  secondcol VARCHAR2(20),
		CONSTRAINT constraint_name_uk
		UNIQUE (secondcol));
		
		Constraint created after table creation:
		ALTER TABLE newtable
		ADD CONSTRAINT constraint_name_uk
		  UNIQUE (secondcol);

CHECK	Ensures that a specified condition is met before the data value is added to the table.	*Constraint created during table creation:*
		CREATE TABLE newtable
		  (firstcol NUMBER,
		  secondcol VARCHAR2(20),
		  thirdcol NUMBER CHECK (BETWEEN 20
		    AND 30));

128

Constraint	Description	Example

	For example, an order's ship date can't be "less than" its order date.	*or*
		`CREATE TABLE newtable`
		`  (firstcol NUMBER,`
		`   secondcol VARCHAR2(20),`
		`   thirdcol NUMBER,`
		`CONSTRAINT constraint_name_ck CHECK`
		`   (thirdcol BETWEEN 20 AND 80));`
		
		Constraint created after table creation:
		`ALTER TABLE newtable`
		`ADD CONSTRAINT constraint_name_ck`
		`   CHECK (thirdcol BETWEEN 20 AND 80);`

NOT NULL	Requires that the specified column can't contain a NULL value. It can be created *only* with the column-level approach to table creation.	*Constraint created during table creation:*
		`CREATE TABLE newtable`
		`  (firstcol NUMBER,`
		`   secondcol VARCHAR2(20),`
		`   thirdcol NUMBER NOT NULL);`
		Constraint created after table creation:
		`ALTER TABLE newtable`
		`MODIFY (thirdcol NOT NULL);`

Review Questions

To answer these questions, refer to the tables in the JustLee Books database.

1. What is the difference between a PRIMARY KEY constraint and a UNIQUE constraint?
2. How can you verify the constraints that exist for a table?
3. A table can have a maximum of how many PRIMARY KEY constraints?
4. Which type of constraint can be used to make certain the category for a book is included when a new book is added to inventory?
5. Which type of constraint should you use to ensure that every book has a profit margin between 15% and 25%?
6. How is adding a NOT NULL constraint to an existing table different from adding other types of constraints?
7. When must you define constraints at the table level rather than the column level?
8. To which table do you add a FOREIGN KEY constraint if you want to make certain every book ordered exists in the BOOKS table?
9. What is the difference between disabling a constraint and dropping a constraint?
10. What is the simplest way to determine whether a particular column can contain NULL values?

Multiple Choice

To answer the following questions, refer to the tables in the JustLee Books database.

1. Which of the following statements is correct?
   a. A PRIMARY KEY constraint allows NULL values in the primary key column(s).
   b. You can enable a dropped constraint if you need it in the future.
   c. Every table must have at least one PRIMARY KEY constraint, or Oracle 11*g* doesn't allow the table to be created.
   d. None of the above statements is correct.

2. Which of the following is *not* a valid constraint type?
   a. PRIMARY KEYS
   b. UNIQUE
   c. CHECK
   d. FOREIGN KEY

3. Which of the following SQL statements is invalid and returns an error message?
   a. `ALTER TABLE books ADD CONSTRAINT books_pubid_uk UNIQUE (pubid);`
   b. `ALTER TABLE books ADD CONSTRAINT books_pubid_pk PRIMARY KEY (pubid);`
   c. `ALTER TABLE books ADD CONSTRAINT books_pubid_nn NOT NULL (pubid);`
   d. `ALTER TABLE books ADD CONSTRAINT books_pubid_fk FOREIGN KEY (pubid)`
      `REFERENCES publisher (pubid);`
   e. All of the above statements are invalid.

4. What is the maximum number of PRIMARY KEY constraints allowed for a table?
   a. 1
   b. 2
   c. 30
   d. 255

5. Which of the following is a valid SQL command?
   a. `ALTER TABLE books ADD CONSTRAINT UNIQUE (pubid);`
   b. `ALTER TABLE books ADD CONSTRAINT PRIMARY KEY (pubid);`
   c. `ALTER TABLE books MODIFY (pubid CONSTRAINT NOT NULL);`
   d. `ALTER TABLE books ADD FOREIGN KEY CONSTRAINT (pubid) REFERENCES`
      `publisher (pubid);`
   e. None of the above commands is valid.

6. How many NOT NULL constraints can be created at the table level by using the CREATE TABLE command?
   a. 0
   b. 1
   c. 12
   d. 30
   e. 255

7. The FOREIGN KEY constraint should be added to which table?

   a. the table representing the "one" side of a one-to-many relationship

   b. the parent table in a parent-child relationship

   c. the child table in a parent-child relationship

   d. the table that doesn't have a primary key

8. What is the maximum number of columns you can define as a primary key when using the column-level approach to creating a table?

   a. 0

   b. 1

   c. 30

   d. 255

9. Which of the following commands can you use to rename a constraint?

   a. RENAME

   b. ALTER CONSTRAINT

   c. MOVE

   d. NEW NAME

   e. None of the above commands can be used.

10. Which of the following is a valid SQL statement?

   a.
```
CREATE TABLE table1
  (col1 NUMBER PRIMARY KEY,
   col2 VARCHAR2(20) PRIMARY KEY,
   col3 DATE DEFAULT SYSDATE,
   col4 VARCHAR2(2));
```

   b.
```
CREATE TABLE table1
  (col1 NUMBER PRIMARY KEY,
   col2 VARCHAR2(20),
   col3 DATE,
   col4 VARCHAR2(2) NOT NULL,
    CONSTRAINT table1_col3_ck
    CHECK (col3 = SYSDATE));
```

   c.
```
CREATE TABLE table1
  (col1 NUMBER,
   col2 VARCHAR2(20),
   col3 DATE,
   col4 VARCHAR2(2),
    PRIMARY KEY (col1));
```

   d.
```
CREATE TABLE table1
  (col1 NUMBER,
   col2 VARCHAR2(20),
   col3 DATE DEFAULT SYSDATE,
   col4 VARCHAR2(2);
```

11. In the initial creation of a table, if a UNIQUE constraint is included for a composite column that requires the combination of entries in the specified columns to be unique, which of the following statements is correct?

   a. The constraint can be created only with the ALTER TABLE command.
   b. The constraint can be created only with the table-level approach.
   c. The constraint can be created only with the column-level approach.
   d. The constraint can be created only with the ALTER TABLE ... MODIFY command.

12. Which type of constraint should you use on a column to allow entering only values above 100?

   a. PRIMARY KEY
   b. UNIQUE
   c. CHECK
   d. NOT NULL

13. Which of the following commands can be used to enable a disabled constraint?

   a. ALTER TABLE ... MODIFY
   b. ALTER TABLE ... ADD
   c. ALTER TABLE ... DISABLE
   d. ALTER TABLE ... ENABLE

14. Which of the following keywords allows the user to delete a record from a table, even if rows in another table reference the record through a FOREIGN KEY constraint?

   a. CASCADE
   b. CASCADE ON DELETE
   c. DELETE ON CASCADE
   d. DROP
   e. ON DELETE CASCADE

15. Which of the following data dictionary objects should be used to view information about the constraints in a database?

   a. USER_TABLES
   b. USER_RULES
   c. USER_COLUMNS
   d. USER_CONSTRAINTS
   e. None of the above objects should be used.

16. Which of the following types of constraints can't be created at the table level?

   a. NOT NULL
   b. PRIMARY KEY
   c. CHECK
   d. FOREIGN KEY
   e. None of the above constraints can be created at the table level.

17. Suppose you created a PRIMARY KEY constraint at the same time you created a table and later decide to name the constraint. Which of the following commands can you use to change the constraint's name?

    a. ALTER TABLE ... MODIFY
    b. ALTER TABLE ... ADD
    c. ALTER TABLE ... DISABLE
    d. None of the above commands can be used.

18. You're creating a new table consisting of three columns: Col1, Col2, and Col3. Col1 should be the primary key and can't have any NULL values, and each entry should be unique. Col3 must not contain any NULL values either. How many total constraints do you have to create?

    a. 1
    b. 2
    c. 3
    d. 4

19. Which of the following types of restrictions can be viewed with the DESCRIBE command?

    a. NOT NULL
    b. FOREIGN KEY
    c. UNIQUE
    d. CHECK

20. Which of the following is the valid syntax for adding a PRIMARY KEY constraint to an existing table?

    a. ALTER TABLE *tablename* ADD CONSTRAINT PRIMARY KEY (*columnname*);
    b. ALTER TABLE *tablename* ADD CONSTRAINT (*columnname*) PRIMARY KEY *constraintname*;
    c. ALTER TABLE *tablename* ADD[CONSTRAINT *constraintname*] PRIMARY KEY;
    d. None of the above is valid syntax.

Hands-On Assignments

JustLee Books has become the exclusive distributor for a number of books. The company now needs to assign sales representatives to retail bookstores to handle the new distribution duties. For these assignments, create new tables to support the following:

1. Modify the following SQL command so that the Rep_ID column is the PRIMARY KEY for the table and the default value of Y is assigned to the Comm column. (The Comm column indicates whether the sales representative earns commission.)

```
CREATE TABLE store_reps
  (rep_ID NUMBER(5),
     last VARCHAR2(15),
     first VARCHAR2(10),
     comm CHAR(1) );
```

2. Change the STORE_REPS table so that NULL values can't be entered in the name columns (First and Last).

3. Change the STORE_REPS table so that only a Y or N can be entered in the Comm column.

4. Add a column named Base_salary with a datatype of NUMBER(7,2) to the STORE_REPS table. Ensure that the amount entered is above zero.

5. Create a table named BOOK_STORES to include the columns listed in the following chart:

6. Add a constraint to make sure the Rep_ID value entered in the BOOK_STORES table is a

Column Name	Datatype	Constraint Comments
Store_ID	NUMBER(8)	PRIMARY KEY column
Name	VARCHAR2(30)	Should be UNIQUE and NOT NULL
Contact	VARCHAR2(30)	
Rep_ID	VARCHAR2(5)	

valid value contained in the STORE_REPS table. The Rep_ID columns of both tables were initially created as different datatypes. Does this cause an error when adding the constraint? Make table modifications as needed so that you can add the required constraint.

7. Change the constraint created in Assignment #6 so that associated rows of the BOOK_STORES table are deleted automatically if a row in the STORE_REPS table is deleted.

8. Create a table named REP_CONTRACTS containing the columns listed in the following chart. A composite PRIMARY KEY constraint including the Rep_ID, Store_ID, and Quarter columns should be assigned. In addition, FOREIGN KEY constraints should be assigned to both the Rep_ID and Store_ID columns.

Column Name	Datatype
Store_ID	NUMBER(8)
Name	NUMBER(5)
Quarter	CHAR(3)
Rep_ID	NUMBER(5)

9. Produce a list of information about all existing constraints on the STORE_REPS table.

10. Issue the commands to disable and then enable the CHECK constraint on the Base_salary column.

Advanced Challenge

Create two tables based on the E-R model shown in Figure 4-41 and the business rules in the following list for a work order tracking database. Include all the constraints in the CREATE TABLE statements. You should have *only* two CREATE TABLE statements and no ALTER TABLE statements. Name all constraints except NOT NULLs.

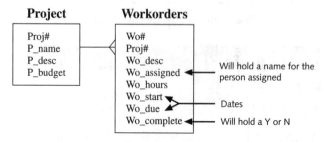

FIGURE 4-41   Workorders E-R model

- Use your judgment for column datatypes and sizes.
- Proj# and Wo# are used to uniquely identify rows in these tables.
- Each project added must be assigned a name, and no duplicate project names are allowed.
- Each work order must be assigned to a valid project when added and be assigned a description and number of hours.
- Each work order added must have a different description.
- The number of hours assigned to a work order should be greater than zero.
- If data is provided for the Wo_complete column, only Y or N are acceptable values.

Create and execute the SQL statements needed to enforce the data relationships among these tables.

Case Study: *City Jail*

In previous chapters, you have designed and created tables for the City Jail database. These tables don't include any constraints. Review the information in Chapters 1 and 3 case studies to determine what constraints you might need for the City Jail database.

First, using the format in the following chart, create a list of constraints needed. Second, create and execute all the SQL statements needed to add these constraints. Follow these steps to create and alter the tables:

1. First, drop the APPEALS, CRIME_OFFICERS, and CRIME_CHARGES tables constructed in Chapter 3. These three tables are to be built last, using a CREATE TABLE command that includes all the necessary constraints.

2. Second, use the ALTER TABLE command to add all constraints to the existing tables. Note that the sequence of constraint addition has an impact. Any tables referenced by FOREIGN KEYs must already have the PRIMARY KEY created.
3. Third, use the CREATE TABLE command, including all constraints, to build the three tables dropped in the first step.

Table Name	Column(s)	Constraint Type	Condition

DATA MANIPULATION AND TRANSACTION CONTROL

LEARNING OBJECTIVES

After completing this chapter, you should be able to do the following:

- Use the INSERT command to add a record to an existing table
- Manage virtual columns in data manipulations
- Use quotes in data values
- Use a subquery to copy records from an existing table
- Use the UPDATE command to modify a table's existing rows
- Use substitution variables with an UPDATE command
- Delete records
- Manage transactions with the transaction control commands COMMIT, ROLLBACK, and SAVEPOINT
- Differentiate between a shared lock and an exclusive lock
- Use the SELECT … FOR UPDATE command to create a shared lock

INTRODUCTION

In Chapters 3 and 4, you issued data definition language (DDL) commands to create, alter, and drop database tables. You also used basic INSERT commands to add table data rows to test table constraints. This chapter delves further into methods for adding new data rows and introduces modifying or deleting existing data rows. All the operations performed in this chapter differ from DDL statements in that they affect the data stored *in* tables, not the actual structure of tables. Commands that modify data are called **data manipulation language (DML)** commands. Commands that save data permanently or undo data changes are referred to as transaction control commands. Table 5-1 lists the commands you use in this chapter.

TABLE 5-1   DML and Transaction Control Commands

Command	Description
INSERT	Adds new rows to a table; can include a subquery to copy rows from an existing table
UPDATE	Adds data to, or modifies data in, existing rows
DELETE	Removes rows from a table
COMMIT	Saves changed data in a table permanently
ROLLBACK	Allows "undoing" uncommitted changes to data
SAVEPOINT	Enables setting markers in a transaction
LOCK TABLE	Prevents other users from making changes to a table
SELECT ... FOR UPDATE	Creates a shared lock on a table to prevent another user from making changes to data in specified columns

DATABASE PREPARATION

Go to the Chapter 5 folder in your data files. Before working through the examples in this chapter, run the JLDB_Build_5.sql script to ensure that all necessary tables and constraints are available. This script removes existing tables and creates a new set of tables. Refer to the steps at the beginning of Chapter 2 for loading and executing a script. Ignore any errors in the DROP TABLE statements at the beginning of the script. An "object does not exist" error merely indicates that the table wasn't created in the schema previously.

NOTE

The JLDB_Build_5.sql script includes creating a table named ACCTMANAGER, as shown in Figure 5-1. Review the table structure, including datatypes, constraints, and DEFAULT options. The ACCTMANAGER table is used throughout this chapter for DML tasks. It currently contains no data rows.

```
CREATE TABLE acctmanager
(amid CHAR(4),
 amfirst VARCHAR2(12)  NOT NULL,
 amlast VARCHAR2(12)  NOT NULL,
 amedate DATE DEFAULT SYSDATE,
 amsal NUMBER(8,2),
 amcomm NUMBER(7,2) DEFAULT 0,
 region CHAR(2),
  CONSTRAINT acctmanager2_amid_pk PRIMARY KEY (amid),
  CONSTRAINT acctmanager2_region_ck
    CHECK (region IN ('N', 'NW', 'NE', 'S', 'SE', 'SW', 'W', 'E')));
```

FIGURE 5-1   The ACCTMANAGER table creation

INSERTING NEW ROWS

As discussed in Chapter 3, the management of JustLee Books is implementing a new commission policy for regional account managers. The ACCTMANAGER table was created to store data about account managers. Now that the table has been created and all necessary constraints for the table are in place, it's time to add data to the table. The data shown in Table 5-2 is for account managers you add to the ACCTMANAGER table in the next section. (Blank spaces indicate that data hasn't been provided yet.)

TABLE 5-2   Data for Account Managers

ID	Name	Employment Date	Salary	Commission	Region
T500	Nick Taylor	September 5, 2009	$42,000.00	$3,500.00	NE
L500	Mandy Lopez	October 1, 2009	$47,000.00	$1,500.00	
J500	Sammie Jones	Today	$39,500.00	$2,000.00	NW

Using the INSERT Command

An **INSERT** command is used to add new rows of data to a table. The syntax of this command is shown in Figure 5-2.

```
INSERT INTO tablename [(columnname, ...)]
VALUES (datavalue, ...);
```

FIGURE 5-2   Syntax of the INSERT command

Note the following syntax elements in Figure 5-2:

- The keywords **INSERT INTO** are followed by the name of the table into which rows will be entered. The table name is followed by a list of the columns containing the data. The square brackets surrounding this list indicate that using a column list is optional.

- The **VALUES** clause identifies the data values to be inserted in the table. You list the data values in parentheses after the VALUES keyword.
- If the data entered in the VALUES clause contains a value for every column and is in the same order as columns in the table, column names can be omitted in the INSERT INTO clause.
- If you enter data for only some columns, or if columns are listed in a different order than they're listed in the table, the column names *must* be provided in the INSERT INTO clause in the same order as they're given in the VALUES clause. You must list the column names inside parentheses after the table name in the INSERT INTO clause.
- If more than one column is listed, column names must be separated by commas.
- If more than one data value is entered, the values must be separated by commas.
- You must use single quotes to enclose nonnumeric data inserted in a column. (That is, the column's datatype is not NUMBER.)

To insert the first account manager's data (refer to Table 5-2) in the ACCTMANAGER table, use the command shown in Figure 5-3.

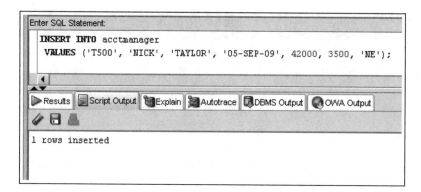

FIGURE 5-3   The INSERT command for Nick Taylor

The INSERT INTO clause shown in Figure 5-3 doesn't contain a list of column names because the VALUES clause contains a valid entry for every column in the ACCTMANAGER table, and the data is given in the same order as columns are listed in the table.

The character data in the VALUES clause is entered in all uppercase characters, as for all tables in the JustLee Books database. When you enter character data, it retains the case you use in the INSERT INTO command. For example, if an account manager's name is entered in mixed case (uppercase and lowercase letters), the table stores the name in mixed case. Mixed case can make future record searches difficult if the data doesn't follow a consistent format or case because character data matching is case sensitive.

The date value uses the default date format for Oracle11g: two-digit day, three-character-month, and two-digit year, separated by hyphens. (You can also provide a four-digit year value.) Notice that single quotes aren't required for numeric values in the INSERT INTO statement, such as the salary and commission values. Also, no formatting characters, such as a comma or dollar sign, should be included in values because they raise an error. A NUMBER column stores only digits. You add formatting characters when querying numeric values in Chapter 10.

NOTE

Methods to contend with case sensitivity issues and different date formats are addressed in Chapter 10.

After you execute the INSERT INTO command, the message "1 rows inserted" is displayed, indicating that data has been inserted in the table. To verify that the row was added, you can use a SELECT statement to view the table's contents. Figure 5-4 shows a query to confirm that data for the first account manager, Nick Taylor, has been added to the ACCTMANAGER table.

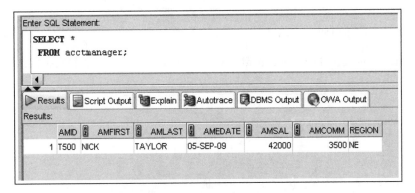

FIGURE 5-4   Verify that data was inserted

TIP

When inserting table data, the most common error is forgetting to enclose data for nonnumeric columns in single quotes. When this occurs, Oracle 11g displays the message "ERROR:ORA-01756: quoted string not properly terminated." You can correct the problem by simply reissuing the command with the required single quotes.

The next record to enter in the ACCTMANAGER table contains data about Mandy Lopez. However, as shown in Table 5-2, she hasn't yet been assigned a marketing region, so that column is left blank. You can take one of the following approaches to indicate that the Region column contains a NULL value:

- List all columns *except* the Region column in the INSERT INTO clause, and provide data for the listed columns in the VALUES clause.
- In the VALUES clause, substitute *two single quotes* in the position that should contain the account manager's assigned region. Oracle 11g interprets the two

single quotes to mean that a NULL value should be stored in the column. Be sure you don't add a blank space between these single quotes, however. Doing so adds a blank space value rather than a NULL value.

• In the VALUES clause, include the keyword NULL in the position where the region should be listed. As long as the keyword NULL isn't enclosed in single quotes, Oracle 11g leaves the column blank. However, if the keyword is mistakenly entered as 'NULL', Oracle 11g tries to store the word "NULL" in the column.

Figure 5-5 shows the INSERT INTO commands for entering NULL values with these three methods.

```
1.  INSERT INTO acctmanager (amid, amfirst, amlast, amedate, amsal, amcomm)
       VALUES ('L500', 'MANDY', 'LOPEZ', '01-OCT-09', 47000, 1500);
2.  INSERT INTO acctmanager
       VALUES ('L500', 'MANDY', 'LOPEZ', '01-OCT-09', 47000, 1500, '');
3.  INSERT INTO acctmanager
       VALUES ('L500', 'MANDY', 'LOPEZ', '01-OCT-09', 47000, 1500, NULL);
```

FIGURE 5-5   Methods for entering NULL values

Execute the third method, and Oracle 11g adds Mandy Lopez to the ACCTMANAGER table, as shown in Figure 5-6.

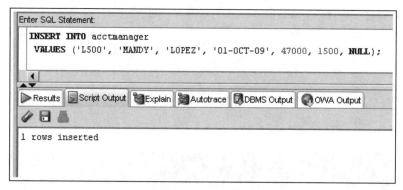

FIGURE 5-6   The INSERT INTO command for Mandy Lopez

NOTE

Many developers consider using a column list in an INSERT command a good practice. First, reading statements that include a column list is easier because you don't need to reference a column structure listing to recall the columns in the table. Second, if columns are added to the table later, an INSERT statement using the original column list still executes successfully.

Next, you need to enter the record for Sammie Jones in the ACCTMANAGER table. Note that Sammie's employment date should be set to the current date. You could include the value SYSDATE in the VALUES clause to have Oracle 11g insert the current date in the Amedate column, as shown in Figure 5-7.

```
INSERT INTO acctmanager (amid, amfirst, amlast, amedate, amsal, amcomm, region)
  VALUES ('J500', 'SAMMIE' 'JONES', 'SYSDATE', 39500, 2000, 'NW');
```

FIGURE 5-7   Using SYSDATE as a data value

However, in this case, the Amedate column has a DEFAULT option set to SYSDATE. To instruct Oracle 11g to use the DEFAULT option, you can use one of two methods:

- Include a column list in the INSERT INTO clause that omits the Amedate column.
- Use the keyword DEFAULT for the column value in the VALUES clause.

Figure 5-8 shows the commands for both methods.

```
1. INSERT INTO acctmanager (amid, amfirst, amlast, amsal, amcomm, region)
     VALUES ('J500', 'SAMMIE', 'JONES', 39500, 2000, 'NW');

2. INSERT INTO acctmanager (amid, amfirst, amlast, amedate, amsal, amcomm, region)
     VALUES ('J500', 'SAMMIE', 'JONES', 'DEFAULT', 39500, 2000, 'NW');
```

FIGURE 5-8   Methods for using a DEFAULT option

Use Figure 5-9 as a guide to execute the first method shown in Figure 5-8 for adding the Sammie Jones record. The column list in the INSERT INTO clause lists all columns of the ACCTMANAGER table except Amedate. Although the columns are listed in the same sequence as in the actual table, this order isn't a requirement. What *is* required is that the data listed in the VALUES clause matches the *exact order* of the columns listed in the INSERT INTO clause. The order of the list dictates which data value is assigned to which column.

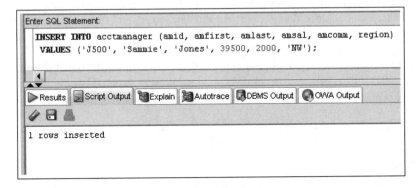

FIGURE 5-9   Use an INSERT statement that applies a DEFAULT column option

T I P

What if the employment date for Sammie Jones is supposed to be NULL? Because the column has a DEFAULT option setting, using a column list and excluding the Amedate column activates the DEFAULT option and enters the current date for the column. To prevent this result, you must supply a NULL value in the INSERT command for the Amedate column.

As shown in Figure 5-10, a display of the ACCTMANAGER table's current contents confirms the addition of the three records. Because the current date was used for Sammie Jones's employment date, your results will vary from what's shown in Figure 5-10. Note that the name values for Sammie Jones are in mixed case, which illustrates that character values are stored in the case that's used when the data is inserted.

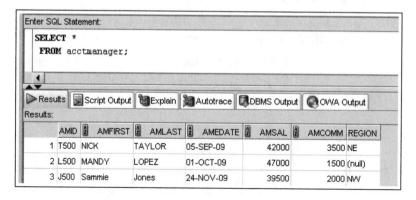

FIGURE 5-10 Contents of the ACCTMANAGER table

NOTE

The order of rows shown in Figure 5-10 might also vary from yours because no sorting operation is used.

Handling Virtual Columns

In Chapter 3, you created an Amearn column for the ACCTMANAGER table. This column is a virtual column, which is a column that generates values based on other column values. In other words, the database system generates the value for the column automatically based on the manipulation (in this case, a calculation) defined for the column.

What happens if a virtual column is included in an INSERT command? A virtual column must be ignored in an INSERT command, or an error occurs. To add a virtual column to the ACCTMANAGER table, execute the ALTER TABLE statement shown in Figure 5-11.

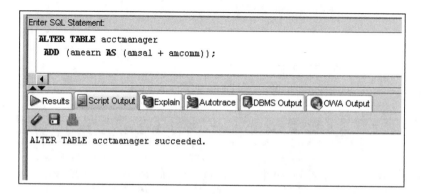

FIGURE 5-11 Add a virtual column

The new column, Amearn, contains the total earnings for an account manager, which is the salary plus commission.

Next, perform the query shown in Figure 5-12 to view the data values generated by the virtual column for existing rows. Keep in mind that the values for Amearn are generated during the query, using the calculation defined for the column.

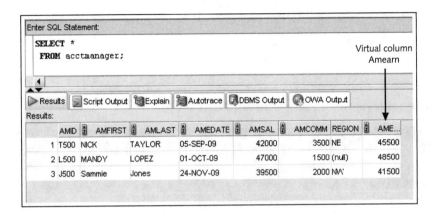

FIGURE 5-12   View data generated by a virtual column

Figure 5-13 shows an INSERT attempt of a fourth account manager. This command includes a value for the virtual column Amearn. Note that the error message clearly states a value can't be inserted in a virtual column.

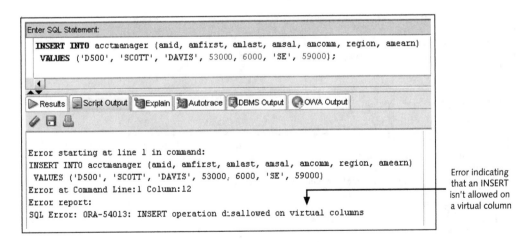

FIGURE 5-13   Error caused by using a virtual column in an INSERT statement

To add this row, you must use an INSERT command with a column list excluding the virtual column, as shown in Figure 5-14.

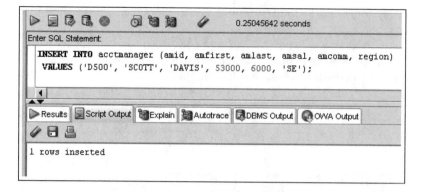

FIGURE 5-14 Successful INSERT on a table with a virtual column

> ### NOTE
>
> Any calculations involving a NULL value always result in a NULL value. Therefore, virtual columns that reference NULL value columns for calculations generate a NULL value. You discover how to optionally substitute values for a NULL value for calculations in Chapter 10.

Handling Single Quotes in an INSERT Value

Inserting values containing single quotes raises an error because they're confused with the single quotes used to enclose character or string values. For example, what if another account manager named Peg O'hara needs to be added to the ACCTMANAGER table? Figure 5-15 shows the INSERT statement, including the single quote in the account manager's last name.

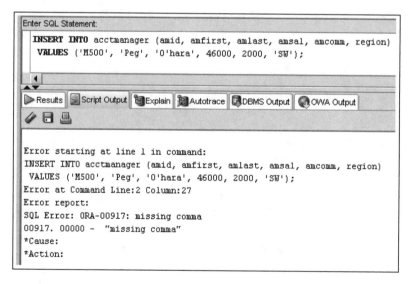

FIGURE 5-15 Error with a single quote in the value

The error states that the issue is a missing comma. The single quote in the last name value is treated as the closing quote for the string value 'O' for the last name. When the INSERT command attempts to read values following this value, an error is raised.

To instruct Oracle 11g to treat a single quote as part of a string value, enter two single quotes together in the value. Don't use the double quote (") because the command actually inserts this character in the string value, and the result would be O"hara. Figure 5-16 shows the successful INSERT command for account manager Peg O'hara. The command in Figure 5-17 displays the data in the ACCTMANAGER table to confirm that a single quote is inserted in the last name value O'hara.

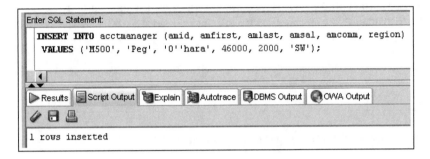

FIGURE 5-16   Use two single quotes in the INSERT value

Enter SQL Statement:
```
SELECT *
FROM acctmanager;
```

Results:

	AMID	AMFIRST	AMLAST	AMEDATE	AMSAL	AMCOMM	REGION	AME...
1	T500	NICK	TAYLOR	05-SEP-09	42000	3500	NE	45500
2	L500	MANDY	LOPEZ	01-OCT-09	47000	1500	(null)	48500
3	J500	Sammie	Jones	24-NOV-09	39500	2000	NW	41500
4	D500	SCOTT	DAVIS	24-NOV-09	53000	6000	SE	59000
5	M500	Peg	O'hara	24-NOV-09	46000	2000	SW	48000

Single quote in value inserted successfully

FIGURE 5-17   Query to confirm that a single quote was inserted successfully

Results might vary, depending on your computer's system date, because SYSDATE was used for employment dates for three account managers.

NOTE

The INSERT commands covered so far add one row per statement. Oracle 11*g* also includes an INSERT ALL command, which enables you to add multiple rows with a single statement.

Inserting Data from an Existing Table

In Chapter 3, you learned how to use the CREATE TABLE command with a subquery to create and populate a new table based on an existing table's structure and content. However, what if a table already exists, and you need to add copies of records stored in another table? In this case, you can't use the CREATE TABLE command. Because the table already exists, you need to use the INSERT INTO command with a subquery. Figure 5-18 shows the syntax for combining an INSERT INTO command with a subquery.

```
INSERT INTO tablename [(columnname, …)]
subquery;
```

FIGURE 5-18 Syntax of the INSERT INTO command with a subquery

Note the following elements in Figure 5-18:

- The main difference between using the INSERT INTO command with data values and with a subquery is that *the VALUES clause isn't included with a subquery*. The keyword VALUES indicates that the clause contains data values that must be inserted in the indicated table. However, the user isn't entering data values; the data is derived from the subquery's results.
- Also, unlike the CREATE TABLE command, the INSERT INTO command doesn't require enclosing the subquery in parentheses, although including parentheses doesn't generate an error message.

The management of JustLee Books is exploring a new bonus policy for account managers. A new table called ACCTBONUS has been created to test the new bonus calculations, and this table should contain the ID, salary, and region for all account managers. The ACCTMANAGER table currently contains all account manager data, and the requested account manager data must be copied to the existing ACCTBONUS table. Use a DESCRIBE statement and SELECT * query to view the structure and data in the ACCTBONUS table. Note that the table is currently empty.

After the requested data has been copied to the ACCTBONUS table, SQL statements from the Budget Department's program that calculates commissions can be tested. Moving this data to a separate table gives the Information Technology Department the flexibility of making changes to the original ACCTMANAGER table without interfering with the Budget Department's program testing. Figure 5-19 shows the command to copy rows from the ACCTMANAGER table to the ACCTBONUS table.

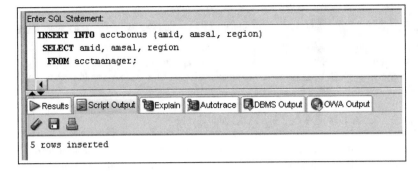

FIGURE 5-19 INSERT INTO command with a subquery

Note the following elements in Figure 5-19:

- The SELECT clause of the subquery lists the columns to be copied from the ACCTMANAGER table, which is identified in the FROM clause.
- Even though a column list is used, the INSERT INTO clause isn't required to contain one because in this example, the columns the subquery returns are in the same order as columns in the ACCTBONUS table.

After the command has been executed, query the ACCTBONUS table to confirm that all account managers now exist in the ACCTBONUS table. As shown in the results in Figure 5-20, five account manager rows now exist in the ACCTBONUS table.

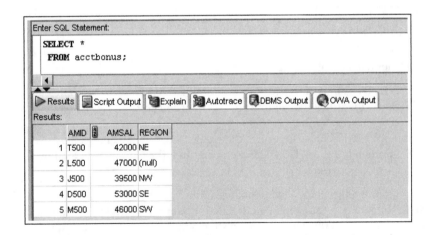

FIGURE 5-20 ACCTBONUS data rows

NOTE

In later chapters, you learn many more options to use in the SELECT statement, such as restricting the rows that are included and performing aggregate calculations to summarize data. All these options can be used in a subquery.

NOTE

Oracle 11*g* has several utilities for importing and exporting data to and from external files, such as SQL*Loader and Data Pump. This topic is beyond the scope of this textbook, so it isn't addressed in this chapter. Appendix D offers an introduction to SQL*Loader to give you an overview of a bulk import operation.

MODIFYING EXISTING ROWS

Often you need to change column data values. For example, when customers move, their mailing addresses need to be updated; when the wholesale costs of books change, retail prices need to be changed. Because the INSERT INTO command can be used only to add new rows to a table, you can't use it to modify existing data. To alter existing table data, you use the UPDATE command. In this section, you learn how to perform updates and create interactive update scripts by using substitution variables.

Using the UPDATE Command

You change the contents of existing rows with the **UPDATE** command. The syntax of this command is shown in Figure 5-21.

```
UPDATE tablename
SET columnname = new_datavalue, …
[WHERE condition];
```

FIGURE 5-21 Syntax of the UPDATE command

Note the following elements in Figure 5-21:

- The UPDATE clause identifies the table containing the records to be changed.
- The **SET** clause identifies the columns to be changed and the new values to be assigned to these columns.
- The optional **WHERE** clause identifies the exact records to be changed by the UPDATE command. If the WHERE clause is omitted, the column specified in the SET clause is updated for *all records* in the table.

Next, you use this command to make several changes that have been requested for data in the ACCTMANAGER table. First, the employment date for Sammie Jones is incorrect and needs to be changed to August 1, 2009. The command shown in Figure 5-22 can be issued to correct the employment date. Notice the SET clause, which indicates only the new value to be entered in the Amedate column. The original value is simply overwritten, so including it isn't necessary.

The WHERE clause identifies exactly which record should be altered. In this case, the easiest way to specify that only Sammie Jones's employment date should be changed is to include the condition that the Amid column must be equal to J500. Because the Amid column is the primary key for the ACCTMANAGER table, no two records can have the same Amid; therefore, only the record for Sammie Jones is affected by the update.

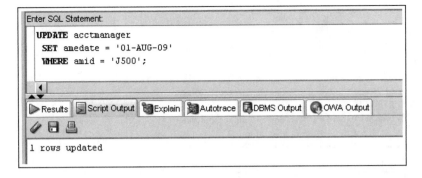

FIGURE 5-22 Command to correct the Amedate column value

Next, the northern regions are being closed, and the account managers assigned to these regions need to be reassigned to the western region. You can make this change by using the command shown in Figure 5-23.

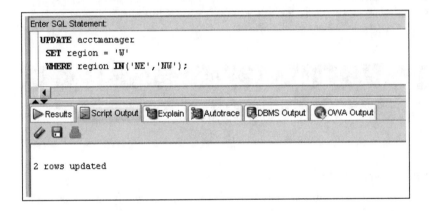

FIGURE 5-23 UPDATE command to reassign regions

Query the table to verify the data changes. Notice that in this case, two of the existing rows in the table were modified.

What if you need to modify more than one column of a row? You can do this by adding multiple columns in the SET clause of the UPDATE command. For example, the employment date for Mandy Lopez needs to be changed to be October 10, 2009, and she needs to be assigned to the southern region. Execute the statement in Figure 5-24.

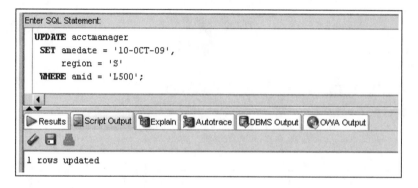

FIGURE 5-24   Updating multiple columns

CAUTION

In Chapter 4, you executed INSERT statements that raised constraint violations. Keep in mind that all data manipulations must meet constraint requirements to complete successfully. Otherwise, the DML action is stopped, and a constraint violation error message is displayed.

NOTE

A MERGE statement is also available in Oracle 11*g* to perform DML operations. This command, covered in Chapter 12, allows a sequence of conditional INSERT and UPDATE commands in a single statement.

Using Substitution Variables

Sometimes just adding a record to a table seems like a lot of effort, and modifying an existing record seems to take even more effort, especially if you need to add or modify 10 or 20 records. For example, say the Region column has just been added to the CUSTOMERS table with the ALTER TABLE command. Every customer's record needs to be updated with the value for this new column. Depending on the strategy you use, the UPDATE command must be reissued several times—at least once for every identified region. Instead of typing the same command again and again for the few values that differ, using a substitution variable is much simpler.

NOTE

Keep in mind that SQL commands can be embedded in applications to perform database interaction activities. If so, values from user input in screen elements, such as text boxes, are passed to the SQL statement via variables. The use of SQL*Plus substitution variables is similar to this process.

A **substitution variable** in an SQL command instructs Oracle 11g to substitute a value in place of the variable at the time the command is actually executed. To include a substitution variable in an SQL command, simply enter an ampersand (&) followed by the name used for the variable.

First, clear all current values stored in the Region column of the CUSTOMERS table with the UPDATE statement shown in Figure 5-25. Be sure the single quotes for the SET clause value don't contain a blank space because that sets the column to a NULL value. A WHERE clause isn't included in this UPDATE statement, so all rows are modified. Next, perform a query to view all rows in the CUSTOMERS table to confirm that the Region value is NULL for all rows.

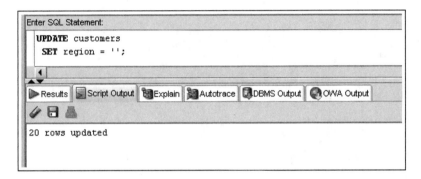

FIGURE 5-25 Clear the Region column

Now take a look at modifying the Region column for customers in a specific state. For example, you want to modify the records of all customers residing in California so that the Region column contains the value W, representing the western region. The command is shown in Figure 5-26.

```
UPDATE customers
  SET region = 'W'
  WHERE state = 'CA';
```

FIGURE 5-26 Command to set the Region value for California customers

To alter the command shown in Figure 5-26 so that you can reuse it for each state in which JustLee Books has customers, you need to enter a substitution variable in place of the value for the State column. Furthermore, the SET clause can contain a substitution variable, so the same command could be used for every region to be updated. Figure 5-27 shows the new command containing substitution variables.

When Oracle 11*g* executes the command shown in Figure 5-27, the user is first prompted to enter a value for the substitution variable named Region. The name of a substitution variable doesn't need to be the same as an existing column name; however, it should clearly indicate the data being requested from the user. Because the SET clause needs the user to enter a value for the region, the variable is named Region.

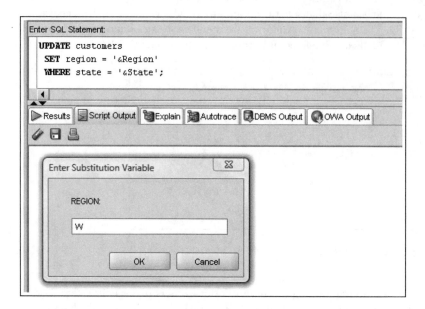

FIGURE 5-27   Prompt for substitution variable input

Enter the value W at the Region variable prompt. After entering a value for the Region column, you're asked for the value of the second substitution variable in the WHERE clause. This substitution variable, named State, is used to define exactly which rows to update. Enter the value CA at the State variable prompt. Then execute the query shown in Figure 5-28 to verify that customer rows with a State value of CA now have a region value of W.

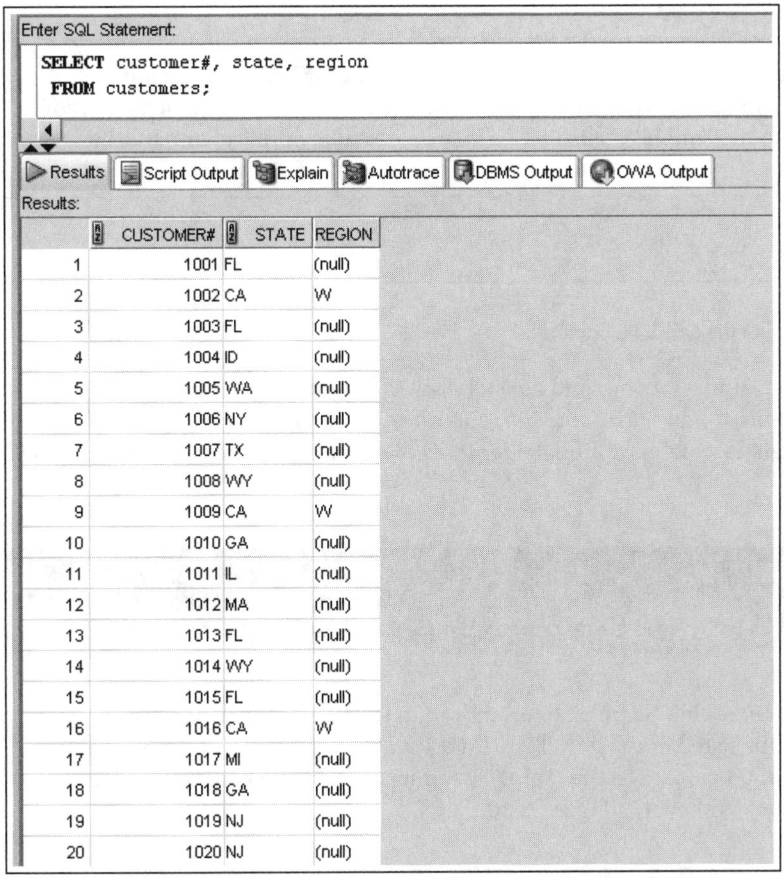
Enter SQL Statement:
SELECT customer#, state, region
FROM customers;
Results
Script Output
Explain
Autotrace
DBMS Output
OWA Output
Results:
CUSTOMER# STATE REGION
1 1001 FL (null)
2 1002 CA W
3 1003 FL (null)
4 1004 ID (null)
5 1005 WA (null)
6 1006 NY (null)
7 1007 TX (null)
8 1008 WY (null)
9 1009 CA W
10 1010 GA (null)
11 1011 IL (null)
12 1012 MA (null)
13 1013 FL (null)
14 1014 WY (null)
15 1015 FL (null)
16 1016 CA W
17 1017 MI (null)
18 1018 GA (null)
19 1019 NJ (null)
20 1020 NJ (null)

FIGURE 5-28 Verify UPDATE results

The statement can now be reexecuted easily. By using substitution variables, the statement becomes interactive, and the user can continue to update the Region column for as many states as necessary.

If a user can't complete all the customer record updates during one session, the command can be stored permanently in a script to be executed later. A script is a text file containing one or more SQL statements. You can create a script containing the UPDATE statement with substitution variables by selecting File, Save from the main menu in SQL Developer. This menu command saves the statements currently in the SQL statement pane. By default, files are saved with an .sql extension to associate the file with SQL tools. To use the saved script, click File, Open from the menu and select the file.

TIP

If you're using the SQL*Plus client tool, research the SAVE and SPOOL commands to store an SQL statement in a script and the @ and START commands to execute a script.

DELETING ROWS

When you need to remove rows from database tables, use the **DELETE** command. Compared to other commands covered in this chapter, the DELETE command is incredibly simple—perhaps even too simple! Figure 5-29 shows the syntax of this command.

```
DELETE FROM tablename
[WHERE condition];
```

FIGURE 5-29   Syntax of the DELETE command

The syntax of the DELETE command doesn't allow specifying any column names because DELETE *applies to an entire row and can't be applied to specific columns in a row*. The WHERE clause, which is optional, identifies the rows to be deleted from the specified table.

CAUTION

If you omit the WHERE clause, all rows are deleted from the specified table.

Suppose you're notified that Sammie Jones moved to the Customer Service Department and should no longer be listed in the ACCTMANAGER table. Because her information has already been inserted, you must use the DELETE command to remove her row from the ACCTMANAGER table, as shown in Figure 5-30.

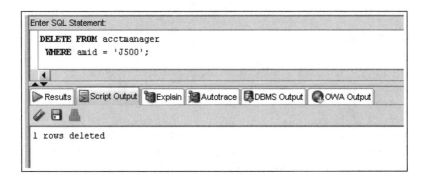

FIGURE 5-30   DELETE command to remove a row from the ACCTMANAGER table

The WHERE clause identifies the exact record—where Amid is equal to J500—to be removed from the ACCTMANAGER table. After the record is deleted, the row for Sammie Jones no longer exists in the table, as shown in Figure 5-31.

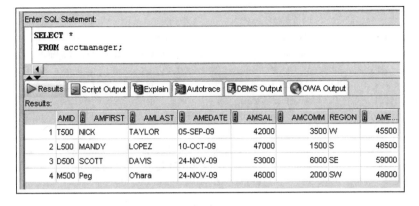

Enter SQL Statement:

```
SELECT *
  FROM acctmanager;
```

Results | Script Output | Explain | Autotrace | DBMS Output | OWA Output

Results:

	AMID	AMFIRST	AMLAST	AMEDATE	AMSAL	AMCOMM	REGION	AME...
1	T500	NICK	TAYLOR	05-SEP-09	42000	3500	W	45500
2	L500	MANDY	LOPEZ	10-OCT-09	47000	1500	S	48500
3	D500	SCOTT	DAVIS	24-NOV-09	53000	6000	SE	59000
4	M500	Peg	O'hara	24-NOV-09	46000	2000	SW	48000

FIGURE 5-31   The row for Sammie Jones has been removed

The WHERE clause in this DELETE statement causes only a single row to be deleted. As with an UPDATE statement, however, a group of rows could be deleted as well. Be aware that a DELETE command without a WHERE clause, as shown in Figure 5-32, deletes all rows in the table because no specific record is identified for deletion.

```
DELETE FROM acctmanager;
```

FIGURE 5-32   DELETE command without the WHERE clause

USING TRANSACTION CONTROL STATEMENTS

Changes to data made by DML commands aren't saved permanently to the table when you execute the SQL statement. Therefore, you have the flexibility of issuing **transaction control** statements to save the modified data or undo the changes if they were made in error. Until the data has been saved permanently to the table, no other users can view any changes you have made. All DML commands you execute remain in a transaction queue until you save the actions permanently or undo the actions.

A **transaction** is a term used to describe DML statements representing data actions that should logically be performed together. A common example is a bank transaction. For example, you withdraw $500 from a savings account and want to put half this amount in your checking account and the other half in a mutual fund account. If the system crashes after the withdrawal is carried out but before the deposits to the other two accounts are accomplished, would you lose $500? Not if transaction control is in effect. Transaction control statements determine at which points DML activity is saved permanently. In the bank transaction example, the save doesn't occur until all three actions are committed.

COMMIT and ROLLBACK Commands

A **COMMIT** command issued implicitly or explicitly permanently saves the DML statements issued previously. An **explicit COMMIT** occurs when you enter a COMMIT statement. By default, an **implicit COMMIT** occurs when you exit client tools, such as SQL Developer. It also occurs if a DDL command, such as CREATE or ALTER TABLE, is issued. In other words, if a user adds several records to a table and then creates a new table, the records added before the DDL command is issued are committed automatically (implicitly).

In Oracle 11g, a transaction consists of a series of statements that have been issued and not committed. A transaction could consist of one SQL statement or 2000 SQL statements issued over an extended period. The duration of a transaction is defined by when a commit occurs implicitly or explicitly.

A program can manage the statements that should be grouped together in a transaction by determining when to issue a commit based on an event. For example, if a purchase order has been entered containing the purchase of several items, this transaction involves an insert into the ORDERS table and multiple inserts into the ORDERITEMS table. A commit covering all these inserts might not be issued until the user clicks a Finalize Order button and the credit card approval code is returned. At this point, the commit saves all the inserts permanently as a group.

> **N O T E**
>
> You can verify that data hasn't been saved permanently by issuing an UPDATE statement in one session and then logging in to a second session to view the table and see that the changes aren't visible. Then try it again, but in the first session, issue a COMMIT statement. Now check the data in the second session to see the changes.

Unless a DML operation is committed, it can be undone by issuing the **ROLLBACK** command. For example, if you haven't exited Oracle 11g since beginning to work through the examples in this chapter, executing a ROLLBACK statement reverses all the rows you entered or altered during your work in this chapter. Similarly, in the purchase order described previously, a Cancel button, for example, allows the data entry user to undo all inserts issued during the order entry. Therefore, the ROLLBACK command reverses all DML operations performed since the last commit was issued.

By contrast, commands such as CREATE TABLE, TRUNCATE TABLE, and ALTER TABLE *can't be rolled back* because they are DDL commands, and a commit occurs automatically when they're executed. Note, however, that if the system crashes, a rollback occurs automatically after Oracle 11g restarts, and any operations not committed previously are undone.

To ensure that all operations performed so far in this chapter are safe from being reversed accidentally, issue the command shown in Figure 5-33 before continuing with the remaining examples in this chapter. Next, query the ACCTMANAGER table and verify that the data matches the listing shown earlier in Figure 5-31.

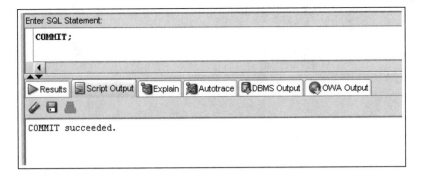

FIGURE 5-33   Command to save data changes permanently

159

NOTE

Be sure to issue the COMMIT statement in Figure 5-33 before completing the exercises in this section. If your data doesn't match the listing in Figure 5-31, modify it as needed, and then issue a COMMIT statement again.

TIP

If you're using the SQL Developer tool, verify that the Autocommit option isn't on. This option is located in the preference settings. On the menu, select Tools, Preferences, Database and Worksheet Parameters.

SAVEPOINT Command

Along with COMMIT commands, developers sometimes use the **SAVEPOINT** command to create a type of bookmark in a transaction. This command is commonly used in the banking industry. For example, a customer is making both a deposit and a withdrawal through an ATM. If the customer first makes a deposit and then requests a withdrawal, but cancels the withdrawal before the money is dispensed, is the entire customer transaction canceled?

To address this issue, a program can be designed to commit the deposit as one transaction and then begin the withdrawal as a separate transaction. However, some designers have the program issue a command with the syntax SAVEPOINT *name*; after the deposit is completed to identify a particular "point" in a transaction—a potential ROLLBACK point. If a subsequent portion of the transaction is canceled, the program simply issues a command with the syntax ROLLBACK TO SAVEPOINT *name*;—and any SQL statements issued after the SAVEPOINT command aren't permanently updated to the database. A COMMIT command still needs to be executed to update the database with any data added or changed by the first part of the transaction.

To see how this command works, make several changes to the ACCTMANAGER table by using transaction control statements. First, execute the series of UPDATE statements shown in Figure 5-34. This series include a COMMIT following the first UPDATE and a SAVEPOINT before the last UPDATE.

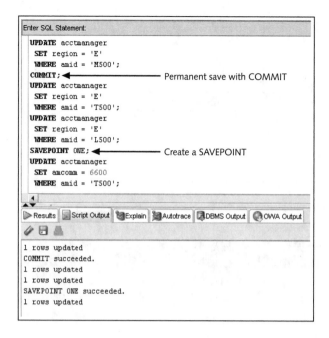

FIGURE 5-34 Establishing a SAVEPOINT

Second, query the data as shown in Figure 5-35. The query verifies that the region has been changed for three account manager records, and the commission value has been changed for one account manager record.

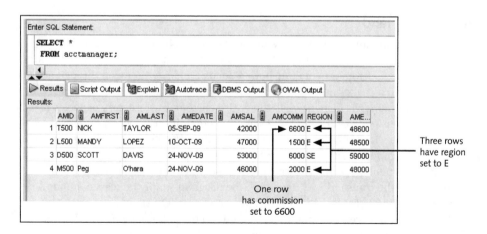

FIGURE 5-35 Verify that modifications were made

What would happen if you issued a ROLLBACK command at this point? All UPDATES except the first one in the series are undone, and the associated region and commission values return to their original values before the UPDATE statements. The region change in the first UPDATE (for Amid M500), however, doesn't revert because a COMMIT was issued following this UPDATE.

What if you want to undo only the last UPDATE? Issue a ROLLBACK TO SAVEPOINT statement, as shown in Figure 5-36. Then query the data as shown in Figure 5-37 to confirm the result.

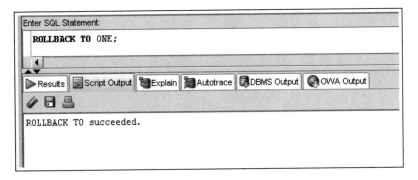

FIGURE 5-36   Undo changes to the SAVEPOINT

Figure 5-37 shows that the last UPDATE was undone; however, the three previous UPDATE statements (before the SAVEPOINT) are intact. If you now want to permanently save the two UPDATE statements before the SAVEPOINT, you could issue a COMMIT statement. However, what if you need to undo these modifications as well? Issue a ROLLBACK command to undo the two UPDATE statements, as shown in Figure 5-38. Verify the results by querying the table again, and compare your results to Figure 5-39.

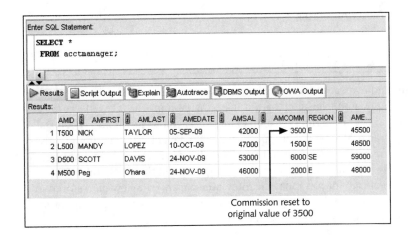

FIGURE 5-37   Verify the results of ROLLBACK TO SAVEPOINT

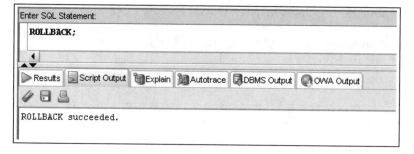

FIGURE 5-38   Undo all pending changes

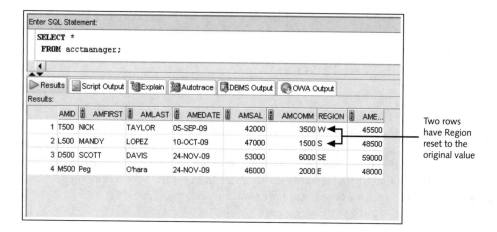

Two rows
have Region
reset to the
original value

FIGURE 5-39   Verify the ROLLBACK results

USING TABLE LOCKS

The discussion of DML and transaction control statements needs to address the fact that most database systems have numerous concurrent users. What happens if two users try to change the same record at the same time? Which change is saved to the table? When DML commands are issued, Oracle 11g, by default, performs a row-level lock, which implicitly "locks" the rows being affected so that no other user can change these rows. In addition, a lock is placed on the table so that other users can't attempt to lock the whole table while the row lock is active. The lock is a **shared lock**, meaning other users can view data stored in the table, but they can't alter the table structure or perform other types of DDL operations, in addition to not being able to change the specific rows that are locked.

LOCK TABLE Command

Although rarely used outside a program, a user can explicitly lock a table in SHARE mode by issuing the **LOCK TABLE** command. Figure 5-40 shows the syntax of this command.

```
LOCK TABLE tablename IN SHARE MODE;
```

FIGURE 5-40   Syntax of LOCK TABLE in SHARE mode

When DDL operations are performed, Oracle 11g places an **exclusive lock** on the table so that no other user can alter the table or attempt adding to or updating the table's contents. If an exclusive lock exists on a table, no other user can place an exclusive lock or a shared lock on the same table. In addition, if a user has a shared lock on a table, no other user can place an exclusive lock on the same table. If necessary, the user can instruct Oracle 11g to lock a table in EXCLUSIVE mode, using the command syntax shown in Figure 5-41.

```
LOCK TABLE tablename IN EXCLUSIVE MODE;
```

FIGURE 5-41   Syntax of LOCK TABLE in EXCLUSIVE mode

CAUTION

Always be careful when explicitly locking a table. If one user locks a portion of a table in SHARE mode, and another user locks a different portion of a table, and the completion of one user's command depends on the portion of a table locked by the other user, a deadlock occurs. Usually, Oracle 11g detects deadlocks automatically and returns an error message to one of the users. When an error message is returned, the lock is also released, so the command issued by the other user is completed. Locks (including exclusive locks) are *released automatically if the user issues a transaction control statement, such as ROLLBACK or COMMIT, or exits the system.*

SELECT ... FOR UPDATE Command

A data consistency issue can occur when a user looks at the contents of a record, makes a decision based on those contents, and then updates the record—only to find out that between the SELECT command and the UPDATE command, the record's contents have changed. For example, you're assigned the task of increasing the retail price of certain books, and you're told to base the new retail price on a percentage of the book's cost. As you begin to update retail prices, you realize someone has updated the cost of the books. Ugh! Now what should you do?

As mentioned, DML operations aren't stored permanently in a table until a COMMIT command is issued. To provide a consistent view for all users accessing the table in a multiuser environment, no changes can be seen by other users until the changes have been committed. This can create major headaches, however, when working with transaction-type tables that are constantly being changed to reflect new orders, account balances, and so on.

To avoid this problem, you can use the **SELECT ... FOR UPDATE** command to view a record's contents when you anticipate that the record will need to be modified. The SELECT ... FOR UPDATE command places a shared lock on the records to be changed and prevents any other user from acquiring a lock on the same records. The syntax is the same as a regular SELECT statement, except the FOR UPDATE clause is added at the end, as shown in Figure 5-42.

```
SELECT columnnames,…
FROM tablename, …
[WHERE condition]
FOR UPDATE;
```

FIGURE 5-42   Syntax of the SELECT ... FOR UPDATE command

If a user decides to update a record, a regular UPDATE command is used to perform the change. However, if the user doesn't change any of the data included in the SELECT ... FOR UPDATE command, a COMMIT or ROLLBACK command must still be issued, or the selected rows remain locked, and no other users can make changes to those rows.

Chapter Summary

- Data manipulation language (DML) includes the INSERT, UPDATE, DELETE, COMMIT, and ROLLBACK commands.
- The INSERT INTO command is used to add new rows to an existing table.
- The column list specified in the INSERT INTO clause must match the order of data entered in the VALUES clause.
- A virtual column must be ignored in all DML actions because the database system generates this column value automatically.
- You can use a NULL value in an INSERT INTO command by including the keyword NULL, omitting the column from the column list of the INSERT INTO clause, or entering two single quotes (without a space) in the position of the NULL value.
- To assign a DEFAULT option value, a column must be excluded from the column list in an INSERT statement, or the keyword DEFAULT must be included as the value for the column.
- In a DML statement, two single quotes together must be used to represent a single quote in a value.
- If rows are copied from a table and entered in an existing table by using a subquery in the INSERT INTO command, the VALUES clause must be omitted because it's irrelevant.
- You can change the contents of a row or group of rows with the UPDATE command.
- You can use substitution variables to allow you to execute the same command several times with different data values.
- DML operations aren't stored permanently in a table until a COMMIT command is issued implicitly or explicitly.
- A transaction consists of a set of DML operations committed as a block.
- Uncommitted DML operations can be undone by issuing the ROLLBACK command.
- A SAVEPOINT serves as a marker for a point in a transaction and allows rolling back only a portion of the transaction.
- Use the DELETE command to remove records from a table. If the WHERE clause is omitted, *all* rows in the table are deleted.
- Table locks can be used to prevent users from mistakenly overwriting changes made by other users.
- Table locks can be in SHARE mode or EXCLUSIVE mode.
- EXCLUSIVE mode is the most restrictive table lock and prevents any other user from placing any locks on the same table.
- A lock is released when a transaction control statement is issued, a DDL statement is executed, or the user exits the system by using the EXIT command.
- SHARE mode allows other users to place shared locks on other portions of the table, but it prevents users from placing an exclusive lock on the table.
- The SELECT ... FOR UPDATE command can be used to place a shared lock for a specific row or rows. The lock isn't released unless a DDL command is issued or the user exits the system.

Chapter 5 Syntax Summary

The following table summarizes the syntax you have learned in this chapter. You can use the table as a study guide and reference.

Syntax Guide

Command	Description	Example
Optional SELECT Clauses		

INSERT	Adds new rows to a table; a subquery can be included to copy rows from an existing table	`INSERT INTO acctmanager`
		`VALUES ('T500', 'NICK TAYLOR',`
		`  '05-SEP-09', 'NE');`
		or
		`INSERT INTO acctmanager`
		`SELECT amid, amname, amedate, region`
		`FROM acctmanager`
		`WHERE amedate <= '01-OCT-09';`

UPDATE	Adds data to, or modifies data in, an existing row	`UPDATE acctmanager`
		`SET amedate = '05-SEP-09'`
		`WHERE amid = 'J500';`

COMMIT	Saves changed data in a table permanently	`COMMIT;`
ROLLBACK	Allows the user to "undo" uncommitted changes to data	`ROLLBACK;`
DELETE	Removes rows from a table	`DELETE FROM acctmanager` `WHERE amid = 'D500';`

LOCK TABLE	Prevents other users from making changes to a table	`LOCK TABLE customers IN SHARE MODE;`
		or
		`LOCK TABLE customers IN EXCLUSIVE MODE;`

SELECT . . . FOR UPDATE	Creates a shared lock on a table to prevent another user from making changes to data in specified columns	`SELECT cost`
		`FROM books`
		`WHERE category = 'COMPUTER'`
		`FOR UPDATE;`

Interactive Operator		

&	Identifies a substitution variable; allows prompting the user to enter a specific value for the substitution variable	`UPDATE customers`
		`SET region = '&Region'`
		`WHERE state = '&State';`

Review Questions

1. Which command should you use to copy data from one table and have it added to an existing table?
2. Which command can you use to change the existing data in a table?
3. When do changes generated by DML operations become stored in database tables permanently?
4. Explain the difference between explicit and implicit locks.
5. If you add a record to the wrong table, what's the simplest way to remove the record from the table?
6. How does Oracle 11g identify a substitution variable in an SQL command?
7. How are NULL values included in a new record being added to a table?
8. When should the VALUES clause be omitted from the INSERT INTO command?
9. What happens if a user attempts to add data to a table, and the addition would cause the record to violate an enabled constraint?
10. What two methods can be used to activate a column's DEFAULT option in an INSERT command?

Multiple Choice

1. Which of the following is a correct statement?
   a. A commit is issued implicitly when a user exits SQL Developer or SQL*Plus.
   b. A commit is issued implicitly when a DDL command is executed.
   c. A commit is issued automatically when a DML command is executed.
   d. All of the above are correct.
   e. Both a and b are correct.
   f. Both a and c are correct.

2. Which of the following is a valid SQL statement?
   a. SELECT * WHERE amid = 'J100' FOR UPDATE;
   b. INSERT INTO homework10 VALUES (SELECT * FROM acctmanager);
   c. DELETE amid FROM acctmanager;
   d. rollback;
   e. all of the above

3. Which of the following commands can be used to add rows to a table?
   a. INSERT INTO
   b. ALTER TABLE ADD
   c. UPDATE
   d. SELECT ... FOR UPDATE

4. Which of the following statements deletes all rows in the HOMEWORK10 table?
   a. DELETE * FROM homework10;
   b. DELETE *.* FROM homework10;

c. DELETE FROM homework10;

d. DELETE FROM homework10 WHERE amid = '*';

e. Both c and d delete all rows in the HOMEWORK10 table.

5. Which of the following statements places a shared lock on at least a portion of a table named HOMEWORK10?

a. SELECT * FROM homework10 WHERE col2 IS NULL FOR UPDATE;

b. INSERT INTO homework10 (col1, col2, col3) VALUES ('A', 'B', 'C');

c. UPDATE homework10 SET col3 = NULL WHERE col1 = 'A';

d. UPDATE homework10 SET col3 = LOWER (col3) WHERE col1 = 'A';

e. all of the above

6. Assuming the HOMEWORK10 table has three columns (Col1, Col2, and Col3, in this order), which of the following commands stores a NULL value in Col3 of the HOMEWORK10 table?

a. INSERT INTO homework10 VALUES ('A', 'B', 'C');

b. INSERT INTO homework10 (col3, col1, col2) VALUES (NULL, 'A', 'B');

c. INSERT INTO homework10 VALUES (NULL, 'A', 'B');

d. UPDATE homework10 SET col1 = col3;

7. Which of the following symbols designates a substitution variable?

a. &

b. $

c. #

d. _

8. Which of the following input values results in a successful INSERT of O'hara?

a. 'O^hara'

b. 'O''hara' (two single quotes following the O)

c. 'O"hara' (a double quote following the O)

d. Data values can't contain quotes.

9. Which of the following commands locks the HOMEWORK10 table in EXCLUSIVE mode?

a. LOCK TABLE homework10 EXCLUSIVELY;

b. LOCK TABLE homework10 IN EXCLUSIVE MODE;

c. LOCK TABLE homework10 TO OTHER USERS;

d. LOCK homework10 IN EXCLUSIVE MODE;

e. Both b and d lock the table in EXCLUSIVE mode.

10. You issue the following command: INSERT INTO homework10 (col1, col2, col3) VALUES ('A', NULL, 'C'). The command will fail if which of the following statements is true?

a. Col1 has a PRIMARY KEY constraint enabled.

b. Col2 has a UNIQUE constraint enabled.

c. Col3 is defined as a DATE column.

d. None of the above would cause the command to fail.

11. Which of the following releases a lock currently held by a user on the HOMEWORK10 table?

    a. A COMMIT command is issued.

    b. A DDL command is issued to end a transaction.

    c. The user exits the system.

    d. A ROLLBACK command is issued.

    e. all of the above

    f. none of the above

12. Assume you have added eight new orders to the ORDERS table. Which of the following is true?

    a. Other users can view the new orders as soon as you execute the INSERT INTO command.

    b. Other users can view the new orders as soon as you issue a ROLLBACK command.

    c. Other users can view the new orders as soon as you exit the system or execute a COMMIT command.

    d. Other users can view the new orders only if they place an exclusive lock on the table.

13. Which of the following commands removes all orders placed before April 1, 2009?

    a. DELETE FROM orders WHERE orderdate < '01-APR-09';

    b. DROP FROM orders WHERE orderdate < '01-APR-09';

    c. REMOVE FROM orders WHERE orderdate < '01-APR-09';

    d. DELETE FROM orders WHERE orderdate > '01-APR-09';

14. How many rows can be added to a table by executing the INSERT INTO ... VALUES command?

    a. 1

    b. 2

    c. 3

    d. unlimited

15. You accidentally deleted all the orders in the ORDERS table. How can the error be corrected after a COMMIT command has been issued?

    a. ROLLBACK;

    b. ROLLBACK COMMIT;

    c. REGENERATE RECORDS orders;

    d. None of the above restores the deleted orders.

16. Which of the following is the standard extension used for a script file?

    a. .spt

    b. .srt

    c. .script

    d. .sql

Data Manipulation and Transaction Control

17. A rollback occurs automatically when:

    a.   A DDL command is executed.

    b.   A DML command is executed.

    c.   The user exits the system.

    d.   none of the above

18. What is the maximum number of rows that can be deleted from a table at one time?

    a.   1

    b.   2

    c.   3

    d.   unlimited

19. Which of the following is a correct statement?

    a.   If you attempt to add a record that violates a constraint for one of the table's columns, only the valid columns for the row are added.

    b.   A subquery nested in the VALUES clause of an INSERT INTO command can return only one value without generating an Oracle 11*g* error message.

    c.   If you attempt to add a record that violates a NOT NULL constraint, a blank space is inserted automatically in the appropriate column so that Oracle 11*g* can complete the DML operation.

    d.   None of the above statements is correct.

20. What is the maximum number of records that can be modified with a single UPDATE command?

    a.   1

    b.   2

    c.   3

    d.   unlimited

Hands-On Assignments

To perform the following assignments, refer to the tables created in the JLDB_Build_5.sql script at the beginning of the chapter.

1. Add a new row in the ORDERS table with the following data: Order# = 1021, Customer# = 1009, and Order date = July 20, 2009.

2. Modify the zip code on order 1017 to 33222.

3. Save the changes permanently to the database.

4. Add a new row in the ORDERS table with the following data: Order# = 1022, Customer# = 2000, and Order date = August 6, 2009. Describe the error raised and what caused the error.

5. Add a new row in the ORDERS table with the following data: Order# = 1023 and Customer# = 1009. Describe the error raised and what caused the error.

6. Create a script using substitution variables that allows a user to set a new cost amount for a book based on the ISBN.

7. Execute the script and set the following values: isbn = 1059831198 and cost = $20.00.

8. Execute a command that undoes the change in Step 7.

9. Delete Order# 1005. You need to address both the master order record and the related detail records.

10. Execute a command that undoes the previous deletion.

Advanced Challenge

Currently, the contents of the Category column in the BOOKS table are the actual name for each category. This structure presents a problem if one user enters COMPUTER for the Computer category and another user enters COMPUTERS. To avoid this and other problems that might occur, the database designers have decided to create a CATEGORY table containing a code and description for each category. The structure for the CATEGORY table should be as follows:

Column Name	Datatype	Width	Constraints
CATCODE	VARCHAR2	3	PRIMARY KEY
CATDESC	VARCHAR2	11	NOT NULL

The data for the CATEGORY table is as follows:

CATCODE	CATDESC
BUS	BUSINESS
CHN	CHILDREN
COK	COOKING
COM	COMPUTER
FAL	FAMILY LIFE
FIT	FITNESS
SEH	SELF HELP
LIT	LITERATURE

Required:

- Create the CATEGORY table and populate it with the given data. Save the changes permanently.
- Add a column to the BOOKS table called Catcode.
- Add a FOREIGN KEY constraint that requires all category codes entered in the BOOKS table to already exist in the CATEGORY table. Set the Catcode values for the existing rows in the BOOKS table, based on each book's current Category value.
- Verify that the correct categories have been assigned in the BOOKS table, and save the changes permanently.
- Delete the Category column from the BOOKS table.

Case Study: *City Jail*

Execute the CityJail_5.sql script to rebuild the CRIMINALS and CRIMES tables of the City Jail database. The statements at the beginning of this script drop existing tables in your schema with the same table names.

NOTE

The CityJail_5.sql script is not included in the data files; it's included with the solution files, so it will be provided by your instructor.

Review the script so that you're familiar with the table structure and constraints, and then do the following:

1. Create and execute statements to perform the following DML activities. Save the changes permanently to the database.

   a. Create a script to allow a user to add new criminals (providing prompts to the user) to the CRIMINALS table.

   b. Add the following criminals, using the script created in the previous step. No value needs to be entered at the prompt if it should be set to the DEFAULT column value. Query the CRIMINALS table to confirm that new rows have been added.

Criminal_ID	Last	First	Street	City	State	Zip	Phone	V_status	P_status
1015	Fenter	Jim		Chesapeake	VA	23320		N	N
1016	Saunder	Bill	11 Apple Rd	Virginia Beach	VA	23455	7678217443	N	N
1017	Painter	Troy	77 Ship Lane	Norfolk	VA	22093	7677655454	N	N

   c. Add a column named Mail_flag to the CRIMINALS table. The column should be assigned a datatype of CHAR(1).

d. Set the Mail_flag column to a value of 'Y' for all criminals.

e. Set the Mail_flag column to 'N' for all criminals who don't have a street address recorded in the database.

f. Change the phone number for criminal 1016 to 7225659032.

g. Remove criminal 1017 from the database.

2. Execute a DML statement to accomplish each of the following actions. Each statement produces a constraint error. Document the error number and message, and briefly explain the cause of the error. If your DML statement generates a syntax error rather than a constraint violation error, revise your statement to correct any syntax errors. You can review the CityJail_5.sql file to identify table constraints.

a. Add a crime record using the following data: Crime_ID = 100, Criminal_ID = 1010, Classification = M, Date_charged = July 15, 2009, Status = PD.

b. Add a crime record using the following data: Crime_ID = 130, Criminal_ID = 1016, Classification = M, Date_charged = July 15, 2009, Status = PD.

c. Add a crime record using the following data: Crime_ID = 130, Criminal_ID = 1016, Classification = P, Date_charged = July 15, 2009, Status = CL.

CHAPTER 6

ADDITIONAL DATABASE OBJECTS

LEARNING OBJECTIVES

After completing this chapter, you should be able to do the following:

- Define the purpose of a sequence and explain how it can be used in a database
- Use the CREATE SEQUENCE command to create a sequence
- Explain why gaps might appear in integers generated by a sequence
- Call and use sequence values
- Identify which options can't be changed by the ALTER SEQUENCE command
- Delete a sequence
- Create indexes with the CREATE INDEX command
- Explain the main index structures: B-tree and bitmap
- Verify index use with the explain plan
- Describe variations on conventional indexes, including a function-based index and an index organized table
- Verify index existence via the data dictionary
- Rename an index with the ALTER INDEX command
- Remove an index with the DELETE INDEX command
- Create and remove a public synonym

INTRODUCTION

The tables and constraints you created in previous chapters are considered database objects. A **database object** is anything with a name and a defined structure. Three other database objects commonly used in Oracle 11*g* are sequences, indexes, and synonyms, which you examine in this chapter. The following list identifies the role of each object:

- A sequence generates sequential integers that organizations can use to assist with internal controls or simply to serve as primary keys for tables.
- A database index serves the same basic purpose as an index in a book, allowing users to locate specific records quickly.
- A synonym is a simpler name, like a nickname, given to an object with a complex name or to provide an alternative name for identifying database objects. Synonyms can simplify referencing objects with complex names and objects that are moved to different schemas.

This chapter explains how to create, maintain, and delete sequences and how to create and delete indexes and synonyms. Table 6-1 is an overview of this chapter's contents.

TABLE 6-1   Overview of Chapter Contents

Description	Command Syntax

Create a sequence to generate a series of integers	CREATE SEQUENCE *sequencename*
	    [INCREMENT BY *value*]
	    [START WITH *value*]
	    [{MAXVALUE *value* | NOMAXVALUE}]
	    [{MINVALUE *value* | NOMINVALUE}]
	    [{CYCLE | NOCYCLE}]
	    [{ORDER | NOORDER}]
	    [{CACHE *value* | NOCACHE}];

Alter a sequence	ALTER SEQUENCE *sequencename*
	    [INCREMENT BY *value*]
	    [{MAXVALUE *value* | NOMAXVALUE}]
	    [{MINVALUE *value* | NOMINVALUE}]
	    [{CYCLE | NOCYCLE}]
	    [{ORDER | NOORDER}]
	    [{CACHE *value* | NOCACHE}] ;

Drop a sequence	DROP SEQUENCE *sequencename*;
Create a B-tree index	CREATE INDEX *indexname* ON *tablename*(*columnname*, . . .);
Create a bitmap index	CREATE BITMAP INDEX *indexname* ON *tablename* (*columnname*, . . .);
Create a function-based index	CREATE INDEX *indexname* ON *tablename* (*expression*);

TABLE 6-1 Overview of Chapter Contents (continued)

Description	Command Syntax

Create an index organized table	CREATE TABLE *tablename*
	   (*columnname datatype*,
	    columnname datatype)
	  ORGANIZATION INDEX;

Rename an index	ALTER INDEX *indexname* RENAME TO *newindexname*;
Drop an index	DROP INDEX *indexname*;
Create a synonym	CREATE [PUBLIC] SYNONYM *synonymname* FOR *objectname*;
Drop a synonym	DROP [PUBLIC] SYNONYM *synonymname*;

DATABASE PREPARATION

This chapter assumes you have executed the JustLee Books database script, JLDB_Build_5.sql, as instructed in Chapter 5.

SEQUENCES

A **sequence** is a database object you can use to generate a series of integers. These integers are most commonly used to generate a unique primary key for each record or for internal control purposes. A brief overview of these two concepts follows.

When you use values generated by a sequence as a primary key, there's no true correlation between the number assigned to a record and the entity it represents. However, depending on the parameters used to create the sequence, database users can be assured that no two records have the same primary key value. Ensuring that primary key values aren't duplicated is especially important if different users are assigned the task of entering records in a database table because they might attempt to assign the same primary key value to different records. For example, if several customer service representatives are entering new customers at the same time, how are customer numbers assigned? Are all the customer service representatives in the same room, asking one another "What number did your last customer receive?" Not likely. Chances are that they're using a sequence, so each customer service representative can be certain that every customer number is unique.

NOTE

When a database-generated value is used as a primary key value, it's often referred to as a surrogate key. If existing data, such as a book ISBN, is used to provide the table's primary key value, it's often referred to as a natural key.

A sequence can also be used to provide business and auditing controls. Every organization should have some control mechanisms to avoid problems with transaction auditing, embezzlement, and accounting errors. Most organizations use sequential numbers to track checks, purchase orders, invoices, and anything else used to record financial events. With sequential numbers, an auditor can determine whether items such as checks or invoices are missing, which can reveal accounting problems—unrecorded transactions or employees obtaining blank checks or invoices for their own use, for example.

Creating a Sequence

You create a sequence with the **CREATE SEQUENCE** command, using the syntax shown in Figure 6-1. Optional commands are shown in square brackets. Curly brackets indicate that one of the two options listed can be used, but not both.

```
CREATE SEQUENCE sequencename
[INCREMENT BY value]
[START WITH value]
[{MAXVALUE value | NOMAXVALUE}]
[{MINVALUE value | NOMINVALUE}]
[{CYCLE | NOCYCLE}]
[{ORDER | NOORDER}]
[{CACHE value | NOCACHE}];
```

FIGURE 6-1   Syntax of the CREATE SEQUENCE command

Notice that most items in a CREATE SEQUENCE command are optional; not much is required to create a basic sequence. The statement shown in Figure 6-2 creates a sequence named CUSTOMERS_CUSTOMER#_SEQ.

```
CREATE SEQUENCE customers_customer#_seq;
```

FIGURE 6-2   Create a sequence

You might have some questions about this sequence. By what amount will the generated numbers be incremented? When will the sequence run out of numbers to generate? What number will be the first one generated? Because the statement in Figure 6-2 doesn't set any options, the default values for options are in effect. This is why understanding all the option settings before creating a sequence is important. Take a closer look at each syntax element in Figure 6-1. Note that some elements, such as CACHE, must indicate a value (how many numbers to generate and store in memory), whereas other options, such as NOCACHE, don't need a value indicated.

The CREATE SEQUENCE keywords are followed by the name for identifying the sequence. A standard naming convention is including _seq at the end of the name to make it easier to identify as a sequence.

The **INCREMENT BY** clause specifies the interval between two sequential values. For checks and invoices, this interval is usually 1. However, for sequences representing credit card or bank account numbers, for example, the interval might be much larger so that no two account numbers are similar. (An interval such as 13,519, for instance, might be more appropriate for this purpose.) If the sequence is incremented by a positive value, the values the sequence generates are in ascending order. However, if a negative value is specified, the values the sequence generates are in descending order. If you need values in descending order, you must include the keywords INCREMENT BY and provide a negative value. If the INCREMENT BY clause isn't included when the sequence is created, the default setting is used, which increases the sequence by one for each integer generated.

The **START WITH** clause establishes the starting value for the sequence. Oracle 11*g* begins each sequence at 1 unless another value is specified in the START WITH clause. For example, if you want all customer numbers to consist of four digits, you can assign the value 1000 to the START WITH clause to avoid assigning account numbers of fewer than four digits to the first 999 customers.

TIP

The START WITH value is critical if you have existing data. If you're importing existing data into a new database, you need to consider what values exist for columns the sequence will populate. For example, if you already have customer numbers from 1000 to 2500, you need to set a START WITH value of 2501 for the sequence used to populate the Customer Number column.

Continuing with the syntax shown in Figure 6-1, the **MINVALUE** and **MAXVALUE** clauses establish a minimum or maximum value for the sequence. If the sequence is incremented with a positive value, using the MINVALUE clause doesn't make sense. In this case, the sequence value won't go below the START value, so the START value *is* the minimum value. By the same logic, you might assume if the sequence is incremented with a negative value (for descending order), a MAXVALUE clause isn't necessary. However, this isn't the case. If a negative increment is used and you set a MINVALUE, the MAXVALUE must also be set. Typically, MAXVALUE is set to the same value as the START value.

By default, if you don't specify minimum and maximum values, Oracle 11*g* assumes the **NOMINVALUE** and **NOMAXVALUE** options are used. When the NOMINVALUE option is assumed—or assigned—the lowest possible value for an increasing sequence is 1, and the lowest possible value for a decreasing sequence is -10^{26}, or -100,000,000,000,000,000,000,000,000. For the NOMAXVALUE option, 10^{27}, or 1,000,000,000,000,000,000,000,000,000, is the highest possible value for an ascending sequence, and -1 is the highest possible value for a descending sequence.

The **CYCLE** and **NOCYCLE** options determine whether Oracle 11*g* should begin reissuing values from the sequence after reaching the minimum or maximum value. If the CYCLE option is specified and Oracle 11*g* reaches the maximum value for an ascending sequence or the minimum value for a descending sequence, the CYCLE option initiates the cycle of numbers again. If the sequence is being used to generate values for a primary key, cycling can cause problems if the sequence tries to assign a value that already exists in the table.

However, some organizations reuse check numbers, order numbers, and so forth after an extended period instead of letting these numbers become astronomically large. Therefore, the sequence must be allowed to reuse the same sequence of numbers. If a user doesn't specify a cycle option, Oracle 11g applies the default NOCYCLE option to the sequence. If the NOCYCLE option is in effect, Oracle 11g doesn't generate any numbers after the minimum or maximum value has been reached, and an error message is returned when the user requests another value from the sequence.

The **ORDER** and **NOORDER** options are used in **application cluster environments**, in which multiple users might be requesting sequence values at the same time during large transactions (as when printing a large quantity of checks or invoices). The ORDER option instructs Oracle 11g to return sequence values in the same order in which requests are received. If this option isn't specified in the CREATE SEQUENCE command, the default NOORDER option is assumed. When the sequence is being used to generate a primary key, the order of sequence values isn't a problem because each value is still unique.

Generating sequence values can slow down processing requests from other users, especially if large volumes of these values are requested in a short period. If the **NOCACHE** option is specified when the sequence is created, each number is generated when the request is received. However, if an organization's transactions require large amounts of sequential numbers throughout a session, the **CACHE** option can be used to have Oracle 11g generate a set of values ahead of time and store them in the server's memory. Then, when a user requests a sequence value, the next available value is assigned—without Oracle 11g having to generate the number. On the other hand, if the CACHE option isn't specified, Oracle 11g assumes a default option of CACHE 20 and stores 20 sequential values in memory automatically for users to access.

When working with sequences and cached values, remember that when a value is generated, it has been assigned and can't be regenerated until the sequence begins a new cycle. Therefore, if Oracle 11g caches 20 values, the values have been generated regardless of whether they're actually used. If the system crashes, or if users don't use the values, these generated values are lost. If the sequence is used for internal control purposes, some gaps in the sequence could be caused by nonuse after a system crash. These gaps can't be documented, which can be a cause for concern. For example, suppose 50 sequential numbers to be assigned as order numbers are cached. After a few orders have been received, the Oracle 11g server crashes and has to be restarted. All unassigned numbers are now lost, and a gap in the order number sequence results. Gaps might also appear if transactions are rolled back because the sequence value is already considered used and can't be returned for "reuse."

Sequences aren't assigned to a specific column or table. They are independent objects and, therefore, different users can use the same sequence to generate values that are inserted into several different tables. In other words, the same sequence could be used to generate order numbers and customer numbers. This use results in gaps in the sequence values appearing in each table. Although gaps aren't a concern if the values are used for a primary key column that requires only unique values, they're a problem if the sequence values are used for internal control purposes. If the sequence's purpose is to provide a means for auditing control and to make certain no checks, invoices, and so on are missing, a sequence should be used for only one table, and the generated values shouldn't be cached.

Next, take a look at an example of how sequences can work. All orders placed by JustLee Books customers are assigned a four-digit number to uniquely identify each order. As the sales volume increases, multiple data entry clerks enter orders in the ORDERS table. Therefore, two clerks could try to enter the same order number if they're assigning order numbers manually. Although the PRIMARY KEY constraint for the ORDERS table prevents two orders from having the same number, it still slows down data entry because one of the clerks must choose a different order number and then reenter the order. To avoid this problem, you need to create a sequence to generate the order numbers used in the ORDERS table, following these guidelines:

- The sequence should be named ORDERS_ORDER#_SEQ to identify it as a sequence object created to generate order numbers for orders in the ORDERS table. (Using this naming scheme isn't an absolute requirement, however. Although the name was assigned to indicate its purpose, the sequence value generated can still be used in any table.)
- The INCREMENT BY clause must instruct Oracle 11g that each generated number should be increased by the value of 1.
- Because the last order number stored in the ORDERS table is 1020, the sequence needs to start at 1021 so that there's no gap in order numbers and no previously assigned value is duplicated.
- Oracle 11g should also be instructed not to cache any values in memory, which means each value is generated only when it's requested.
- The CYCLE option must indicate that the sequence values can't be reused after the maximum value has been reached.

For any clause not included explicitly, the default values are in effect. The command to create this sequence is shown in Figure 6-3.

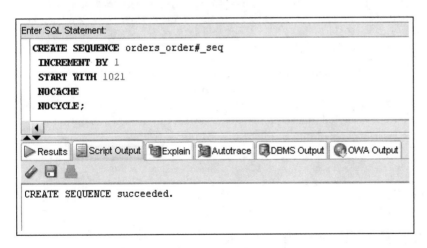

FIGURE 6-3 Generate a sequence for order numbers

You can verify which sequences exist by querying the USER_OBJECTS data dictionary object with a SELECT statement, as shown in Figure 6-4. This statement produces a list of all existing sequences.

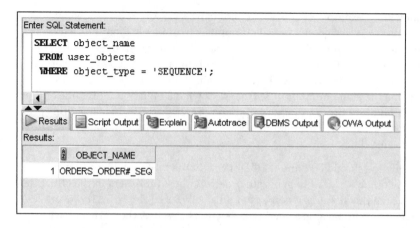

FIGURE 6-4 Query USER_OBJECTS to verify existing sequences

To verify each setting for sequence options, you can query the USER_SEQUENCES data dictionary object, as shown in Figure 6-5. This query is also a quick way to identify which value is the next one to be assigned in the sequence—without generating a number accidentally. The next value to be assigned in a sequence created with the NOCACHE option is indicated in the LAST_NUMBER column in the SELECT query's results.

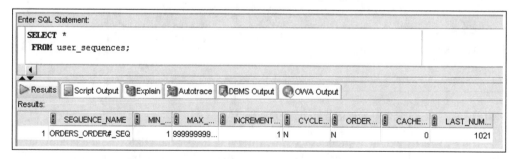

FIGURE 6-5 Verifying sequence option settings

Using Sequence Values

You can access sequence values by using the two pseudocolumns NEXTVAL and CURRVAL. **Pseudocolumns** are data associated with table data, much like columns, but aren't physical columns stored in the database. The pseudocolumn **NEXTVAL** (NEXT VALUE) is used to actually generate the sequence value. In other words, it calls the sequence object and requests the value of the next number in the sequence. After a value is generated, it's stored in the **CURRVAL** (CURRENT VALUE) pseudocolumn so that you can reference it again.

Next, you use the NEXTVAL pseudocolumn to record a new order, using the ORDERS_ORDER#_SEQ sequence created earlier. An order is received from customer 1010 on April 6, 2009, for one copy of *Big Bear and Little Dove* to be shipped to 123 West Main, Atlanta, GA 30418. To process the order, the first step is placing the order information in the ORDERS table. Figure 6-6 shows the command to add the order to the ORDERS table.

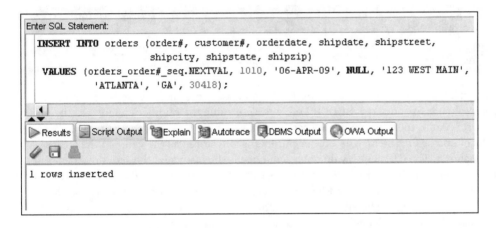

Enter SQL Statement:

```
INSERT INTO orders (order#, customer#, orderdate, shipdate, shipstreet,
                    shipcity, shipstate, shipzip)
   VALUES (orders_order#_seq.NEXTVAL, 1010, '06-APR-09', NULL, '123 WEST MAIN',
           'ATLANTA', 'GA', 30418);
```

Results | Script Output | Explain | Autotrace | DBMS Output | OWA Output

```
1 rows inserted
```

FIGURE 6-6   Inserting a row, using a sequence to provide a PRIMARY KEY value

In Figure 6-6, the orders_order#_seq.NEXTVAL reference in the VALUES clause instructs Oracle 11*g* to generate the next sequential value from the ORDERS_ORDER#_SEQ sequence. Because the reference is listed as the first column of the VALUES clause, the generated value is stored in the first column of the column list, identified in the INSERT INTO clause as order#. The NEXTVAL pseudocolumn is preceded by the sequence name, which identifies the sequence that should generate the value. Figure 6-7 shows the order added to the ORDERS table.

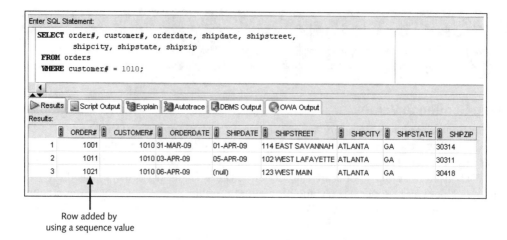

Row added by
using a sequence value

FIGURE 6-7 Order added, using a sequence value for the Order#

After the row has been added to the ORDERS table, you can use the SELECT command to view the order number assigned to the new order. The next step is to add the ordered item, *Big Bear and Little Dove*, to the ORDERITEMS table.

NOTE

If an error message occurs when you try to insert the new order, make certain the order number doesn't exist already. If it does, use the DELETE command to remove the existing row.

When adding a book to the ORDERITEMS table, the order number must be entered. One approach is to query the ORDERS table to determine the number assigned to the order. However, because the user didn't generate another number by referencing the NEXTVAL pseudocolumn again, the assigned order number is still stored as the value in the CURRVAL pseudocolumn. The CURRVAL value of a sequence is the last value the sequence generated in the user session. Therefore, you can use CURRVAL's contents to add the order number to the ORDERITEMS table without having to recall the assigned order number. Use the INSERT INTO command shown in Figure 6-8 to insert the order number, item number, ISBN, quantity, and "paideach" values in the ORDERITEMS table.

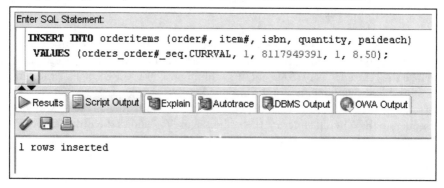

FIGURE 6-8 Using CURRVAL to insert an order detail row

The command shown in Figure 6-8 instructs Oracle 11*g* to place the value stored in the CURRVAL pseudocolumn for the ORDERS_ORDER#_SEQ sequence in the first column of the ORDERITEMS table. Any reference to CURRVAL doesn't cause Oracle 11*g* to generate a new order number. However, if the example in Figure 6-8 had referenced NEXTVAL, a new sequence number would have been generated, and the order number entered in the ORDERITEMS table wouldn't have been the same as the order number already stored in the ORDERS table. Figure 6-9 shows CURRVAL inserted in the ORDERITEMS table.

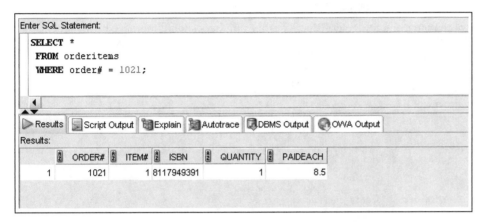

FIGURE 6-9   Verifying the CURRVAL value

> **NOTE**
>
> When a user logs in to Oracle 11*g*, no value is initially stored in the CURRVAL pseudocolumn; the current value is NULL. After a NEXTVAL call has been issued to generate a sequence value, CURRVAL stores that value until the next value is generated. CURRVAL contains *only the last value generated*.

Altering Sequence Definitions

You can change settings for a sequence by using the **ALTER SEQUENCE** command. However, any changes are applied only to values generated *after* the modifications are made. The only restrictions that apply to changing the sequence settings are as follows:

- The START WITH clause can't be changed because the sequence would have to be dropped and re-created to make this change.
- The changes can't make previously issued sequence values invalid. (For example, they can't change the defined MAXVALUE to a number less than a sequence number that has already been generated.)

As shown in Figure 6-10, the ALTER SEQUENCE command follows the same syntax as the CREATE SEQUENCE command. Only the options that need to be added or modified must be included in the ALTER SEQUENCE command. Any options you don't specify in the ALTER SEQUENCE command remain at their current settings.

```
ALTER SEQUENCE sequencename
   [INCREMENT BY value]
   [{MAXVALUE value | NOMAXVALUE}]
   [{MINVALUE value | NOMINVALUE}]
   [{CYCLE | NOCYCLE}]
   [{ORDER | NOORDER}]
   [{CACHE value | NOCACHE}];
```

FIGURE 6-10   Syntax of the ALTER SEQUENCE command

Suppose management decides that all order numbers should increase by a value of 10 rather than 1. Figure 6-11 shows the command to change the INCREMENT BY setting for the ORDERS_ORDER#_SEQ sequence.

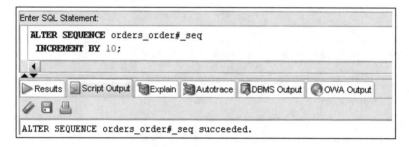

FIGURE 6-11   Command to change the INCREMENT BY setting for a sequence

Because no other settings were changed in this command, their values are unaffected. You can view the new setting for the sequence by using the USER_SEQUENCES data dictionary object, as shown in Figure 6-12. The LAST_NUMBER column now indicates that the next number to issue from the sequence is 1031, which is 10 more than the start value of 1021.

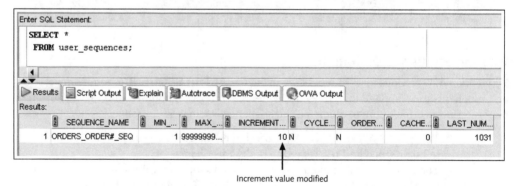

Increment value modified

FIGURE 6-12   New settings for the ORDERS_ORDER#_SEQ sequence

As you begin experimenting with sequences, you can use the Oracle 11g DUAL table to call the NEXTVAL and CURRVAL sequence values. The DUAL table is available for performing queries that aren't retrieving table data. A SELECT statement requires both a SELECT and FROM clause to execute in Oracle 11g. However, the data you retrieve might not be stored in a table. For example, if you want to retrieve the current system date, you can use the DUAL table in a query, as shown in Figure 6-13. If the FROM clause isn't included in the query, the statement generates an error.

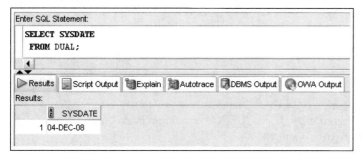

FIGURE 6-13   Using the DUAL table

The DUAL table can assist in experimenting with sequences because you don't need to complete INSERT commands to test the generated sequence values. The statements shown in Figure 6-14 demonstrate testing sequence values by using the DUAL table to complete the sequence call.

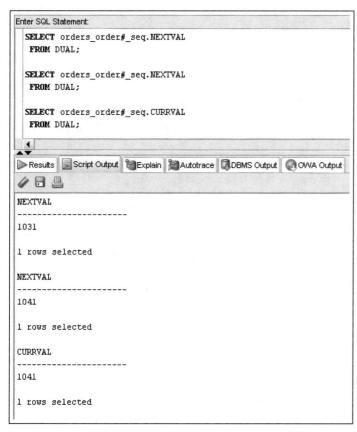

FIGURE 6-14   Testing sequence values with the DUAL table

Removing a Sequence

To delete a sequence, use the **DROP SEQUENCE** command. Figure 6-15 shows the syntax of this command.

```
DROP SEQUENCE sequencename;
```

FIGURE 6-15   Syntax of the DROP SEQUENCE command

When a sequence is dropped, it doesn't affect any values previously generated and stored in a database table. When the DROP SEQUENCE command is executed successfully, the message shown in Figure 6-16 is displayed.

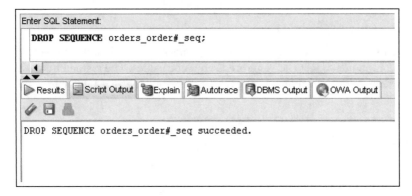

FIGURE 6-16   Dropping the ORDERS_ORDER#_SEQ sequence

To verify that the sequence no longer exists, query the USER_SEQUENCES data dictionary object, as shown in Figure 6-17.

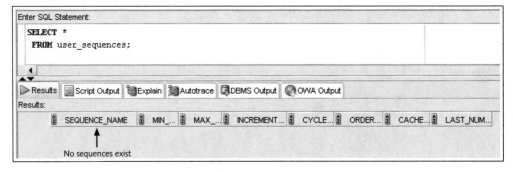

FIGURE 6-17   Verify that the sequence is removed

INDEXES

An Oracle 11g **index** is a database object that stores a map of column values and the ROWIDs of matching table rows. A **ROWID** is the physical address of a table row. A database index is much like the index at the end of this textbook. If you look up a topic such as "primary key," you can scan the handful of alphabetically sorted index pages quickly to determine the page location for this topic. If no index existed, you would need to scan the entire textbook to find references to this topic, which could be time consuming and tedious. In a similar way, database indexes make data retrieval more efficient.

A common challenge in managing databases is improving data retrieval speed. As tables become populated with many rows, processing the operations involved in query searches (WHERE conditions) and sorting (ORDER BY and joins) takes increasing amounts of time. Much of this increase in execution time is caused by disk I/O or disk reads (reading data from physical disk drives).

NOTE

You explore query options for data searching and sorting in more depth in Chapters 8 and 9.

A CREATE TABLE statement in Oracle 11g by default creates a **heap-organized table**, which is an unordered collection of data. As rows of data are inserted, they're physically added to the table in no particular order. As rows are deleted, the space can be reused by new rows. Therefore, if a search condition such as "WHERE zip = 90404" is included in a SELECT statement, a full table scan is performed. In a **full table scan**, each row of the table is read, and the zip value is checked to determine whether it satisfies the condition. Even if the table contains 10 million rows, every single row is read into memory and reviewed.

This example raises several important issues about query performance and the need for indexes:

- First, disk I/O is typically the largest factor in a query's total execution time. Indexes are the primary means of reducing disk I/O.
- Second, if data of the column used in a search condition has high selectivity or **cardinality** (meaning a large number of distinct values), the full table scan results in reading many more rows than needed. For example, if only 100 rows contain a zip value of 90404, more than nine million rows that don't match the condition are read.

NOTE

Much of the Oracle documentation uses the term "cardinality" instead of "selectivity" to describe the level of distinct values in a table column. If a column contains many distinct values, it's described as having high cardinality.

- Third, the amount of buffer pool space used to hold the full table data affects other system users.

The database **buffer pool** serves as the database server's shared cache memory area. As users perform SQL queries, the data is placed in memory and can be reused from memory for quicker retrieval instead of performing disk I/O operations again. As the buffer space fills, previous query data is cleared to make space for more recent query data. A full table scan on a large table could use a lot of buffer space and, therefore, clear a large amount of cached data and cancel the benefits of data caching.

Many of these issues can be addressed by applying indexes to table columns. Oracle 11g provides a number of different indexes. To delve into the functionality and physical structure of an index, in the following sections you learn about the B-tree index, which is the default index structure, and the bitmap index. These indexes are the two main physical index structures used in an Oracle database. Following this discussion, two other index variations based on usage goals are introduced: function-based indexes and index organized tables.

B-Tree Indexes

The **B-tree (balanced-tree) index** is the most common index used in Oracle. You can create this type of index with a basic CREATE INDEX statement. For example, many queries on customer data search for specific zip code values or a range of zip code values. Figure 6-18 shows the SQL statement used to create an index on the Zip column of the CUSTOMERS table.

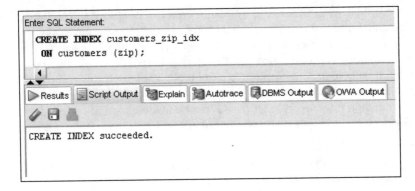

FIGURE 6-18 Creating an index on the Zip column

Now that you have an index, examine the organization of a B-tree index. Figure 6-19 depicts the organization of a B-tree index, using the zip code example.

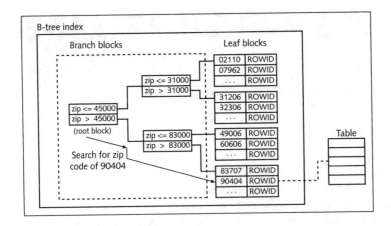

FIGURE 6-19 B-tree index organization

Data in the Oracle database is stored in basic structures called **data blocks**. The amount of data a block can contain is determined by the size setting, defined in the database configuration. Both indexes and table data are stored in data blocks. The index begins with the root node block, which provides the initial breakdown or ranges of column values. The upper blocks (branch blocks) of a B-tree index contain index data that points to lower-level index blocks. The branch blocks continue to provide value breakdowns until the ranges are narrow enough to be divided into blockwide ranges (leaf blocks). The lowest-level index blocks (leaf blocks) contain every indexed data value and a corresponding ROWID for locating the actual table row. The leaf blocks are doubly linked: to a branch block as well as to previous and next leaf nodes to support range value searches and specific (equality) value searches.

So for a query searching for the specific zip code value 90404, branch 1, branch 2, and leaf 4 blocks are read. At this point, the ROWID (pointer to a table row) is used to read the matching data block and the identified row from the CUSTOMERS table. In total, four data blocks are read to resolve the query (assuming a single row matches the specified zip code value).

What if the search finds two matching rows? Two ROWIDs are then identified in the leaf 4 block read and, if the two table rows are stored on two different table data blocks, the read then involves one additional data block, or a total of five data blocks. In contrast, if the CUSTOMERS table consists of 100,000 data blocks, a full table scan requires reading 100,000 rather than four or five data blocks! As you can see, using indexes reduces disk I/O dramatically.

NOTE

A B-tree index is referred to as a "balanced" index because it attempts to make all leaf blocks have equal depth, so every value search should result in an equal number of index block reads.

An index can be created implicitly by Oracle 11g or explicitly by a user (as you did with the previous CREATE INDEX statement). Oracle 11g creates an index automatically when a PRIMARY KEY or UNIQUE constraint is created for a column. Because both these constraints enforce uniqueness, the purpose of adding the index is to allow Oracle 11g to determine whether a value exists in a table without having to perform a full table scan. Oracle 11g accesses the index whenever a value is inserted into or changed in a column designated as a primary key column or a column that can contain only unique values.

TIP

If you're adding a PRIMARY KEY or UNIQUE constraint on a column that's already involved in another constraint, Oracle 11g might not generate an index for the column automatically. Be careful to confirm creation of the index. You learn how to check the data dictionary for index information later in this chapter.

So if the purpose of an index is to improve data retrieval efficiency, why not just create an index for every column? Although indexes can speed up row retrieval, their use isn't always appropriate for the following reasons:

- Because an index is a database object based on table values, Oracle 11*g* must update it *every* time a DML operation is performed on an underlying table. Therefore, if you have a table that's modified (updates, inserts, deletes) frequently, the speed of processing the update slows down because Oracle 11*g* must now update both the table *and* the index. Furthermore, if a table has 10 indexes and a row is added, all 10 indexes must be updated separately.
- B-tree indexes are typically beneficial only if a small percentage of the table is expected to be returned in query results. Because having an index requires Oracle 11*g* to examine the index to identify records meeting the criteria and then retrieve rows from the actual table, large result sets could require Oracle 11*g* to do additional work. If a majority of table blocks need to read to retrieve the requested data, index reads might actually add more block reads than a full table scan. In fact, having an index in this situation might slow down data retrieval. Various users and publications have guidelines on indexes. For example, some guidelines state that indexes should be used if the query condition will return less than 10% of the table rows. These guidelines don't apply to all situations, however; the best way to prove whether an index is beneficial is to test it.
- Typically, small tables don't benefit from indexes. In this case, a full table scan might be as fast as the combined effort of reading index blocks and then the appropriate table blocks.
- More storage space is required for the database because indexes are additional database objects.

Understanding the statement execution plan and the database optimizer is critical in analyzing index usage and applicability. The execution plan or **explain plan** identifies the steps the database system uses to resolve the query, including whether an index scan or full table scan is used. The **optimizer** provides the logic the database system uses in determining the best path of execution, based on available information. The optimizer considers the distribution of data values to determine whether using an index is beneficial. In other words, just because an index exists doesn't mean the system will use it.

NOTE

Performance-tuning topics are beyond the scope of this textbook. The concepts introduced in this section and in Appendix E provide just the fundamentals to give you an overview of performance tuning.

To see that an index might not be used, examine the explain plan of a query using a zip code search on the CUSTOMERS table, shown in Figure 6-20. After issuing a query in SQL Developer, click the Execute Explain Plan button to display the explain plan.

Execute Explain Plan button

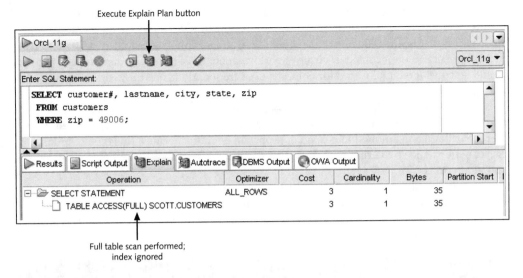

Full table scan performed;
index ignored

FIGURE 6-20 View the explain plan indicating a full table scan

The explain plan indicates that a full table scan is used to search for records with a matching zip code. The CUSTOMERS_ZIP_IDX index created earlier isn't used to perform the query, probably because the table contains very few rows of data. Reading and checking all rows in the table is faster than the two-step process of navigating the index to find the ROWIDs for the rows matching the condition, and then retrieving those rows from the CUSTOMERS table.

What does the explain plan look like if the index *is* used? Figure 6-21 shows the same query with the addition of an optimizer hint. The hint instructs Oracle to use an index if one is available. (Hints are also beyond the scope of this textbook; however, Figure 6-21 shows you what an explain plan indicating an index scan looks like.)

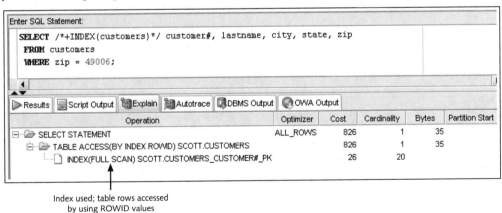

Index used; table rows accessed
by using ROWID values

FIGURE 6-21 View the explain plan indicating an index scan

Given these considerations, you should first weigh the benefits of improving query performance against the decreased performance for data manipulation actions. Determine the

Additional Database Objects

volume of queries and DML statements regularly executed on the database and which type of statement has performance priority. In some circumstances, database query activity is the priority; that is, the database is used mainly to support queries, and little DML activity occurs or is of less importance in terms of performance. Under these circumstances, indexes should be considered and tested for columns used frequently in WHERE conditions or for sorting operations, including table joins.

On the other hand, if database operations involve more DML actions than query actions, and if DML performance is the priority, index creation should be minimized. However, consider using unique indexes on appropriate columns because this type of index assists performance by verifying that duplicate values aren't entered in a column during DML operations.

As index candidate columns are identified, perform tests to determine whether the index actually improves data retrieval. Testing involves measuring query execution time with and without the index and comparing the results. To do this, you can use tools such as the TIMING feature and review the explain plan.

NOTE

The TIMING feature is explained in Appendix E.

NOTE

Sorting operations include using an ORDER BY clause and table joins. Sorting many rows for output can be memory and processor intensive because data rows are manipulated into order in memory, particularly if rows contain large amounts of selected data. Indexes can improve sorting operations by allowing rows to be retrieved in sorted order. Join operations, which link data between two tables, involve sorting operations. These operations assist in matching row values of the two tables by ordering rows according to the columns used to join the tables.

A unique index is typically created automatically when a PRIMARY KEY or UNIQUE constraint is defined on a column. Unique indexes can also be explicitly created by including the UNIQUE keyword in the CREATE INDEX statement, as shown in Figure 6-22. This statement creates a unique index on the Title column of the BOOKS table.

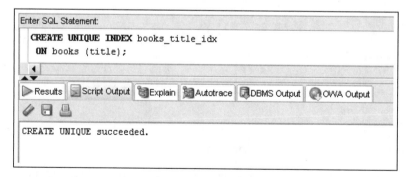

FIGURE 6-22   Explicitly creating a unique index

If an index is created to improve performance on large sorting operations, consider whether the sort order required is ascending or descending. By default, an index is created with an ascending order on the index column value. However, a descending sort can be used in an index by including a DESC sort option when you create the index, as shown in Figure 6-23.

```
CREATE INDEX customers_zip_desc_idx
   ON customers(zip DESC);
```

FIGURE 6-23   Indicating a descending sort for index values

The existence of NULL values in a column must also be considered in creating indexes for a column that's used in search conditions frequently. A B-tree index doesn't include any rows with NULL values in the indexed column. Therefore, if the column contains a lot of NULL values, an index might be useful because it eliminates searching through all the NULL value rows with a full table scan. However, if your search condition is hunting for NULL values (that is, WHERE zip IS NULL), a full table scan is performed regardless of whether a B-tree index exists on the column. In this case, a function-based index, covered later in this chapter, could be useful.

The indexes discussed so far consist of only a single column. However, indexes can include multiple columns of a table. These indexes are called **composite** or **concatenated indexes**. For example, an index on customer name could be created by using the statement shown in Figure 6-24.

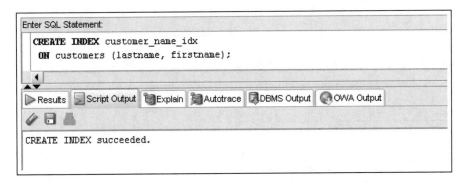

FIGURE 6-24   Creating a composite index

This index could improve the performance of queries that include a search condition on both the Lastname and Firstname columns. Generally, a composite index is more efficient than creating two separate single-column indexes because less I/O is required to read a single index. In addition, the selection results of two separate indexes don't have to be combined. In this case, the indexed key value contains both the first and last name of the customer. Putting the most selective column first typically produces the most efficient result. This index can also be used on search conditions involving only the leading (first) column—Lastname, in this example.

Bitmap Indexes

A **bitmap index** varies in structure and use from a B-tree index. This index is useful for improving queries on columns that have low selectivity (low cardinality, or a small number of distinct values). The index is a two-dimensional array containing one column for each distinct value in the column being indexed. Each row is linked to a ROWID and contains a bit (0 or 1) indicating whether the column value matches this index value. For example, the CUSTO-MERS table contains a Region column that can have one of eight possible values: N, NW, NE, S, SE, SW, W, or E. You can create a bitmap index on the Region column with the command shown in Figure 6-25, which adds the keyword BITMAP to the CREATE INDEX command.

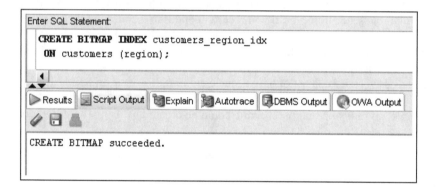

FIGURE 6-25   Creating a bitmap index on the Region column

Figure 6-26 illustrates the organization of a bitmap index, using the Region column example.

Bitmap index
Region

N	NW	NE	S	SE	SW	W	E
0	1	0	0	0	0	0	0
0	0	0	0	1	0	0	0
1	0	0	0	0	0	0	0
0	1	0	0	0	0	0	0
0	0	0	0	0	1	0	0

FIGURE 6-26   Organization of a bitmap index

Notice in Figure 6-26 that each row has only one bit turned on (set to 1), which indicates the Region value for the corresponding row of the CUSTOMERS table. The bitmap index structure can be particularly useful in queries involving compound conditions (using AND and OR operators).

For example, a WHERE clause might attempt to identify all male customers in the NW region (assuming there's a gender column in the CUSTOMERS table). In this type of query, if a bitmap index exists on both columns, the bitmap index information for the two column values (Gender = 'M', Region = 'NW') can be combined to identify the rows to be queried from the table.

When you use bitmap indexes, query performance improvements are offset by less efficient DML statement execution, as with B-tree indexes. Bitmap indexes tend to decrease DML performance when new values are added to the indexed column because a new bitmap column must be added to the index.

Function-Based Indexes

A **function-based index** can be useful if a query search is based on an expression (calculated value) or a function. After an expression or function is used on the column in the search condition, a non-function-based index is ignored. For JustLee Books, one commonly used search criterion is profit. Management might be looking for values that fall above or below a certain dollar value for profit. To speed up the retrieval of rows meeting a given condition, you can create an index based on the calculated profit for each book. The only difference in the CREATE INDEX command is that the expression or function on which the index is based is used in the ON clause, instead of including just the column name. For example, to create an index on the BOOKS table for the dollar profit returned by each book, you could use the command shown in Figure 6-27.

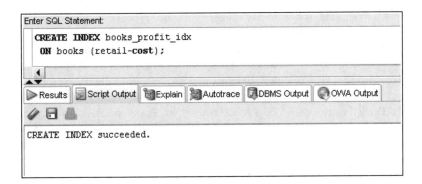

FIGURE 6-27   Creating a function-based index

> **NOTE**
>
> You can create a function-based index in a B-tree structure (default) or a bitmap structure. Add the keyword BITMAP to create the function in a bitmap structure.

A function-based index can also help improve query performance for search conditions on NULL values. By default, NULL values in an indexed column cause that row to not be indexed. To work around this problem, you could use a function-based index with the NVL function (covered in Chapter 10). The NVL function simply instructs the query to substitute a specified value if a NULL value is retrieved.

For example, JustLee Books routinely executes queries to identify orders that haven't been shipped. For this query, the WHERE clause checks for a NULL value in the Shipdate column. A basic B-tree index on this column doesn't include NULL value rows and, therefore, results in a full table scan. However, creating a function-based index, as shown in Figure 6-28, allows NULL values to be indexed.

```
CREATE INDEX orders_shipdate_idx
   ON orders(NVL(shipdate,'null'));
```

FIGURE 6-28   Creating a function-based index for NULL values

In this example, the index is created by including the NULL value rows. The NVL function converts the NULL values to the string value 'null' that's stored in the index for applicable rows. Therefore, the query search must match the same substitution action—WHERE NVL(shipdate, 'null') = 'null'—for this index to be used.

Index Organized Tables

An **index organized table (IOT)** is a variation of the B-tree index structure, used as an alternative to the conventional heap-organized table. This structure stores the entire table's contents in a B-tree index with rows sorted in the primary key value order. It combines the index and table into a single structure. Search and sort operations involving primary key column can be improved with this index. It has an advantage over other types of indexes because only one physical database object is needed to house both the index and data values.

The leaf blocks of an IOT contain the primary key value and the entire row of data. A primary key is required to create an IOT because it's used as the row identifier. A search by ROWID isn't required with this type of index. Because the B-tree index becomes the table structure, an IOT is created at the time of table creation. To create an IOT, add the keywords ORGANIZATION INDEX to the CREATE TABLE statement. For example, if you anticipate performing many searches and sort operations on the ISBN column of the BOOKS table, you could create an IOT, as shown in Figure 6-29.

```
CREATE TABLE books2
(ISBN VARCHAR2(10),
 title VARCHAR2(30),
 pubdate DATE,
 pubID NUMBER (2),
 cost NUMBER (5,2),
 retail NUMBER (5,2),
 category VARCHAR2(12),
   CONSTRAINT books2_isbn_pk PRIMARY KEY(isbn))
 ORGANIZATION INDEX;
```

FIGURE 6-29   Creating an IOT for the BOOKS table

Notice that the table is named BOOKS2 to avoid conflict with the existing BOOKS table. As with other types of indexes, an IOT can affect DML operations because the primary key order must be maintained in this structure.

Verifying an Index

After an index has been created implicitly or explicitly, you can use the USER_INDEXES data dictionary view to verify that the index exists. Figures 6-30 and 6-31 show the commands to verify existing indexes for a specific table.

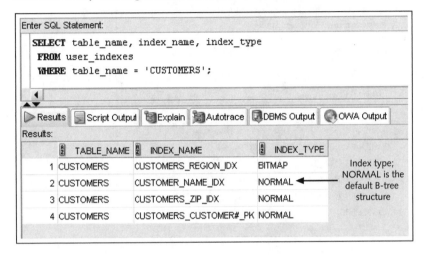

FIGURE 6-30   Identify indexes on the CUSTOMERS table

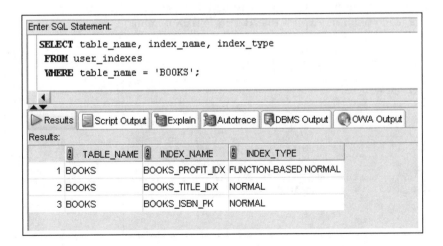

FIGURE 6-31   Identify indexes on the BOOKS table

Querying USER_INDEXES verifies existing indexes; however, it doesn't identify which columns each index includes. The USER_IND_COLUMNS data dictionary view includes information on the columns included in each index. Figure 6-32 shows the command to list index column information for the CUSTOMERS table.

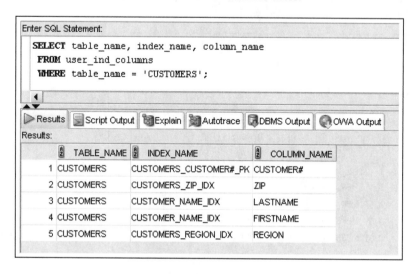

FIGURE 6-32   More index details from USER_IND_COLUMNS

Altering or Removing an Index

The only modification you can perform on an existing index is a name change. If you need to change the name of an index, use the **ALTER INDEX** command, as shown in Figure 6-33.

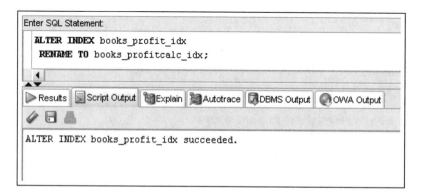

FIGURE 6-33   Rename an index

If an inappropriate index exists (for example, queries on an indexed column tend to return a large number of rows or updates are slow), you can delete an index with the **DROP INDEX** command. As mentioned, except for a name change, an index can't be modified; if you need to change an existing index, you have to delete and then re-create it. Figure 6-34 shows the syntax of the DROP INDEX command.

```
DROP INDEX indexname;
```

FIGURE 6-34   Syntax of the DROP INDEX command

As with other DROP and DDL commands, after the DROP INDEX command is executed, the statement can't be rolled back, and the index is no longer available to Oracle 11g. Figure 6-35 shows the command to drop the BOOKS_PROFITCALC_IDX index.

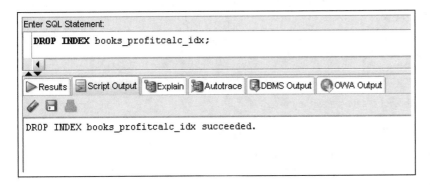

FIGURE 6-35   Dropping the BOOKS_PROFITCALC_IDX index

SYNONYMS

A **synonym** is an alternative name or alias for a database object, such as a table or a sequence. A synonym is used for several reasons:

- A synonym gives you the convenience of not having to use a full object name, including the schema name, if required.
- A synonym hides the actual object name, which can improve security.
- Synonyms help minimize application modification by allowing the alias to be used in application code. When object changes are necessary (object name changes or object schema changes, for example), you don't have to change the application code; instead, you can modify the synonym to identify the correct object.

Simplifying schema references can be an important benefit of using synonyms. When a user creates an object, unless otherwise specified, it belongs to his or her schema. By grouping objects according to the owner, multiple objects with the same name can exist in the same database, but only if each object belongs to a different schema.

For example, a user named Jeff creates a table called PROFITTABLE. Unless Jeff indicates otherwise, the table is an object in the schema called Jeff. If any other user who has permission wants to access Jeff's table, the user must identify the table by using the correct schema name in the FROM clause of the SELECT statement (for example, Jeff.PROFITTABLE). If the table name isn't prefixed by a schema name, Oracle 11g searches for the table only in the schema of the user who issued the SELECT statement. If a table with the same name doesn't exist in the user's schema, Oracle 11g returns an error message indicating that the table doesn't exist.

This situation can cause problems if several users must access a table frequently. In the JustLee Books database, different users (customer service representatives) enter orders in the ORDERS table. Before a user can enter an order into the table, he or she must remember who actually owns the table, and then prefix the table name with the correct schema. To simplify this process, Oracle 11g allows you to create synonyms that serve as a substitute for an object name. The syntax of the CREATE SYNONYM command is shown in Figure 6-36.

```
CREATE [PUBLIC] SYNONYM synonymname
  FOR objectname;
```

FIGURE 6-36   Syntax for creating a synonym

Note the following elements in Figure 6-36:

- The synonym name in the CREATE SYNONYM clause identifies the substitute name, or permanent alias, for the object listed in the FOR clause.
- The optional PUBLIC keyword can be used so that any user in the database can use that synonym to refer to the object.
- The object listed in the FOR clause can be the name of a table, constraint, view, or any other Oracle 11g database object.

Create a synonym for the ORDERS table, using the statement shown in Figure 6-37.

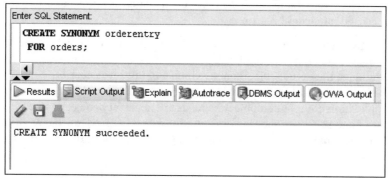

FIGURE 6-37   Create a synonym

A synonym can be a **private synonym**, which users use to reference objects they own, or a **public synonym**, which users use to access another user's database objects. If you add the PUBLIC keyword to the command in Figure 6-37, any user can reference the ORDERS table by using ORDERENTRY as the table name. Users don't need to reference the correct schema when a public synonym is available for an object. Because the PUBLIC keyword allows any database user to use the synonym, only someone with database administrator (DBA) privileges is allowed to delete the synonym. Why? To make certain that one user doesn't delete the synonym, which could affect the work of other users.

Review the SELECT statement in Figure 6-38, which uses the synonym. The data is actually being retrieved from the ORDERS table.

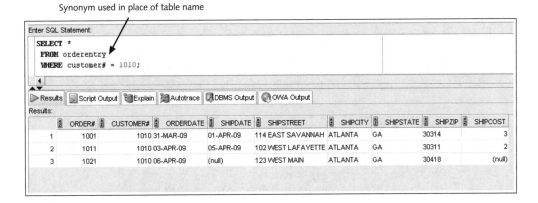

FIGURE 6-38   Using a synonym

As shown in Figure 6-38, after a synonym is created, you can substitute it for the object name in a command. When Oracle 11g tries to find that object in the database, it takes this path:

1. It searches for an object with the same name as the synonym.
2. If no object is found, it searches for private synonyms with the synonym's name.
3. If no private synonym is found, it searches for public synonyms with the synonym's name.
4. If no public synonym is found, Oracle 11g returns an error message, indicating that the object doesn't exist.

Many users like to create synonyms to avoid typing long or complex object names. You can create synonyms for your personal use simply by omitting the PUBLIC keyword from the CREATE SYNONYM clause. If the PUBLIC keyword isn't included, the user who created the private synonym is the *only one* who can use it. However, you can use a private synonym to reference objects in someone else's schema. For example, if user Jane frequently references the PROFITTABLE table in user Jeff's schema, Jane can create a synonym for the object Jeff.PROFITTABLE, and then she doesn't need to remember to include the schema name when she accesses his table.

Deleting a Synonym

The **DROP SYNONYM** command is used to delete both private and public synonyms. However, if the synonym being dropped is a public synonym, you *must* include the keyword PUBLIC. Figure 6-39 shows the syntax of the DROP SYNONYM command.

```
DROP [PUBLIC] SYNONYM synonymname;
```

FIGURE 6-39   Syntax of the DROP SYNONYM command

If you attempt to drop a public synonym and forget to include the PUBLIC keyword, Oracle 11g returns an error message, indicating that a private synonym by that name doesn't exist. To drop the ORDERENTRY synonym you just created, use the command shown in Figure 6-40. After the synonym has been deleted, any reference to ORDERENTRY in an SQL statement returns an error message indicating that the object doesn't exist.

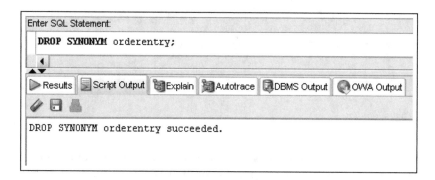

FIGURE 6-40   Command to delete the ORDERENTRY synonym

NOTE

You must have DBA privileges to drop public synonyms.

Chapter Summary

- A sequence can be created to generate a series of integers.
- The values generated by a sequence can be stored in any table.
- A sequence is created with the CREATE SEQUENCE command.
- Gaps in sequences might occur if the values are stored in various tables, if numbers are cached but not used, or if a rollback occurs.
- A value is generated by using the NEXTVAL pseudocolumn.
- The CURRVAL pseudocolumn is NULL until a value is generated by NEXTVAL.
- The USER_OBJECTS data dictionary object can be used to confirm the existence of all schema objects.
- The USER_SEQUENCES data dictionary object is used to view sequence settings.
- The ALTER SEQUENCE command is used to modify an existing sequence. The only settings that can't be modified are the START WITH option and any option that would be invalid because of previously generated values.
- The DUAL table is helpful for testing sequence value generation.
- The DROP SEQUENCE command deletes an existing sequence.
- An index can be created to speed up the query process.
- DML operations are always slower when indexes exist.
- Oracle 11g creates an index for PRIMARY KEY and UNIQUE constraints automatically.
- An explicit index is created with the CREATE INDEX command.
- An index can be used by Oracle 11g automatically if a query criterion or sort operation is based on a column or an expression used to create the index.
- The two main structures for indexes are B-tree and bitmap.
- The explain plan can verify whether an index is used in a query.
- Function-based indexes are used to index an expression or the use of functions on a column or columns.
- An index organized table is a table stored in a B-tree structure to combine the index and table into one database object.
- Information about an index can be retrieved from the USER_INDEXES and USER_IND_COLUMNS views.
- An index can be dropped with the DROP INDEX command.
- An index can be renamed with the ALTER INDEX command.
- Except for a name change, an index can't be modified. It must be deleted and then re-created.
- A synonym provides a permanent alias for a database object.
- A public synonym is available to any database user.
- A private synonym is available only to the user who created it.
- A synonym is created by using the CREATE SYNONYM command.
- A synonym is deleted by using the DROP SYNONYM command.
- Only a user with DBA privileges can drop a public synonym.

Chapter 6 Syntax Summary

The following table summarizes the syntax you have learned in this chapter. You can use the table as a study guide and reference.

Syntax Guide

Description	Command Syntax	Example

Create a sequence to generate a series of integers	CREATE SEQUENCE *sequencename*
	    [INCREMENT BY *value*]
	    [START WITH *value*]
	    [{MAXVALUE *value* |
	        NOMAXVALUE}]
	    [{MINVALUE *value* |
	        NOMINVALUE}]
	    [{CYCLE | NOCYCLE}]
	    [{ORDER | NOORDER}]
	    [{CACHE *value* | NOCACHE}] ;
		CREATE SEQUENCE
		    orders_order#_seq
		    INCREMENT BY 1
		    START WITH 1021
		    NOCACHE NOCYCLE ;

Alter a sequence	ALTER SEQUENCE *sequencename*
	    [INCREMENT BY *value*]
	    [{MAXVALUE *value* |
	        NOMAXVALUE}]
	    [{MINVALUE *value* |
	        NOMINVALUE}]
	    [{CYCLE | NOCYCLE}]
	    [{ORDER | NOORDER}]
	    [{CACHE *value* | NOCACHE}] ;
		ALTER SEQUENCE
		    orders_order#_seq
		    INCREMENT BY 10 ;

Drop a sequence	DROP SEQUENCE *sequencename*;	DROP SEQUENCE orders_order#_seq;

Create a B-tree index	CREATE INDEX *indexname*
	ON *tablename*
	    (*columnname*, ...);
		CREATE INDEX
		    customers_lastname_idx
		ON customers(lastname) ;

Create a bitmap index	CREATE BITMAP INDEX *indexname*
	    ON *tablename*
	        (*columnname*, ...);
		CREATE BITMAP INDEX
		    customers_region_idx
		ON customers (region) ;

Create a function-based index	CREATE INDEX *indexname* ON *tablename* (*expression*);	CREATE INDEX books_profit_idx ON books (retail-cost) ;

Create an index organized table	CREATE TABLE *tablename*
	    (*columnname datatype*,
	        columnname datatype)
	ORGANIZATION INDEX;
		CREATE TABLE
		    books(isbn NUMBER(10),
		    title VARCHAR2(30))
		ORGANIZATION INDEX;

Description	Command Syntax	Example
Rename an index	ALTER INDEX *indexname* RENAME TO *newindexname*;	**ALTER INDEX books_profit_idx RENAME TO books_profitcalc_idx;**
Drop an index	DROP INDEX *indexname*;	**DROP INDEX books_profit_idx;**
Create a synonym	CREATE [PUBLIC] SYNONYM *synonymname* FOR *objectname*;	**CREATE PUBLIC SYNONYM orderentry FOR orders;**
Drop a synonym	DROP [PUBLIC] SYNONYM *synonymname*;	**DROP PUBLIC SYNONYM orderentry;**

Data Dictionary View Prefix	Description
ALL_	Displays objects accessible by the user
DBA_	Displays all objects in the database
USER_	Displays objects owned by the user
V$	Displays dynamic database statistics

Review Questions

1. How can a sequence be used in a database?
2. How can gaps appear in values generated by a sequence?
3. How can you indicate that the values generated by a sequence should be in descending order?
4. When is an index appropriate for a table?
5. What is the difference between the B-tree and bitmap index structures?
6. When does Oracle 11g automatically create an index for a table?
7. Under what circumstances should you not create an index for a table?
8. What is an IOT and under what circumstances might it be useful?
9. What command is used to modify an index?
10. What is the purpose of a synonym?

Multiple Choice

To answer the following questions, refer to the tables in the JustLee Books database.

1. Which of the following generates a series of integers that can be stored in a database?
    a. a number generator
    b. a view
    c. a sequence
    d. an index
    e. a synonym

2. Which syntax is correct for removing a public synonym?
    a. DROP SYNONYM *synonymname*;
    b. DELETE PUBLIC SYNONYM *synonymname*;
    c. DROP PUBLIC SYNONYM *synonymname*;
    d. DELETE SYNONYM *synonymname*;

3. Which of the following commands can you use to modify an index?
    a. ALTER SESSION
    b. ALTER TABLE
    c. MODIFY INDEX
    d. ALTER INDEX
    e. none of the above

4. Which of the following generates an integer in a sequence?
    a. NEXTVAL
    b. CURVAL
    c. NEXT_VALUE
    d. CURR_VALUE
    e. NEXT_VAL
    f. CUR_VAL

5. Which of the following is a valid SQL statement?
    a. ```
        INSERT INTO publisher
            VALUES (pubsequence.nextvalue, 'HAPPY
            PRINTING', 'LAZY LARRY', NULL);
        ```
    b.  ```
        CREATE INDEX a_new_index
            ON (firstcolumn*.02);
        ```
    c. ```
        CREATE SYNONYM pub
            FOR publisher;
        ```
    d.  all of the above
    e.  only a and c
    f.  none of the above
        ```

6. Suppose the user Juan creates a table called MYTABLE with four columns. The first column has a PRIMARY KEY constraint, the second column has a NOT NULL constraint, the third column has a CHECK constraint, and the fourth column has a FOREIGN KEY constraint. Given this information, how many indexes does Oracle 11*g* create automatically when the table and constraints are created?

    a.   0

    b.   1

    c.   2

    d.   3

    e.   4

209

7. Given the table created in Question 6, which of the following commands can Juan use to create a synonym that allows anyone to access the table without having to identify his schema in the table reference?

    a.   `CREATE SYNONYM thetable`
        `FOR juan.mytable;`

    b.   `CREATE PUBLIC SYNONYM thetable`
        `FOR mytable;`

    c.   `CREATE SYNONYM juan`
        `FOR mytable;`

    d.   none of the above

8. Which of the following statements is true?

    a.   A gap can appear in a sequence created with the NOCACHE option if the system crashes before a user can commit a transaction.

    b.   Any unassigned sequence values appears in the USER_SEQUENCE data dictionary table as unassigned.

    c.   Only the user who creates a sequence is allowed to delete it.

    d.   Only the user who created a sequence is allowed to use the value generated by the sequence.

9. When is creating an index manually inappropriate?

    a.   when queries return a large percentage of rows in the results

    b.   when the table is small

    c.   when the majority of table operations are updates

    d.   all of the above

    e.   only a and c

10. If a column has high selectivity or cardinality, which index type is most appropriate?

    a.   IOT

    b.   B-tree

    c.   bitmap

    d.   function-based index

11. If a column has low selectivity, this means:

    a. The column contains many distinct values.

    b. The column contains a small number of distinct values.

    c. A WHERE clause is always used in a query on the column.

    d. The selectivity of a column can't be determined.

12. Oracle 11*g* automatically creates an index for which type of constraints?

    a. NOT NULL

    b. PRIMARY KEY

    c. FOREIGN KEY

    d. UNIQUE KEY

    e. none of the above

    f. only a and b

    g. only b and d

13. Which of the following settings can't be modified with the ALTER SEQUENCE command?

    a. INCREMENT BY

    b. MAXVALUE

    c. START WITH

    d. MINVALUE

    e. CACHE

14. Which node of the b-tree index contains ROWIDs?

    a. branch blocks

    b. root block

    c. leaf blocks

    d. None of the above because the primary key is used to identify rows.

15. If the CACHE or NOCACHE options aren't included in the CREATE SEQUENCE command, which of the following statements is correct?

    a. Oracle 11*g* generates 20 integers automatically and stores them in memory.

    b. No integers are cached by default.

    c. Only one integer is cached at a time.

    d. The command will fail.

    e. Oracle 11*g* generates 20 three-digit decimal numbers automatically and stores them in memory.

16. Which of the following is a valid command?

    a.
```
CREATE INDEX book_profit_idx
   ON (retail-cost) WHERE (retail-cost) > 10;
```

    b.
```
CREATE INDEX book_profit_idx
   ON (retail-cost);
```

c. `CREATE FUNCTION INDEX book_profit_idx`
   `ON books WHERE (retail-cost) > 10;`

d. both a and c

e. none of the above

17. Which of the following can be used to determine whether an index exists?

    a. DESCRIBE *indexname*;

    b. the USER_INDEXES view

    c. the INDEXES table

    d. the USER_INDEX view

    e. all of the above

    f. none of the above

18. Which of the following isn't a valid option for the CREATE SEQUENCE command?

    a. ORDER

    b. NOCYCLE

    c. MINIMUMVAL

    d. NOCACHE

    e. All of the above are valid options.

19. What can be referenced to determine whether an index is used to perform a query?

    a. USER_INDEXES view

    b. query source code

    c. explain plan

    d. database access plan

20. Which of the following commands creates a private synonym?

    a. CREATE PRIVATE SYNONYM

    b. CREATE NONPUBLIC SYNONYM

    c. CREATE SYNONYM

    d. CREATE PUBLIC SYNONYM

Hands-On Assignments

To perform the following assignments, refer to the tables in the JustLee Books database.

1. Create a sequence for populating the Customer# column of the CUSTOMERS table. When setting the start and increment values, keep in mind that data already exists in this table. The options should be set to not cycle the values and not cache any values, and no minimum or maximum values should be declared.

2. Add a new customer row by using the sequence created in Question 1. The only data currently available for the customer is as follows: last name = Shoulders, first name = Frank, and zip = 23567.

3. Create a sequence that generates integers starting with the value 5. Each value should be three less than the previous value generated. The lowest possible value should be 0, and the sequence shouldn't be allowed to cycle. Name the sequence MY_FIRST_SEQ.

4. Issue a SELECT statement that displays NEXTVAL for MY_FIRST_SEQ three times. Because the value isn't being placed in a table, use the DUAL table in the FROM clause of the SELECT statement. What causes the error on the third SELECT?

5. Change the setting of MY_FIRST_SEQ so that the minimum value that can be generated is -1000.

6. Create a private synonym that enables you to reference the MY_FIRST_SEQ object as NUMGEN.

7. Use a SELECT statement to view the CURRVAL of NUMGEN. Delete the NUMGEN synonym and MY_FIRST_SEQ.

8. Create a bitmap index on the CUSTOMERS table to speed up queries that search for customers based on their state of residence. Verify that the index exists, and then delete the index.

9. Create a B-tree index on the customer's Lastname column. Verify that the index exists by querying the data dictionary. Remove the index from the database.

10. Many queries search by the number of days to ship (number of days between the order and shipping dates). Create an index that might improve the performance of these queries.

Advanced Challenge

To perform the following activity, refer to the tables in the JustLee Books database.

Using the training you have received and speculating on query needs, determine appropriate uses for indexes and sequences in the JustLee Books database. Assume all tables will grow quite large in the number of rows. Identify at least three sequences and three indexes that can address needed functionality for the JustLee Books database. In a memo to management, you should identify each sequence and index that you propose and the rationale supporting your suggestions. You should also state any drawbacks that might affect database performance if the changes are implemented.

Case Study: *City Jail*

1. The head DBA has requested the creation of a sequence for the primary key columns of the Criminals and Crimes tables. After creating the sequences, add a new criminal named Johnny Capps to the Criminals table by using the correct sequence. (Use any values for the remainder of columns.) A crime needs to be added for the criminal, too. Add a row to the Crimes table, referencing the sequence value already generated for the Criminal_ID and using the correct sequence to generate the Crime_ID value. (Use any values for the remainder of columns.)

2. The last name, street, and phone number columns of the Criminals table are used quite often in the WHERE clause condition of queries. Create objects that might improve data retrieval for these queries.

3. Would a bitmap index be appropriate for any columns in the City Jail database (assuming the columns are used in search and/or sort operations)? If so, identify the columns and explain why a bitmap index is appropriate for them.

4. Would using the City Jail database be any easier with the creation of synonyms? Explain why or why not.

USER CREATION AND MANAGEMENT

INTRODUCTION

Previous chapters of this textbook have introduced the creation of database objects. After you've created a database, access to database objects must be assigned to users. This chapter focuses on setting up database users, controlling their access to database objects, and defining what actions they can perform on those objects. For example, if an employee's main job for JustLee Books is entering new orders, she needs access only to the customer and order tables, and the actions she can perform on these tables should be limited to selecting and inserting items. This chapter explains how to accomplish this type of access assignment. You examine creating, maintaining, and dropping user accounts; granting and revoking privileges; and simplifying the administration of privileges with roles.

Typically, these duties are performed by a database administrator (DBA) or security officer. However, everyone involved with a database should understand the basic user access principles covered in this chapter. For example, if you're involved in application development, you'll most likely participate in determining the privileges users need to work with applications.

Table 7-1 provides an overview of the commands covered in this chapter. The commands are grouped by category rather than the order in which they're discussed in this chapter.

TABLE 7-1   Overview of Chapter Contents

Description	Command Syntax
Creating, Maintaining, and Dropping User Accounts	
Create a user	CREATE USER *username* IDENTIFIED BY *password*;
Change or expire a password	ALTER USER *username* [IDENTIFIED BY *newpassword*] [PASSWORD EXPIRE];
Drop a user	DROP USER *username*;
Granting and Revoking Privileges	
Grant object privileges to users or roles	GRANT {*objectprivilege*|ALL} [(*columnname*), *objectprivilege* (*columnname*)] ON *objectname* TO {*username*|*rolename*|PUBLIC} [WITH GRANT OPTION];
Grant system privileges to users or roles	GRANT *systemprivilege* [, *systemprivilege*, ...] TO *username*|*rolename* [, *username*|*rolename*, ...] [WITH ADMIN OPTION];
Revoke object privileges	REVOKE *objectprivilege* [, ...*objectprivilege*] ON *objectname* FROM *username*|*rolename*;

TABLE 7-1 Overview of Chapter Contents (continued)

Description	Command Syntax
Granting and Revoking Roles	
Create a role	CREATE ROLE *rolename*;
Grant a role to a user	GRANT *rolename* [, *rolename*] TO *username* [, *username*];
Assign a default role to a user	ALTER USER *username* DEFAULT ROLE *rolename*;
Set or enable a role	SET ROLE *rolename*;
Add a password to a role	ALTER ROLE *rolename* IDENTIFIED BY *password*;
Revoke a role	REVOKE *rolename* FROM *username*|*rolename*;
Drop a role	DROP ROLE *rolename*;

215

DATABASE PREPARATION

If you haven't already run the JLDB_Build_5.sql script from Chapter 5, execute the script before attempting to work through the examples in this chapter.

CAUTION

Most commands in this chapter require database administrator privileges. If you're using an account to access a server and get an "insufficient privileges" error when the command executes, check with your instructor. If you're using your own installation of Oracle, use the SYSTEM user, which has DBA privileges. The SYSTEM user is created automatically during Oracle installation.

DATA SECURITY

Most organizations store data in some type of electronic database. For example, a bank typically manages all customer financial accounts by using a computer database. The news is full of stories about problems caused by unauthorized access to bank account data, ranging from major financial losses to identity theft. Company data needs to be protected from a variety of threats. Some threats are posed by natural disasters, such as floods, fires, and tornadoes. However, the biggest threat to an organization's data often comes in the form of people—computer criminals and the organization's own employees. Computer criminals, sometimes referred to as attackers or cyber criminals, attempt to access a system illegally and then copy, manipulate, or delete its data. Employees can also create havoc with data—often more easily because they're assigned legitimate system access. Disgruntled employees might justify their actions as a way of "getting even" for a promotion they didn't get or retaliating if they feel their employer doesn't value them. However, even employees with good intentions can damage an organization's data by deleting records accidentally or inserting records in the wrong table.

Data security is a wide-ranging topic that covers many facets of information technology, including network access and authentication, operating system manipulation, application security, database account management, physical protection, and user monitoring. This chapter addresses only one area of data security—database account management—and more specifically, it addresses creating database user accounts with passwords and managing the specific database privileges assigned to users. The tasks described in this chapter are important for limiting what a user can do after being logged in to a database and minimizing any intentional or unintentional damage, yet still granting enough rights for users to do their jobs.

Database security involves a two-stage process:

- Authentication to identify the user
- Authorization to allow user access to objects

Authentication is the process of identifying a user attempting to connect to a system, typically based on a username and password. This process doesn't identify the objects the user can access. In this chapter, you address authentication by creating a user who can connect to the database via SQL*Plus. **Authorization** is granting object privileges to users based on their identities, which is addressed by issuing GRANT commands for specific privileges.

Now take a look at database user management in the context of a newly hired employee at JustLee Books. This chapter traces the steps for creating a database account for that employee and then granting the user privileges to access database data.

CREATING A USER

When an employee is hired at JustLee Books, his or her supervisor notifies the DBA of the new employee's name and requests an Oracle 11g database account. The supervisor also tells the DBA the kind of duties the new employee needs to perform. Based on the employee's job responsibilities, the DBA determines what database objects he or she can access.

Creating Usernames and Passwords

The first step in creating a new user account is determining the username and password. By default, a username can contain up to 30 characters, including numbers, letters, and the underscore (_), dollar sign ($), and number (#) symbols. To create a user account, you use the **CREATE USER** command with the syntax shown in Figure 7-1.

```
CREATE USER username
 IDENTIFIED BY password;
```

FIGURE 7-1   Syntax of the CREATE USER command

NOTE

For user creation commands to execute successfully, you must use an account with DBA privileges.

NOTE

The CREATE USER command has additional options to perform tasks such as assigning tablespaces and disk space quotas. This chapter covers only the options associated with enabling login and access to database objects.

The DBA has been notified that a new data entry clerk, Ron Thomas, has been hired by JustLee Books and needs an Oracle 11*g* user account. In most cases, the DBA creates the account by using a coding scheme for the user's account name, such as the user's first initial followed by his last name. Therefore, for the new employee Ron Thomas, you assign the account name "rthomas."

NOTE

Many organizations use a coding scheme for assigning account names to simplify user tracking and produce unique names.

As shown in Figure 7-1, a password is entered in the IDENTIFIED BY clause when the account is created. Usually, a temporary password is assigned by using the PASSWORD EXPIRE option, and the user is allowed to change the password after logging in to the database. This option allows the user to create a password that's easier to remember than a randomly generated password but is still difficult for others to guess. However, the PASSWORD EXPIRE option isn't required to create an account. To create the account for the new employee, use the CREATE USER command shown in Figure 7-2.

```
CREATE USER rthomas
 IDENTIFIED BY little25car
 PASSWORD EXPIRE;
```

FIGURE 7-2 Command to create an account for a new employee

The CREATE USER command in Figure 7-2 creates a new account with the username RTHOMAS and the password little25car. Note that Oracle 11*g* account passwords are case sensitive. After the command is executed, a message indicating that the user was created is displayed.

Even though an account now exists for Ron Thomas, he can't log in to the database yet. If he attempted to log in at this point, the system would prompt him to change his password but then issue a "login denied" message. A user requires some minimum privileges to connect to the database and establish a session; however, no privileges have been granted to this user's account yet. The next section covers granting privileges.

ASSIGNING USER PRIVILEGES

Privileges allow Oracle 11*g* users to execute certain SQL statements. Two types of privileges exist: system privileges and object privileges. **System privileges** allow access to the Oracle 11*g* database and let users perform DDL operations, such as CREATE, ALTER, and DROP, on database objects (for example, tables and views). **Object privileges** allow users to perform DML operations, such as INSERT and UPDATE, on the data contained in database objects. The following sections describe both types of privileges.

System Privileges

Almost 200 system privileges are available in Oracle 11*g*. The ability to create, alter, and drop database objects, such as tables and sequences, is an example of a system privilege. Other system privileges apply to database access and user accounts. For example, to connect to Oracle 11*g*, a user must have the CREATE SESSION privilege. To create new user accounts, a user must have the CREATE USER privilege.

You can use the optional **ANY** keyword when granting a system privilege to allow the user to perform the privilege systemwide. For example, the DROP ANY TABLE command allows a user to delete or truncate any table in the database—not just tables in the user's own schema. You can view all available system privileges in Oracle 11*g* with the data dictionary view SYSTEM_PRIVILEGE_MAP. Figure 7-3 shows a portion of the output from querying this view.

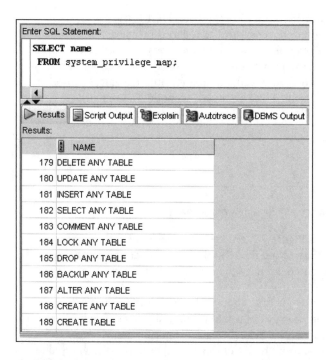

FIGURE 7-3   A partial list of available system privileges

Notice that DML operations, such as INSERT and DELETE, are included in the system privileges list with the ANY keyword. DML operations are considered object rather than system privileges unless the ANY keyword is assigned to the privilege. For example, the INSERT ANY TABLE command gives the user the ability to add rows to any table, regardless of whether he or she owns the table or has explicit permission to access that particular table. Because the privilege is effective systemwide (that is, the privilege can be used on any schema), it's reclassified from an object privilege to a system privilege.

N O T E

System privileges affect all objects; therefore, they aren't assigned by specifying database objects, such as a table name. Object privileges, on the other hand, are assigned to specific database objects.

Granting System Privileges

System privileges are assigned or granted to users with the GRANT command, which uses the syntax shown in Figure 7-4.

```
GRANT systemprivilege [, systemprivilege, …]
 TO username|rolename [, username|rolename, …]
 [WITH ADMIN OPTION];
```

FIGURE 7-4   Syntax of the GRANT command for system privileges

Take a closer look at elements of the statement in Figure 7-4:

- The system privilege being assigned is identified in the **GRANT** clause. If more than one system privilege is being granted, commas separate them. As mentioned, system privileges aren't granted for a specific database object; therefore, no database objects are referenced in this command.
- The users or roles receiving the system privileges are listed in the **TO** clause. Roles define a group of users and are discussed later in this chapter.
- **WITH ADMIN OPTION** allows any user or role identified in the TO clause to grant the system privilege to any other database users.

Now return to the account creation for the new employee, Ron Thomas. He needs to be granted the system privilege CREATE SESSION to connect to the Oracle 11g database. Figure 7-5 shows the command to accomplish this task.

```
GRANT CREATE SESSION
 TO rthomas;
```

FIGURE 7-5   Command to grant the CREATE SESSION privilege

The only privilege assigned to the new user is the ability to log in to Oracle 11g. Log in as user RTHOMAS to test the account. Notice that because the account was created with the PASSWORD EXPIRE option, you're prompted to change the password. However, after being logged in to the database, you can't perform any actions because no other privileges have been granted. For example, if you attempt to execute a SELECT statement to view the contents of the BOOKS table, an error message is returned.

> **NOTE**
>
> The password expiration prompt operates correctly with SQL*Plus; however, currently SQL Developer doesn't manage this prompt correctly and won't enable the login. As of version 1.5 of SQL Developer, a bug has been logged to fix this issue in a future release.

> **NOTE**
>
> The BOOKS table exists in another schema, so the new user, RTHOMAS, must qualify the table name with the schema name. For example, if the BOOKS table has been created in the SCOTT schema, the query's FROM clause must reference the table as SCOTT.BOOKS.

Object Privileges

When a user creates an object, he or she automatically has all object privileges associated with that object. However, if other users need access to the data contained in a database object or the ability to manipulate the data, they must be granted the privilege to do so. Object privileges in Oracle 11g include the following:

- SELECT: Allows users to display data contained in a table, view, or sequence; also allows generating the next sequence value by using NEXTVAL.
- INSERT: Allows users to insert data in a table or view.
- UPDATE: Allows users to modify data in a table or view.
- DELETE: Allows users to delete data in a table or view.
- INDEX: Allows users to create an index for a table.
- ALTER: Allows altering the definition of a table or sequence.
- REFERENCES: Allows users to reference a table when creating a FOREIGN KEY constraint. This privilege can be granted only to a user, not to a role.

> **NOTE**
>
> Additional object privileges, such as EXECUTE, are available that allow users to run a stored function or procedure. See the "Oracle Database SQL Language Reference" in the Oracle documentation library for a complete list of object privileges.

Granting Object Privileges

Object privileges are also granted to users with the GRANT command, which uses the syntax shown in Figure 7-6.

```
GRANT {objectprivilege|ALL} [(columnname),
       objectprivilege (columnname)]
 ON objectname
 TO {username|rolename|PUBLIC}
 [WITH GRANT OPTION];
```

FIGURE 7-6   Syntax of the GRANT command for object privileges

Examine the clauses in Figure 7-6:

- The **GRANT** clause identifies the object privileges being assigned. INSERT, UPDATE, and REFERENCES privileges can also be assigned to specific columns in a table or view. If the object privilege is assigned to a specific column, the column name should be included in the GRANT clause, inside parentheses, after the privilege name. Instead of listing object privileges separately, you can substitute the **ALL** keyword to indicate granting all object privileges. Either an object privilege or the ALL keyword *must be used after the GRANT keyword*. Be careful when granting users all available object privileges because this level of access makes it possible for them to perform any DML operation on the named object.
- The **ON** clause identifies the object (for example, table, view, sequence) to which the privilege applies.
- The **TO** clause identifies the user or role (discussed later in this chapter) receiving the privilege. You can assign privileges to multiple users or roles in the same GRANT command by entering the names in a list, separated by commas. If all database users should be assigned the privilege, the **PUBLIC** keyword can be used in the TO clause instead of a list of names.
- **WITH GRANT OPTION** enables the user to grant the same object privileges to other users.

Now take a look at some different examples of granting object privileges to Ron Thomas, along with their associated commands, as shown in Table 7-2.

TABLE 7-2   Examples of Granting Object Privileges to a User

Example	GRANT Command

Ron Thomas needs the ability to select rows from and insert rows into the CUSTOMERS table, which is in the SCOTT schema.	`GRANT select, insert`
	`   ON scott.customers`
	`   TO rthomas;`

Ron Thomas needs to be able to select any data from the CUSTOMERS table but be able to modify only the Lastname and Firstname columns.	`GRANT select,`
	`      update(lastname, firstname)`
	`   ON scott.customers`
	`   TO rthomas;`

TABLE 7-2   Examples of Granting Object Privileges to a User (continued)

Example	GRANT Command
Ron Thomas is assigned full responsibility for the CUSTOMERS table and, therefore, needs the ability to perform any activity on the CUSTOMERS table and the right to grant these privileges to other users.	```GRANT ALL ON scott.customers TO rthomas WITH GRANT OPTION;```

N O T E

WITH GRANT OPTION can't be used when granting object privileges to roles; it applies only to users.

Now assume Ron Thomas just needs to view the BOOKS table's contents and doesn't need to add, delete, or change anything. In this case, the GRANT command should be issued to assign the necessary object privilege, as shown in Figure 7-7. Remember to replace the SCOTT schema name with the correct schema name containing the JustLee Books database.

```
GRANT SELECT
 ON scott.books
 TO rthomas;
```

FIGURE 7-7   Command for assigning the SELECT privilege

T I P

Creating a synonym for the BOOKS table, as you learned in Chapter 6, eliminates the need to include the schema name in this command.

Because the command shown in Figure 7-7 doesn't include WITH GRANT OPTION, the user RTHOMAS isn't allowed to assign the SELECT privilege for the BOOKS table to any other user. However, Ron Thomas can now view the BOOKS table's contents by using a SELECT command, even though he can't perform any DML or DDL operations. Execute the GRANT command from a DBA account. Then log in as RTHOMAS and test the privilege by executing a SELECT statement on the BOOKS table.

Ron Thomas has now been assigned additional duties that require making corrections to book titles and published dates. The GRANT command in Figure 7-8 assigns the UPDATE privilege needed to modify these two columns of the BOOKS table.

Ron Thomas can now execute UPDATE commands affecting the book title and published date. However, what happens if he tries to modify any other columns of the BOOKS table? Figure 7-9 shows an attempt to modify the publisher ID as well as the published date of a specific book. The error message indicates that Ron Thomas doesn't have the privileges needed to make this change.

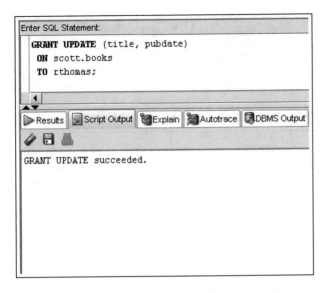

FIGURE 7-8 Assign the UPDATE privilege on specific columns

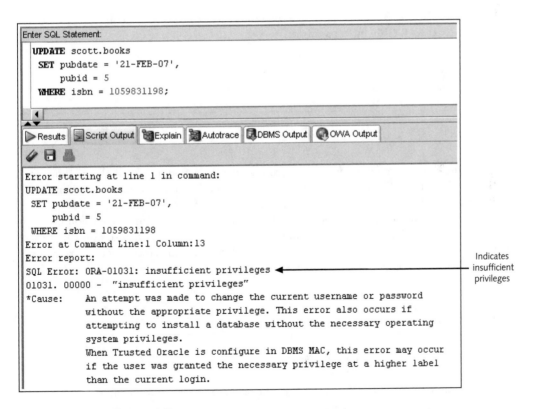

FIGURE 7-9 Insufficient privileges error

MANAGING PASSWORDS

Users often forget their passwords, so DBAs routinely get requests to reset account passwords. After an account has been created, the simplest way to change the password is with the **ALTER USER** command. The DBA can reset the current password and mark the password as "expired," which forces the user to set up a new password immediately after attempting to connect to the database. Figure 7-10 shows the syntax of the ALTER USER command.

```
ALTER USER username
 [IDENTIFIED BY newpassword]
 [PASSWORD EXPIRE];
```

FIGURE 7-10   Syntax of the ALTER USER command

The **IDENTIFIED BY** clause specifies the new password. Figure 7-11 shows the command for resetting RTHOMAS's password.

```
ALTER USER rthomas
 IDENTIFIED by rxy22b
  PASSWORD EXPIRE;
```

FIGURE 7-11   ALTER USER command for resetting a password

Execute the command in Figure 7-11 from your DBA account. To test the modification, log in as RTHOMAS with the new password rxy22b. You should be prompted to change the password, and then the login should be completed successfully.

After logging in to an account successfully, the user can also change the account's password by using the ALTER USER command with the IDENTIFIED BY clause. After Ron logs in to his account and decides that the new password he selected is too difficult to remember, he could issue the command shown in Figure 7-12.

```
ALTER USER rthomas
 IDENTIFIED BY monster42truck;
```

FIGURE 7-12   Command to change a password

The command shown in Figure 7-12 can be issued by the account owner or anyone with the ALTER USER system privilege.

In addition, some interfaces provide mechanisms for users to change their passwords after logging in to an account. The mechanism depends on which interface you're using. For example, the SQL*Plus client tool offers a PASSWORD command that can be issued at the SQL prompt. The user is prompted for the old password and then the new password.

From a DBA's perspective, the previous discussion on logins and passwords just touches the surface of ways you can control user logins. For example, Oracle 11*g* supports a variety of authentication methods beyond the database authentication addressed in this section. Methods include externally processed authentication, authentication processed globally by Secure Sockets Layer (SSL), and authentication processed by proxy. Creating user profiles is another feature for more complex password management, as are password aging rules and verification complexity (for example, setting requirements on the content and length of passwords).

Authentication also raises the topic of **encryption**, which refers to scrambling data to make it unreadable to anyone other than the sender and receiver. Encryption plays a role in protecting data as it's transmitted via network communication and storing data in an encrypted form. Oracle provides encryption for passwords during login transmission and stores users' passwords in the data dictionary in an encrypted format. However, you should consider encrypting other data, too. For example, if customers are providing credit card numbers via the Internet, how can this information be protected? It calls for securing transmissions with a protocol such as SSL and, if the credit card numbers are to be stored, encrypting data columns with Oracle encryption tools. These topics are beyond the scope of this textbook. However, the basic concepts of user account creation and passwords are fundamental to understanding user access.

USING ROLES

In most cases, a user needs more privileges than just the CREATE SESSION system privilege and the SELECT object privilege for one table. For example, as a data entry clerk, Ron Thomas probably needs to enter the ship date for orders, update information in the CUSTOMERS table, and so on. In fact, all JustLee Books data entry clerks need to perform these tasks. Assigning each of the necessary privileges to all data entry clerks separately could be quite cumbersome. A simpler approach is assigning a group of privileges to a role, and then assigning the role to relevant users.

A **role** is a group, or collection, of privileges. In most organizations, roles correlate to users' job duties. For example, customer service representatives might need to view all the data in each database table. However, they probably don't need the privilege of updating data in the BOOKS table (ISBN, cost, retail price, and so on). By grouping employees based on the tasks they need to perform, you can create roles that have been assigned the privileges each group needs. So instead of assigning each user several privileges, you can just assign a collection of privileges—a role—to users.

A role is typically assigned to multiple users representing a workgroup when a new application is introduced. As new employees join the group, the role can be assigned to each user as he or she is added. This process not only simplifies assigning and modifying privileges, but also ensures consistency of privileges for users in a workgroup. In addition, if you have some employees with job duties covering multiple areas (for example, branch supervisors), you can assign multiple roles to them, as shown in Figure 7-13.

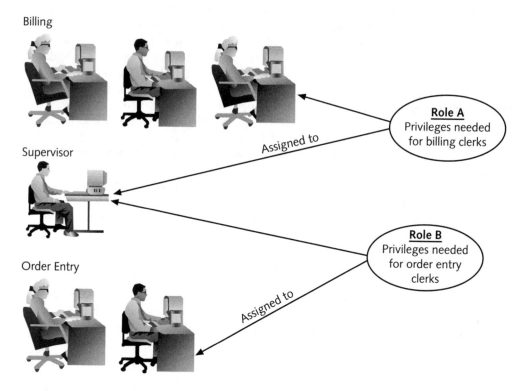

FIGURE 7-13   Assigning multiple roles to users

Creating and Assigning Roles

Before you can assign privileges to a role, you must create the role object by using the **CREATE ROLE** command. Figure 7-14 shows the syntax of this command.

```
CREATE ROLE rolename;
```

FIGURE 7-14   Syntax of the CREATE ROLE command

After the role has been created, you can grant system and/or object privileges to the role, using the same syntax as for granting privileges directly to users. The *only exception* is that an object privilege can't be granted to a role with the WITH GRANT OPTION. After all privileges have been assigned to a role, the role can then be assigned to all relevant users with the GRANT command. The syntax for using the GRANT command to grant a role to a user is shown in Figure 7-15.

```
GRANT rolename [, rolename]
  TO username [, username];
```

FIGURE 7-15   Syntax for granting a role to a user

For example, each order entry clerk should be allowed to issue SELECT, INSERT, and UPDATE commands for the CUSTOMERS, ORDERS, and ORDERITEMS tables in the JustLee Books database. Rather than remember exactly which privileges are required every time a new order entry clerk is hired, a DBA could create a role called ORDERENTRY to be assigned in lieu of multiple privileges. To create the role, connect as the DBA user and issue the command shown in Figure 7-16.

```
CREATE ROLE orderentry;
```

FIGURE 7-16   Command for creating the ORDERENTRY role

Now that you've created the ORDERENTRY role, you can assign the necessary privileges to it by using the commands shown in Figure 7-17. Keep in mind that you should substitute the correct schema for SCOTT—it should be the schema you're using that contains the JustLee Books database tables.

```
GRANT SELECT, INSERT, UPDATE
  ON scott.customers
  TO orderentry;
GRANT SELECT, INSERT, UPDATE
  ON scott.orders
  TO orderentry;
GRANT SELECT, INSERT, UPDATE
  ON scott.orderitems
  TO orderentry;
```

FIGURE 7-17   Commands for granting privileges to the ORDERENTRY role

After assigning privileges to the ORDERENTRY role, you can assign the role to any new order entry clerk by using the GRANT command. Grant the role to Ron Thomas by issuing the command shown in Figure 7-18. Now he can execute SELECT, UPDATE, and INSERT commands on any table in the JustLee Books database.

```
GRANT orderentry
  TO rthomas;
```

FIGURE 7-18   Command for granting the ORDERENTRY role to RTHOMAS

Users can be assigned several roles, based on the different types of tasks they usually perform. For example, suppose you create an ORDERENTRY role and a BILLING role to address both employee groups shown previously in Figure 7-13. Then you create the username SDAVIS for the supervisor named Scott Davis. How can you grant both roles to this supervisor? The command shown in Figure 7-19 accomplishes this task.

```
GRANT orderentry, billing
  TO sdavis;
```

FIGURE 7-19   Command for assigning multiple roles to a user

You can also define a role by including a group of previously defined roles. For example, JustLee Books might have several supervisors who need the same set of roles assigned. To simplify this assignment, you could create a role that includes both the ORDERENTRY and BILLING roles and then assign it to each supervisor who needs both roles. The commands in Figure 7-20 create a role named SUPERVISOR, which includes two other roles.

```
CREATE ROLE supervisor;
GRANT orderentry, billing
  TO supervisor;
```

FIGURE 7-20   Command for creating a role that includes two roles

Using Predefined Roles

You don't necessarily have to create roles from scratch. Oracle has a set of predefined roles available for assigning user privileges. Table 7-3 lists some of these predefined roles.

TABLE 7-3   Some Predefined Roles in Oracle 11*g*

Role Name	Privileges Included
CONNECT	CREATE SESSION
RESOURCE	CREATE CLUSTER, CREATE INDEXTYPE, CREATE OPERATOR, CREATE PROCEDURE, CREATE SEQUENCE, CREATE TABLE, CREATE TRIGGER, CREATE TYPE
DBA	All system privileges as well as WITH ADMIN OPTION

N O T E

The "Oracle Database Security Guide" lists all predefined roles.

You can grant these roles to users in the same way you grant roles that you create. However, predefined roles aren't usually applicable to assigning privileges to application users.

These roles are more closely associated with privileges that developers and database administrators require. In addition, Oracle recommends that organizations create their own roles rather than depend on predefined roles.

Using Default Roles

A user who has been assigned several different roles doesn't have to have all these roles activated at login. Users can be assigned a **default role** that's enabled automatically whenever they log in to the database. The default role should consist of only the privileges the user needs frequently. Privileges that are rarely needed (and could cause problems in the database if used incorrectly) should be assigned to other roles the user can assume when needed. For example, a user is assigned a role containing ALTER TABLE privileges, which aren't used often. In this case, having this role activated only when needed might be better to prevent any unintended table modifications.

You use the ALTER USER and SET ROLE commands to control how roles are activated for a user. After the user account has been created, you can issue an ALTER USER command with the **DEFAULT ROLE** option to assign a default role to a user, using the syntax shown in Figure 7-21.

```
ALTER USER username
   DEFAULT ROLE
(rolename|ALL [EXCEPT role1, role2, ...]|NONE);
```

FIGURE 7-21   Syntax for assigning a default role to a user

As shown in Figure 7-21, users can have none, one, or many roles set as the default role. Including a specific role name enables only that role when the user logs in. The **ALL** option enables all roles assigned to the user. If the **EXCEPT** option is used, all roles except those listed in this clause are enabled. The **NONE** option disables all assigned roles, requiring the user to enable a role after logging in. Table 7-4 lists examples of using default roles for different situations.

TABLE 7-4   Examples of Setting Default Roles

Example	GRANT Command
Ron Thomas needs only the ORDERENTRY role enabled at login.	ALTER USER rthomas DEFAULT ROLE orderentry;
Scott Davis, the supervisor, has been assigned three roles: ORDERENTRY, BILLING, and MOD_TABLES. All these privileges are needed routinely, so all roles should be enabled at login.	ALTER USER sdavis DEFAULT ROLE ALL;

Scott Davis, the supervisor, has been assigned three roles: ORDERENTRY, BILLING, and MOD_TABLES. The privileges for the MOD_TABLES role are needed only periodically, so all roles except this one should be enabled at login.	ALTER USER sdavis
	     DEFAULT ROLE
	     ALL EXCEPT mod_tables;

Enabling Roles After Login

After connecting to the database, a user might need to assume or enable a role or set of privileges other than those assigned as the default role. The user can issue a **SET ROLE** command to do this, using the syntax shown in Figure 7-22.

```
SET ROLE rolename;
```

FIGURE 7-22 Syntax of the SET ROLE command

Note that users can't set their roles to a role that hasn't already been assigned to them. For example, they can't issue the command **SET ROLE DBA** unless they have already been assigned that role by an administrator.

As a safety precaution, some database administrators add a password to a role. For example, the employee Scott has been assigned special privileges through the DBA role. However, he uses this role only when he needs to perform certain operations. One day, he's logged in to the Oracle 11g database, performing day-to-day activities that require only his normal privileges. He's summoned away unexpectedly for a meeting and doesn't remember to log out of the server. Because no special privileges are currently available, forgetting to log out might not be a problem. However, if a disgruntled employee knows Scott has access to the DBA role, that employee can simply sit down at Scott's computer and set the role to DBA with the SET ROLE command. With these privileges available, the disgruntled employee could do a lot of damage in a short time.

To avoid this problem, administrators add passwords to roles that have important and potentially risky privileges. To add a password to a role, simply use the **ALTER ROLE** command with the syntax shown in Figure 7-23.

```
ALTER ROLE rolename
 IDENTIFIED BY password;
```

FIGURE 7-23 Syntax of the ALTER ROLE command

After the ALTER ROLE command has been used to add a password to a role, any user attempting to use the role is required to enter the password, or the role isn't enabled.

NOTE

A password isn't required for roles assigned as a user's default role. It's required only when the SET ROLE command is issued to enable a role.

VIEWING PRIVILEGE INFORMATION

You can query various data dictionary views to determine the privileges currently assigned to a user or role. Table 7-5 describes some commonly used data dictionary views. A Y in the DBA column indicates that DBA privileges are required to see that data dictionary view.

TABLE 7-5 Views for Displaying Privilege Information

DBA	Data Dictionary View	Information
Y	DBA_ROLES	All roles defined in the database
Y	ROLE_TAB_PRIVS	Table privileges assigned to roles
Y	ROLE_SYS_PRIVS	System privileges assigned to roles
Y	DBA_ROLE_PRIVS	Roles assigned to users
Y	DBA_TAB_PRIVS	All table privileges assigned
N	USER_SYS_PRIVS	System privileges granted to the current user
N	USER_TAB_PRIVS	Object privileges granted to the current user
N	USER_ROLE_PRIVS	Roles assigned to the current user
N	ROLE_TAB_PRIVS	Table privileges assigned to the current user via roles
N	SESSION_ROLES	Roles currently enabled for the current user
N	SESSION_PRIVS	Privileges active for the current user that haven't been assigned via a role

The DBA at JustLee Books might need to confirm which table privileges are assigned to the ORDERENTRY role. The statement in Figure 7-24 accomplishes this task.

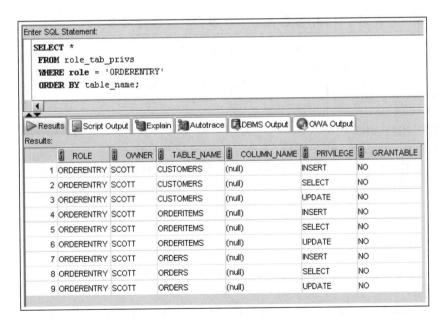

FIGURE 7-24 Verifying privileges assigned to a role

After being logged in, the user RTHOMAS might want to check which system privileges and roles are currently enabled. Figure 7-25 shows the statements for performing this task.

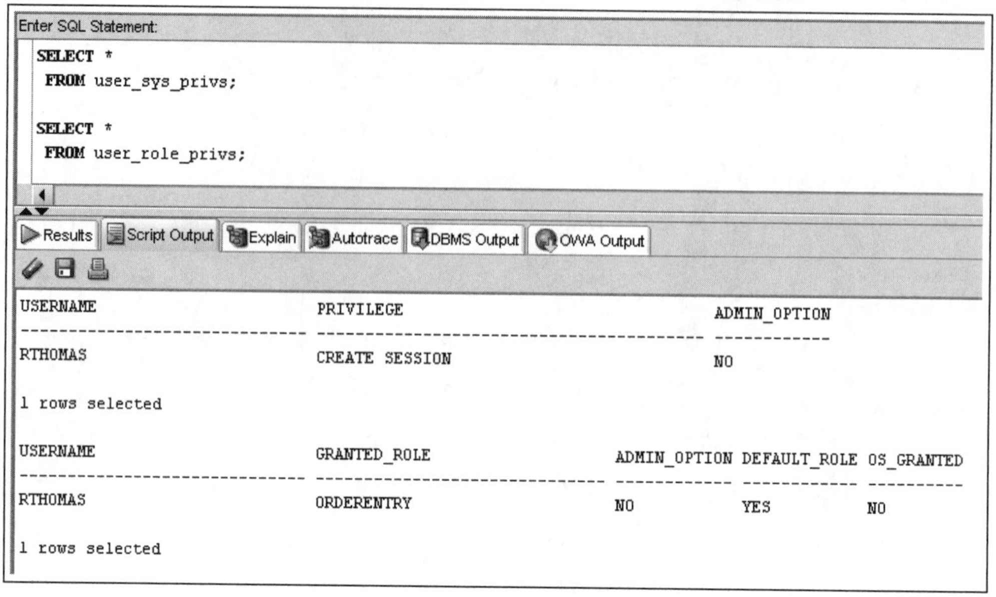

```
Enter SQL Statement:
  SELECT *
  FROM user_sys_privs;

  SELECT *
  FROM user_role_privs;
```

Results | Script Output | Explain | Autotrace | DBMS Output | OWA Output

```
USERNAME                          PRIVILEGE                                  ADMIN_OPTION
------------------------------    ----------------------------------------   -------------
RTHOMAS                           CREATE SESSION                             NO

1 rows selected

USERNAME                          GRANTED_ROLE                  ADMIN_OPTION DEFAULT_ROLE OS_GRANTED
------------------------------    ----------------------------  ------------ ------------ ----------
RTHOMAS                           ORDERENTRY                    NO           YES          NO

1 rows selected
```

FIGURE 7-25   Verifying active system privileges and roles

REMOVING PRIVILEGES AND USERS

Privileges and roles can be removed or dropped as easily as they can be assigned. The REVOKE command is used to remove privileges and roles assigned to a user. The DROP command is used to eliminate roles or users from the database system. The following sections describe these commands.

Revoking Privileges and Roles

Privileges granted to a user or role can be removed by using the **REVOKE** command. Figure 7-26 shows the syntax of the REVOKE command used to remove a system privilege from a user or role.

```
REVOKE systemprivilege [,… systemprivilege]
  FROM username|rolename;
```

FIGURE 7-26   Syntax for revoking a system privilege

The REVOKE command can also be used to revoke object privileges, using the syntax shown in Figure 7-27.

```
REVOKE objectprivilege [,… objectprivilege]
 ON objectname
 FROM username|rolename;
```

FIGURE 7-27   Syntax for revoking an object privilege

In addition, the REVOKE command can be used to remove a role from an account, using the syntax shown in Figure 7-28.

```
REVOKE rolename
 FROM username|rolename;
```

FIGURE 7-28   Syntax for removing a role from an account

For example, to remove the DELETE privilege on the CUSTOMERS table from the ORDERENTRY role, the DBA could issue the command shown in Figure 7-29.

```
REVOKE delete
 ON customers
 FROM orderentry;
```

FIGURE 7-29   Command for removing an object privilege from a role

To remove the ORDERENTRY role from Ron Thomas's user account, the DBA can issue the command shown in Figure 7-30.

```
REVOKE orderentry
FROM rthomas;
```

FIGURE 7-30   Command for removing the ORDERENTRY role from a user account

NOTE

Now that the ORDERENTRY role isn't available to Ron Thomas, he no longer has access to the CUSTO-MERS, ORDERS, and ORDERITEMS tables.

CAUTION

When revoking an object privilege that was originally granted by using WITH GRANT OPTION, the privilege is revoked not only from the specified user, but also from any other users to whom the user might have subsequently granted the privilege. On the other hand, revoking a *system* privilege that was originally granted by using WITH ADMIN OPTION has no cascading effect on other users.

Dropping a Role

A role can be deleted from the Oracle 11g database with the **DROP ROLE** command, which uses the syntax shown in Figure 7-31.

```
DROP ROLE rolename;
```

FIGURE 7-31   Syntax of the DROP ROLE command

When a role is removed from the database, users lose all privileges derived from that role. The only way they can use the privileges previously assigned by the role is to be assigned these privileges again, either by direct grants of the privileges or by creating another role.

For example, you decide the ORDERENTRY role should be more restrictive and specify exactly which columns can be updated in certain tables. The simplest solution is to drop the existing role and re-create it with a new ORDERENTRY role. To drop the role, use the command shown in Figure 7-32.

```
DROP ROLE orderentry;
```

FIGURE 7-32   Command for dropping the ORDERENTRY role

Dropping a User

At times, accounts need to be removed from the system for various reasons, the most common being employees leaving because of events such as retirement. As an example of another reason, suppose Ron's supervisor has just informed you that the correct spelling of Ron's last name is Tomas, not Thomas. There's no ALTER USER option available for changing an account's username. Instead, you must delete Ron's existing account and re-create it with the correct spelling. To remove a user account from an Oracle 11g database, use the **DROP USER** command with the syntax shown in Figure 7-33.

```
DROP USER username;
```

FIGURE 7-33   Syntax of the DROP USER command

The command DROP USER rthomas; drops the existing account named "rthomas." You then need to create a new account with the correct "rtomas" spelling. If the user has any objects in his or her schema, you must use the CASCADE option in the DROP USER command. This option is listed after the username and eliminates all objects in the schema so that the user can be deleted.

Chapter Summary

- Database account management is only one facet of data security.
- A new user account is created with the CREATE USER command. The IDENTIFIED BY clause contains the password for the account.
- System privileges are used to grant access to the database and to create, alter, and drop database objects.
- The CREATE SESSION system privilege is required before users can access their accounts on the Oracle server.
- The system privileges available in Oracle 11*g* can be viewed with the SYSTEM_PRIVILEGE_MAP data dictionary view.
- Object privileges allow users to manipulate data in specific database objects.
- Privileges are assigned with the GRANT command.
- The ALTER USER command, combined with the PASSWORD EXPIRE clause, can be used to force users to change their passwords at the next attempted login.
- The ALTER USER command, combined with the IDENTIFIED BY clause, can be used to change a user's password.
- Privileges can be assigned to roles to make administration of privileges easier.
- Roles are collections of privileges.
- The ALTER USER command, combined with the DEFAULT ROLE keywords, can be used to assign a default role to a user.
- A role can be enabled in a session with the SET ROLE command.
- Privileges can be removed from users and roles by using the REVOKE command.
- Roles can be removed from users by using the REVOKE command.
- A role can be deleted with the DROP ROLE command.
- A user account can be deleted with the DROP USER command.

Chapter 7 Syntax Summary

The following table summarizes the syntax you have learned in this chapter. You can use the table as a study guide and reference.

Syntax Guide

Command Description	Command Syntax	Example
Creating, Maintaining, and Dropping User Accounts		

Create a user	CREATE USER *username*
	    IDENTIFIED BY *password*;
		CREATE USER rthomas
		    IDENTIFIED BY
		        little25car;

Change or expire a password	ALTER USER *username*
	    [IDENTIFIED BY
	        newpassword]
	    [PASSWORD EXPIRE];
		ALTER USER rthomas
		    IDENTIFIED BY
		        monster42truck;

Drop a user	DROP USER *username*;	DROP USER rthomas;

Command Description	Command Syntax	Example
Granting and Revoking Privileges		

Grant object privileges to users or roles	GRANT {*objectprivilege*|ALL}
	  [(*columnname*),
	   objectprivilege
	   (*columnname*)]
	  ON *objectname*
	  TO {*username*|*rolename*|PUBLIC}
	  [WITH GRANT OPTION] ;
		GRANT select, insert
		  ON customers
		  TO rthomas
		  WITH GRANT OPTION;

Grant system privileges to users or roles	GRANT *systemprivilege*
	  [, *systemprivilege*, ...]
	  TO *username*|*rolename*
	  [, *username*|*rolename*,
	  ...]
	  [WITH ADMIN OPTION];
		GRANT CREATE SESSION
		  TO rthomas;

Revoke object privileges	REVOKE *objectprivilege*
	  [,...*objectprivilege*]
	  ON *objectname*
	  FROM *username*|*rolename*;
		REVOKE INSERT
		  ON customers
		  FROM rthomas;

Revoke system privileges	REVOKE *systemprivilege*
	  [,...*systemprivilege*]
	  FROM *username*|*rolename*;
		REVOKE CREATE SESSION
		  FROM rthomas;

Granting and Revoking Roles		
Create a role	CREATE ROLE *rolename*;	**CREATE ROLE orderentry;**

Grant a role to a user	GRANT *rolename*[, *rolename*]
	  TO *username*[, *username*] ;
		GRANT orderentry
		  TO rthomas;

Assign a default role to a user	ALTER USER *username*
	  DEFAULT ROLE *rolename*;
		ALTER USER rthomas
		  DEFAULT ROLE orderentry;

Set or enable a role	SET ROLE *rolename*;	**SET ROLE DBA;**

Add a password to a role	ALTER ROLE *rolename*
	  IDENTIFIED BY *password*;
		ALTER ROLE orderentry
		  IDENTIFIED BY apassword;

Revoke a role	REVOKE *rolename*
	  FROM *username*|*rolename*;
		REVOKE orderentry
		  FROM rthomas;

Drop a role	DROP ROLE *rolename*;	**DROP ROLE orderentry;**

Review Questions

1. What is the purpose of data security?
2. What does a database account with the CREATE SESSION privilege allow the user to do?
3. How is a user password assigned in Oracle 11*g*?
4. What is a privilege?
5. If you're logged in to Oracle 11*g*, how can you determine which privileges are currently available to your account?
6. What types of privileges are available in Oracle 11*g*? Define each type.
7. What is the purpose of a role in Oracle 11*g*?
8. How can you assign a password to a role?
9. What happens if you revoke an object privilege that was granted with the WITH GRANT OPTION? What if the privilege is removed from a user who had granted the same object privilege to three other users?
10. How can you remove a user account from Oracle 11*g*?

Multiple Choice

To answer the following questions, refer to the JustLee Books database.

1. Which of the following commands can be used to change a password for a user account?
   a. ALTER PASSWORD
   b. CHANGE PASSWORD
   c. MODIFY USER PASSWORD
   d. ALTER USER ... PASSWORD
   e. none of the above

2. Which of the following statements assigns the role CUSTOMERREP as the default role for Maurice Cain?
   a. `ALTER ROLE mcain`
      `  DEFAULT ROLE customerrep;`
   b. `ALTER USER mcain`
      `  TO customerrep;`
   c. `SET DEFAULT ROLE customerrep`
      `  FOR mcain;`
   d. `ALTER USER mcain`
      `  DEFAULT ROLE customerrep;`
   e. `SET ROLE customerrep`
      `  FOR mcain;`

3. Which of the following statements is most accurate?

    a. Authentication procedures prevent any data stored in the Oracle 11*g* database from being stolen or damaged.

    b. Authentication procedures are used to limit unauthorized access to the Oracle 11*g* database.

    c. Oracle 11*g* authentication doesn't prevent users from accessing data in the database if they have a valid operating system account.

    d. Authentication procedures restrict the type of data manipulation operations that a user can perform.

4. Which of the following statements creates a user account named DeptHead?

    a.
```
CREATE ROLE depthead
    IDENTIFIED BY apassword;
```

    b.
```
CREATE USER depthead
    IDENTIFIED BY apassword;
```

    c.
```
CREATE ACCOUNT depthead;
```

    d.
```
GRANT ACCOUNT depthead;
```

5. Which of the following privileges must be granted to a user's account before the user can connect to the Oracle 11*g* database?

    a. CONNECT

    b. CREATE SESSION

    c. CONNECT ANY DATABASE

    d. CREATE ANY TABLE

6. Which of the following privileges allows a user to truncate tables in a database?

    a. DROP ANY TABLE

    b. TRUNCATE ANY TABLE

    c. CREATE TABLE

    d. TRUNC TABLE

7. Which of the following tables or views displays the current enabled privileges for a user?

    a. SESSION_PRIVS

    b. SYSTEM_PRIVILEGE_MAP

    c. USER_ASSIGNED_PRIVS

    d. V$ENABLED_PRIVILEGES

8. Which of the following commands eliminates only the user ELOPEZ's ability to enter new books in the BOOKS table?

    a.
```
REVOKE insert
   ON books
   FROM elopez;
```

    b.
```
REVOKE insert
   FROM elopez;
```

c. REVOKE INSERT INTO
FROM elopez;

d. DROP insert
INTO books
FROM elopez;

9. Which of the following commands is used to assign a privilege to a role?

a. CREATE ROLE

b. CREATE PRIVILEGE

c. GRANT

d. ALTER PRIVILEGE

10. Which of the following options requires a user to change his or her password at the next login?

a. CREATE USER

b. ALTER USER

c. IDENTIFIED BY

d. PASSWORD EXPIRE

11. Which of the following options allows a user to grant system privileges to other users?

a. WITH ADMIN OPTION

b. WITH GRANT OPTION

c. DBA

d. ASSIGN ROLES

e. SET ROLE

12. Which of the following is an object privilege?

a. CREATE SESSION

b. DROP USER

c. INSERT ANY TABLE

d. UPDATE

13. Which of the following privileges can be granted only to a user, not to a role?

a. SELECT

b. CREATE ANY

c. REFERENCES

d. READ

e. WRITE

14. Which of the following is used to grant all object privileges for an object to a specified user?

a. ALL

b. PUBLIC

c. ANY

d. OBJECT

15. Which of the following identifies a collection of privileges?

    a.   an object privilege

    b.   a system privilege

    c.   DEFAULT privilege

    d.   a role

16. Which of the following is true?

    a.   If the DBA changes the password for a user while the user is connected to the database, the connection terminates automatically.

    b.   If the DBA revokes the CREATE SESSION privilege from a user account, the user can't connect to the database.

    c.   If a user is granted the privilege to create a table and the privilege is revoked after the user creates a table, the table is dropped from the system automatically.

    d.   all of the above

17. Which of the following commands can be used to eliminate the RECEPTIONIST role?

    a.   `DELETE ROLE receptionist;`

    b.   `DROP receptionist;`

    c.   `DROP ANY ROLE;`

    d.   none of the above

18. Which of the following displays a list of all system privileges available in Oracle 11*g*?

    a.   SESSION_PRIVS

    b.   SYS_PRIVILEGE_MAP

    c.   V$SYSTEM_PRIVILEGES

    d.   SYSTEM_PRIVILEGE_MAP

19. Which of the following can be used to change the role that's currently enabled for a user?

    a.   SET DEFAULT ROLE

    b.   ALTER ROLE

    c.   ALTER SESSION

    d.   SET ROLE

20. Which of the following is an object privilege?

    a.   DELETE ANY

    b.   INSERT ANY

    c.   UPDATE ANY

    d.   REFERENCES

Hands-On Assignments

Create and execute SQL statements to perform the following actions, using the JustLee Books database:

1. Create a new user account. The account name should be a combination of your first initial and your last name.

2. Attempt to log in to Oracle 11*g* with the newly created account.

3. Assign privileges to the new account that allow connecting to the database, creating new tables, and altering an existing table.

4. Using an account with the required privileges, create a role named CUSTOMERREP that allows inserting new rows in the ORDERS and ORDERITEMS tables and deleting rows from these tables.

5. Assign the account created in Assignment 1 the CUSTOMERREP role.

6. Log in to Oracle 11*g* with the new account created in Assignment 1. Determine the privileges currently available to the account.

7. Revoke the privilege to delete rows in the ORDERS and ORDERITEMS tables from the CUSTOMERREP role.

8. Remove the CUSTOMERREP role from the account created in Assignment 1.

9. Delete the CUSTOMERREP role from the Oracle 11*g* database.

10. Delete the user account created in Assignment 1.

Advanced Challenge

Use the JustLee Books database to perform the following activity:

There are three major classifications for employees who don't work for the Information Systems Department of JustLee Books: account managers, who are responsible for the company's marketing activities (for example, promotions based on customers' previous purchases or for specific books); data entry clerks, who enter inventory updates (for example, add new books and publishers, change prices, and so on); and customer service representatives, who are responsible for adding new customers and entering orders in the database. Each employee group has different tasks to perform and, therefore, needs different privileges for various tables in the database. To simplify administration of system and object privileges, a role should be created for each employee group.

Create a document for your supervisor that contains the following information:

- List the tables that each group of employees needs to access from these tables: BOOKS, CUSTOMERS, ORDERS, ORDERITEMS, AUTHOR, BOOKAUTHOR, PUBLISHER, and PROMOTION.
- Name the privileges each group of employees needs.
- For each group of employees, name a role containing the necessary privileges for that group.
- For all groups of employees, list the exact commands for creating and assigning specific privileges to their roles.
- Explain your rationale for the privileges granted to each role.

Case Study: *City Jail*

The City Jail organization is preparing to deploy the new database to four departments. The departments and associated duties for the database are described in the following chart:

Department	Number of Employees	Duties
Criminal Records	8	1. Add new criminals and crime charges. 2. Make changes to criminal and crime charge data as needed for corrections or updates. 3. Keep the police officer information up to date. 4. Maintain the crime codes list.
Court Recording	7	1. Enter and modify all court appeals information. 2. Enter and maintain all probation information. 3. Maintain the probation officer list.
Crimes Analysis	4	1. Analyze all criminal and court data to identify trends. 2. Query all crimes data as needed to prepare federal and state reports.
Data Officer	1	1. Remove crimes, court, and probation data based on approved requests from any of the other departments.

Based on the department duties outlined in the table, develop a plan to assign privileges to employees in all four departments. The plan should include the following:

- A description of what types of objects are required
- A list of commands needed to address user creation for each department

RESTRICTING ROWS AND SORTING DATA

INTRODUCTION

Chapters 3 through 7 covered the creation of database objects, including tables, constraints, indexes, sequences, and users. This chapter shifts the focus back to querying a database. In Chapter 2, you learned how to retrieve specific fields from a table. However, unless you use the DISTINCT or UNIQUE keyword, your results include every record. Sometimes you want to see only records meeting certain conditions—a process referred to as **selection**. Because selection reduces the number of records a query returns, locating a specific record in the output is usually easier. In addition, if data is displayed in a sorted order, identifying trends can be easier. This chapter explains how to perform queries with search conditions and sorting methods. In particular, you see how to use the WHERE clause of the SELECT statement as a search condition and the ORDER BY clause to display results in a specific sequence. You have already seen the WHERE clause used in UPDATE and DELETE DML statements to limit the number of rows affected by the modification. Table 8-1 gives you an overview of this chapter's topics.

TABLE 8-1   Keywords and Operators Used to Restrict and Sort Rows

Element	Description
WHERE clause	Used to specify conditions that must be true for a record to be included in query results
ORDER BY clause	Used to specify the sorted order for displaying query results
Mathematical comparison operators ($=, <, >, <=, >=, <>, !=, \wedge=$)	Used to indicate how a record should relate to a specific search value
Other comparison operators (BETWEEN … AND, IN, LIKE, IS NULL)	Used in conditions with search values that include patterns, ranges, or NULL values
Logical operators (AND, OR, NOT)	Used to join multiple search conditions (AND, OR) or reverse the meaning of a search condition (NOT)

DATABASE PREPARATION

Before working through this chapter's examples, run the JLDB_Build_8.sql script in the Chapter 8 folder of your data files to ensure that all necessary tables and constraints are available. This script removes your existing tables and creates a new set of tables. Refer to the steps at the beginning of Chapter 2 for loading and running a script. Ignore any errors in the DROP TABLE statements at the beginning of the script. An "object does not exist" error indicates merely that the table wasn't created in the schema previously.

WHERE CLAUSE SYNTAX

To retrieve records in an Oracle 11*g* database based on a given condition, add the WHERE clause to the SELECT statement. As indicated by square brackets ([ ]) in the syntax shown in Figure 8-1, the WHERE clause is optional. When used, it should be listed after the FROM clause.

```
SELECT  [DISTINCT | UNIQUE] (*, columnname [ AS alias], ...)
        FROM      tablename
        [WHERE    condition]
        [GROUP BY group_by_expression]
        [HAVING   group_condition]
        [ORDER BY columnname];
```

FIGURE 8-1   Syntax of the SELECT statement

> ## N O T E
>
> Figure 8-1 shows the full structure of a query, as addressed in this textbook. You learn about the GROUP BY and HAVING clauses in Chapter 11, and the ORDER BY clause is explained in "ORDER BY Clause Syntax" later in this chapter.

A **condition** identifies what must exist or a requirement that must be met for a record to be included in the results. Oracle 11*g* searches through each record to determine whether the condition is TRUE. If a record meets the condition, it's returned in the query results. For a simple search of a table, the condition portion of the WHERE clause follows this format:

<column name> <comparison operator> <another named column or a value>

For example, to display a list containing the last name of every customer living in Florida, you use the SQL statement shown in Figure 8-2.

```
SELECT lastname, state
  FROM customers
  WHERE state = 'FL';
```

FIGURE 8-2   Query to perform a simple search based on a given condition

As shown in Figure 8-2, the query specifies the Lastname and State columns stored in the CUSTOMERS table as the data to list in the output. Recall from Chapter 1 that limiting the output to specific columns is referred to as projection. However, you want to see only the records of customers who have the letters FL stored in the State field. Therefore, in the WHERE clause, WHERE is the keyword, State is the name of the column to be searched, the comparison operator "equal to" (=) means the record must contain the exact value specified, and the specified value is FL.

Notice the single quotation marks around FL, which designate it as a string literal. Also, note that the value FL is in uppercase letters to match the format in which data is entered in the State field. The process of limiting output to specific rows is called selection, as mentioned in the beginning of this chapter. Figure 8-3 shows the output of this SELECT statement.

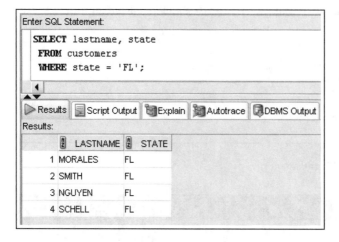

FIGURE 8-3   Results of a state-based search

N O T E

If you get an error message, verify that the FL value is enclosed in single, not double, quotation marks. (Double quotation marks, used in Chapter 2, are for column aliases.) If no rows are returned, make sure the letters FL are capitalized. The data for JustLee Books was originally entered in uppercase characters and is, therefore, stored in the database tables in that format. Any value entered in a string literal (that is, inside single quotation marks) is evaluated *exactly* as entered—both in spacing and letter case. Therefore, if a string literal is entered for a search condition, it must be in the same case as the data being searched, or no rows are returned in the results. You can verify the case of data stored in a table by querying the table and reviewing the output. Although Oracle 11*g* isn't case sensitive when evaluating keywords, table names, and column names, evaluation of data contained in a record *is* case sensitive.

The query results in Figure 8-3 list the last name and state for each customer living in Florida. As you can see, only four rows are returned, even though the table contains 20 customers. The WHERE clause restricts the number of records returned in the results to only those meeting the condition `state = 'FL'`.

Rules for Character Strings

When you use a string literal, such as FL, as part of a search condition, the value must be enclosed in single quotation marks and, as a result, is interpreted exactly as listed. By contrast, if the field referenced in a condition consists only of *numbers*, single quotation marks aren't required. For example, suppose you want to see all data stored in the CUSTOMERS table for customer 1010. The Customer# field has a numeric data-type. Issue the SQL statement to accomplish this task, and compare it to the output shown in Figure 8-4.

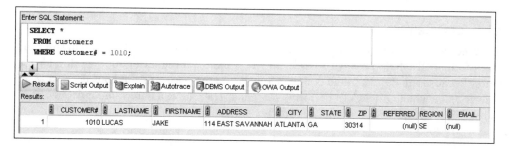

FIGURE 8-4 Search results for Customer 1010

In this example, the value 1010 for the Customer# column isn't enclosed in single quo-tation marks because this column has been defined to store only numbers. Therefore, single quotation marks aren't necessary.

Next, use a WHERE statement to search for a book with the ISBN 1915762492. Figure 8-5 shows the input and the output of this query.

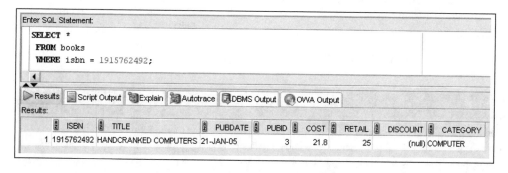

FIGURE 8-5 Search results for ISBN 1915762492

The ISBN column of the BOOKS table is defined as a character (text) field instead of a numeric field because some ISBNs contain letters. In this instance, however, no values stored in the ISBN column contain letters. Therefore, you can search the field by using a numeric value without any single quotation marks. However, if the table has even one record containing a letter in the ISBN field, Oracle 11g returns an error message. In other words, omitting the single quotation marks works in this case, but it might not always work. Using single quotation marks ultimately depends on whether the field is defined to hold text or numeric data. Therefore, *always* use single quotation marks if the column is defined with anything other than a numeric datatype.

Rules for Dates

Sometimes you need to use a date as a search condition. Oracle 11*g* displays dates in the default format DD-MON-YY, with MON being the standard three-letter abbreviation for the month. Because the Pubdate field contains letters and hyphens, it's not considered a numeric value when Oracle 11*g* performs searches. Therefore, a date value must be enclosed in single quotation marks. Figure 8-6 shows a query for books published on January 21, 2005.

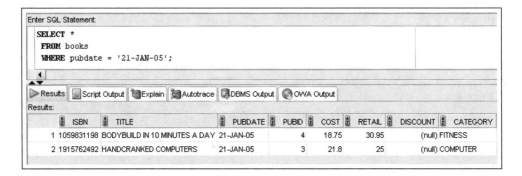

FIGURE 8-6   Querying with a date condition

COMPARISON OPERATORS

So far in this chapter, you have used an equal sign, or **equality operator**, to evaluate search conditions; basically, you instruct Oracle 11*g* to return only results containing the *exact* value you provide. However, many searches aren't based on an "equal to" condition. For example, management needs a list of books for a proposed marketing campaign. The Marketing Department wants to include a gift with the purchase of any book that has a retail price of more than $55.00. Management wants to know which books can be mentioned in an advertisement of this marketing campaign. The equality operator isn't suitable in this situation, so you need a different comparison operator. A **comparison operator** indicates how data should relate to the search value, such as "greater than" or "less than." In this case, you need a comparison operator meaning "greater than" (>) to determine which books meet the "more than $55.00" requirement. Execute the statement in Figure 8-7 to perform this task.

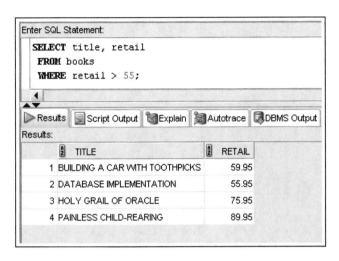

FIGURE 8-7   Searching for books with a retail price greater than $55

Based on these results, you know that four books meet the condition for this sales promotion. Notice that 55 is entered as the value for the Retail condition. This value could have also been entered as 55.00. Oracle 11g accepts a period to indicate decimal positions without considering the entry to be a character value rather than a numeric value. However, if you enter the dollar sign ($) or a comma (to indicate a thousands position), you get an error message indicating that the Retail field is numeric and the value entered is an "invalid character." (The dollar sign is treated as a formatting character, so $55.00 is not equivalent to 55.00.) Unlike some other database management systems, Oracle 11g doesn't have a currency datatype, so it regards the comma and dollar sign as characters.

The "greater than" (>) comparison operator can also be used with text and date fields. Suppose you're about to take a physical inventory of all books in stock. The procedure JustLee Books uses is to give each employee a list of books, and then have the person record the quantity on hand; each person is responsible for a portion of the alphabet. For example, one person might be responsible for all books with titles falling in the A through D range.

Figure 8-8 shows how to create the list of books for the person assigned to inventory all books with a title occurring alphabetically after the letters HO. All book titles with additional characters following HO or beginning with the letters HP and the letters that follow are listed.

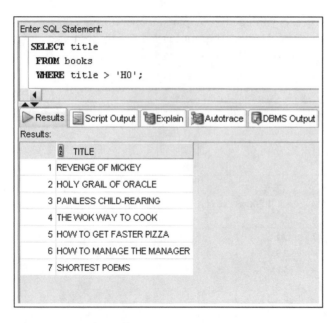

FIGURE 8-8   Searching for book titles with letters greater than HO

Now that you've examined "equal to" and "greater than" comparison operators, take a look at Table 8-2, which lists comparison operators commonly used in Oracle 11*g*. Even though the first group lists comparison operators considered to be mathematical operators, they can be used with a variety of datatypes, including characters and dates.

TABLE 8-2   Comparison Operators

Mathematical Comparison Operators	
=	Equality or "equal to"—for example, $cost = 55.95$
>	Greater than—for example, $cost > 20$
<	Less than—for example, $cost < 20$
<>, !=, or ^=	Not equal to—for example, $cost <> 55.95$ or $cost != 55.95$ or $cost ^= 55.95$
<=	Less than or equal to—for example, $cost <= 20$
>=	Greater than or equal to—for example, $cost >= 20$
Other Comparison Operators	
[NOT] BETWEEN x AND y	Used to express a range—for example, searching for numbers BETWEEN 5 AND 10. The optional NOT is used when searching for numbers that are NOT BETWEEN 5 AND 10.

TABLE 8-2   Comparison Operators (continued)

Other Comparison Operators	
[NOT] IN (x,y,...)	Similar to the OR logical operator. Can search for records meeting at least one condition inside the parentheses—for example, `Pubid IN (1, 4, 5)` returns only books with a publisher ID of 1, 4, or 5. The optional NOT keyword instructs Oracle to return books not published by Publisher 1, 4, or 5.
[NOT] LIKE	Used when searching for patterns if you aren't certain how something is spelled—for example, `title LIKE 'TH%'`. Using the optional NOT means records that *do* contain the specified pattern shouldn't be included in the results.
IS [NOT] NULL	Used to search for records that don't have an entry in the specified field—for example, `Shipdate IS NULL`. Include the optional NOT to find records that *do* have an entry in the field—for example, `Shipdate IS NOT NULL`.

In contrast to the "greater than" (>) operator that returns only rows with a value higher than the value in the condition, the "less than" (<) operator returns only values that are less than the condition. For example, the management of JustLee Books wants a list of all books having a profit of less than 20% of the book's cost. Because profit is determined by subtracting the cost from the retail price, this calculated value can be compared against the book's cost multiplied by .20 (the decimal version of 20%). As shown in Figure 8-9, only one book generates less than a 20% profit margin.

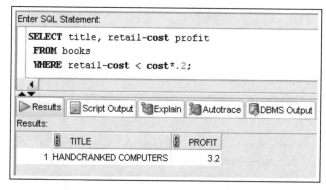

FIGURE 8-9   Searching Profit with the "less than" operator

TIP

You might be tempted to use the column alias Profit in the WHERE clause in Figure 8-9 instead of repeating the calculation `retail-cost`. Using a column alias isn't allowed in the WHERE clause, however, and raises an error. The only clause that allows using column aliases is the sorting clause of ORDER BY, explained in "ORDER BY Clause Syntax" later in this chapter.

The "greater than" and "less than" operators don't include values that exactly match the condition. For example, if you want the results in Figure 8-9 to include books returning exactly 20% profit, the comparison operator must be changed to the "less than or equal to" operator (<=). If you substitute the <= operator for the < operator in Figure 8-9, any book returning exactly 20% profit is also included in the results.

As another example, the Marketing Department is sorting paper files and requests a list of all customers who live in Georgia or in a state listed alphabetically before Georgia (that is, A through GA). The simplest way to identify these customers is to search for all customers by using the condition state <= 'GA', as shown in Figure 8-10.

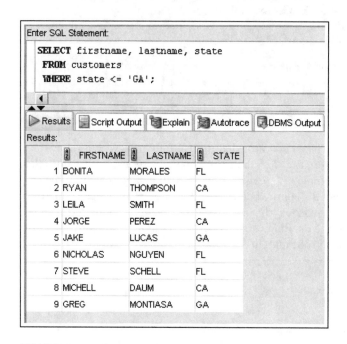

FIGURE 8-10   Searching State with the "less than or equal to" operator

Later, the Marketing Department asks you to identify all customers who live in Georgia or in a state with a state abbreviation "greater than" GA. Although you might think this request is a little unusual because the previous list already includes customers living in Georgia, you nevertheless create the list by using the condition state >= 'GA' and retrieve the names of 13 customers meeting this condition, as shown in Figure 8-11.

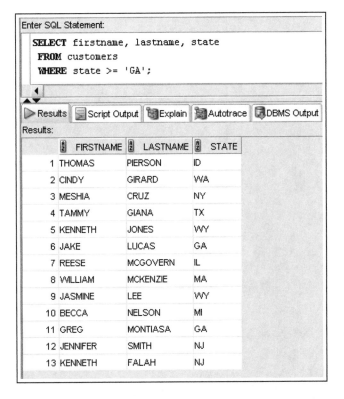

Enter SQL Statement:

```
SELECT firstname, lastname, state
FROM customers
WHERE state >= 'GA';
```

Results | Script Output | Explain | Autotrace | DBMS Output

Results:

	FIRSTNAME	LASTNAME	STATE
1	THOMAS	PIERSON	ID
2	CINDY	GIRARD	WA
3	MESHIA	CRUZ	NY
4	TAMMY	GIANA	TX
5	KENNETH	JONES	WY
6	JAKE	LUCAS	GA
7	REESE	MCGOVERN	IL
8	WILLIAM	MCKENZIE	MA
9	JASMINE	LEE	WY
10	BECCA	NELSON	MI
11	GREG	MONTIASA	GA
12	JENNIFER	SMITH	NJ
13	KENNETH	FALAH	NJ

FIGURE 8-11   Searching State with the "greater than or equal to" operator

For another marketing analysis, the Marketing Department requests a list of all customers who *do not* live in the state of Georgia. You can generate this list by using the condition `state <> 'GA'`, as shown in Figure 8-12.

TIP

Using != or ^= for the "not equal to" operator in Figure 8-12 returns the same results as using <>.

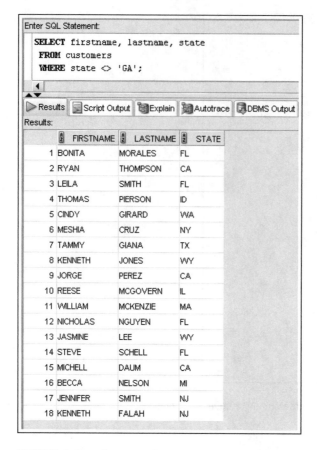

FIGURE 8-12 Searching State with the "not equal to" operator

To see how comparison operators are used with dates, suppose you need to produce a list of all orders placed before April 2009. You could use the WHERE clause with a condition on the Orderdate column, as shown in Figure 8-13. The "less than" comparison translates to "dates before" in this query, as the database system recognizes you're working with a date column.

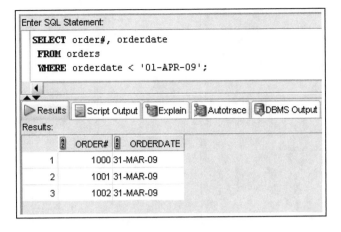

FIGURE 8-13 Searching a date value

BETWEEN ... AND Operator

The **BETWEEN ... AND** comparison operator is used when searching a field for values falling within a specified range. Figure 8-14 shows a query to find any book with a publisher that has an assigned ID between 1 and 3. Notice that the range is inclusive, so it includes any publisher with the ID 1, 2, or 3.

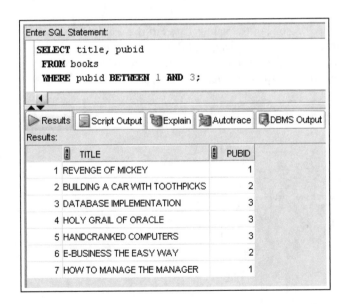

FIGURE 8-14 Searching Pubid with the BETWEEN ... AND operator

Returning to the book inventory example, an alphabetical range of titles could be queried with the BETWEEN ... AND comparison operator. The statement shown in Figure 8-15 identifies all book titles falling in the A through D range.

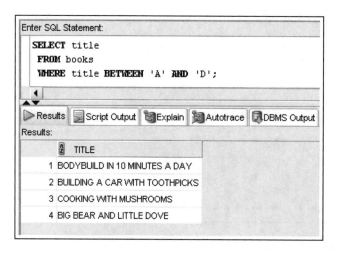

FIGURE 8-15 Searching for a character range with the BETWEEN … AND operator

You could also search for a range of dates with the BETWEEN … AND comparison operator. For example, if JustLee management needs to identify all orders placed from April 1, 2009 to April 4, 2009, a query using `WHERE orderdate BETWEEN '01-APR-09' AND '04-APR-09'` accomplishes this task.

IN Operator

The **IN** operator returns records matching one of the values listed in the condition. Oracle 11g syntax requires *separating list items with commas*, and the entire list *must be enclosed in parentheses*. For example, the output of the query in Figure 8-16 shows that seven books currently in inventory are published by Publisher 1, 2, or 5.

```
Enter SQL Statement:
SELECT title, pubid
FROM books
WHERE pubid IN (1,2,5);
```

Results:

	TITLE	PUBID
1	REVENGE OF MICKEY	1
2	BUILDING A CAR WITH TOOTHPICKS	2
3	E-BUSINESS THE EASY WAY	2
4	PAINLESS CHILD-REARING	5
5	BIG BEAR AND LITTLE DOVE	5
6	HOW TO MANAGE THE MANAGER	1
7	SHORTEST POEMS	5

FIGURE 8-16 Searching Pubid with the IN operator

If a list of all customers residing in California or Texas is needed, you should include the state abbreviations in single quotation marks because they are string literals. The query in Figure 8-17 shows using the IN operator to produce this list.

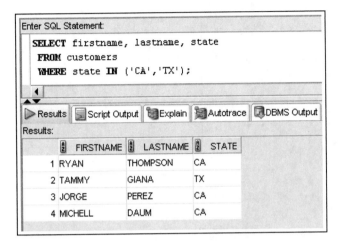

FIGURE 8-17   Searching State with the IN operator

Keep in mind that the **NOT** option can be used with all comparison operators to reverse the operation. For example, if you want to list all customers in states other than California or Texas, you can add the NOT option to the same query, as shown in Figure 8-18. The list now includes all customers in any states except California and Texas.

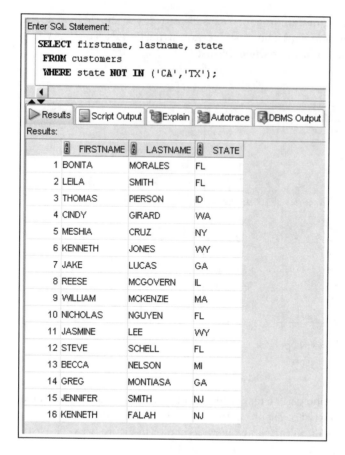

FIGURE 8-18   Using the NOT option

LIKE Operator

The **LIKE** operator is unique, in that it's used with wildcard characters to search for patterns. **Wildcard characters** are used to represent one or more alphanumeric characters. The wildcard characters available for pattern searches in Oracle 11*g* are the percent sign (%) and the underscore symbol (_). The percent sign represents *any number of characters* (zero, one, or more), and the underscore symbol represents *exactly one character*. For example, if you're trying to find any customer whose last name starts with P and don't care about the remaining letters of the last name, you can enter the SQL statement shown in Figure 8-19.

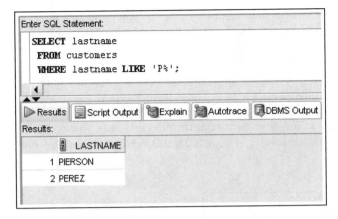

FIGURE 8-19   Searching with the LIKE operator and the % wildcard character

The results include two customers whose last names begin with P. If, however, you're searching for customers whose last names contain a P in any position, you change the search pattern to '%P%'. Oracle interprets this pattern as "It doesn't matter what's before or after the letter P, but a P must be somewhere in the Lastname column."

Suppose you're having difficulty reading the printout of a customer's order because someone spilled coffee on the form. You can tell that the first two digits of the Customer# are 1 and 0, and the last digit is 9. However, you can't read the third number. In this case, you could use the _ wildcard character to represent the missing digit, as shown in Figure 8-20.

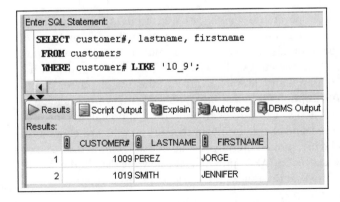

FIGURE 8-20   Searching with the LIKE operator and the _ wildcard character

Oracle 11g interprets the search condition in Figure 8-20 as "Look for any customer number that begins with 10, is followed by any character, and ends with 9." The results return two customers: 1009 and 1019.

The percent sign and underscore symbol can also be combined in the same search condition to create more complex search patterns. Suppose you need to identify every book ISBN that has 4 as the second numeral and ends with 0. The pattern you're trying to identify can be stated as '_4%0' because you know just one number comes before the 4 and there will be additional numbers after the 4. You don't care what the numbers are or how many there are, as long as the last digit is 0. As shown in Figure 8-21, this search pattern identifies two books from the BOOKS table.

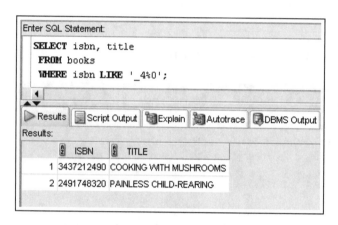

FIGURE 8-21   Searching with the LIKE operator and a combination of wildcard characters

NOTE

The regular expression REGEXP_LIKE extends the pattern-matching capabilities of the LIKE operator. Regular expressions are covered in Chapter 10.

What if you need to use the LIKE operator to search for patterns but also need to search for a wildcard character as a literal in your value? For example, you need to search for a value that starts with the % symbol, contains an uppercase A as the fourth character, and ends with an uppercase T. In this query, you need to use the wildcard characters _ and % with the LIKE operator but also need to search for a literal % symbol as the first character. The LIKE operator includes the **ESCAPE** option for indicating when wildcard symbols should be used as literals rather than translated as wildcard characters. This option allows the user to select the escape character. The escape character must precede any wildcard characters in the search pattern that should be interpreted literally, not as wildcard characters. To perform the search in this example with the ESCAPE option, first review the data in the TESTING table, shown in Figure 8-22. Note that the row with the ID 1 contains a Tvalue of %ccAccT that can be used for this example.

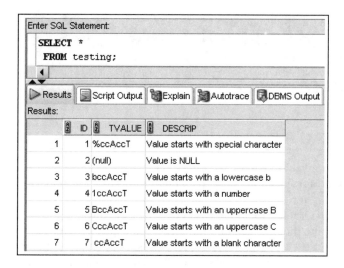

FIGURE 8-22   Review TESTING table data

NOTE

The TESTING table is created with the Chapter 8 database script. This table isn't part of the JustLee database; it's used only to demonstrate LIKE and sorting examples in this chapter. If you don't have this table in your schema, be sure to run the JLDB_Build_8.sql script as instructed previously.

Figure 8-23 shows the query with the pattern identified by using the ESCAPE keyword followed by the escape character \, which must be enclosed in single quotation marks ('\'). This escape character must be placed immediately before any wildcard symbols that should be treated as literal characters. In this case, the first character must be a % symbol, so the escape character is placed immediately before it to instruct the LIKE operator to not treat this character as a wildcard character.

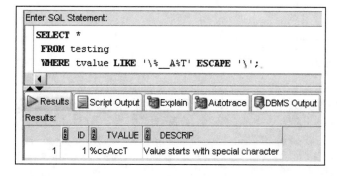

FIGURE 8-23   Using the ESCAPE option with the LIKE operator

LOGICAL OPERATORS

At times, you need to search for records based on two or more conditions. In these situations, you can use **logical operators** to combine search conditions. The logical operators **AND** and **OR** are commonly used for this purpose. (The NOT operator mentioned in Table 8-2 is also a logical operator in Oracle 11*g*, but it's used to reverse the meaning of search conditions rather than combine them.) Keep in mind that when a query executes, records can be filtered with WHERE clause conditions. In other words, each record in the table is compared with the stated condition. If the condition is TRUE when compared with a record, the record is included in the results.

When the AND operator is used in the WHERE clause, *both* conditions combined by the AND operator must be evaluated as TRUE, or the record isn't included in the results. For example, Figure 8-24 shows a query for titles of books that are published by Publisher 3 *and* in the Computer category. Because the search is for books meeting both conditions, the conditions are combined with the AND operator.

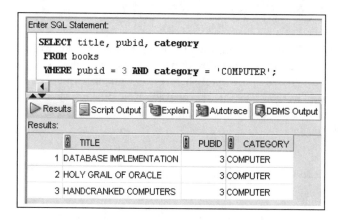

FIGURE 8-24 Searching with multiple conditions and the AND logical operator

On the other hand, if you want a list of books that are published by Publisher 3 *or* in the Computer category, you can use the OR operator, as shown in Figure 8-25. With the OR operator, only one of the conditions must be TRUE to have the record included in the results. As shown in the results, the first three records pass both conditions. The last record passes only the category = 'COMPUTER' condition; the pubid = 3 condition evaluates as FALSE for this record. Because the OR operator has been used, however, only one condition has to be TRUE, so even though this book isn't published by Publisher 3, it's included in the results.

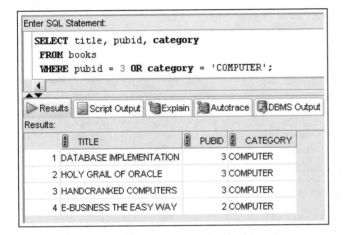

FIGURE 8-25   Searching with multiple conditions and the OR logical operator

---

TIP

Using a series of OR logical operators to join conditions based on the same column is identical to using the IN comparison operator. For example, state = 'GA' OR state = 'CA' is the same as IN ('GA', 'CA'). Keep in mind that each condition must be a full comparison. You can't use WHERE state = 'GA' OR 'CA', or an error occurs. The OR condition must be followed by a complete comparison, such as state = 'CA'.

Next, take a look at the order of logical operators. Because the WHERE clause can contain multiple types of operators, you need to understand the order in which they're resolved:

- Arithmetic operations are solved first.
- Comparison operators (<, >, =, LIKE, and so forth) are solved next.
- Logical operators have a lower precedence and are evaluated last—in the order NOT, AND, and OR.

If you need to change the order of evaluation, simply use parentheses to indicate the operators to be resolved first. To see how this method works, look at the query results in Figure 8-26. The list includes books that are published by Publisher 4 *and* cost more than

$15.00. The list also includes any book from the Family Life category. Although the OR operator is actually listed first in the WHERE clause, Oracle 11*g* first evaluates the Pubid and Cost conditions combined with the AND logical operator. After this operation is solved, the Category condition preceding the OR logical operator is evaluated.

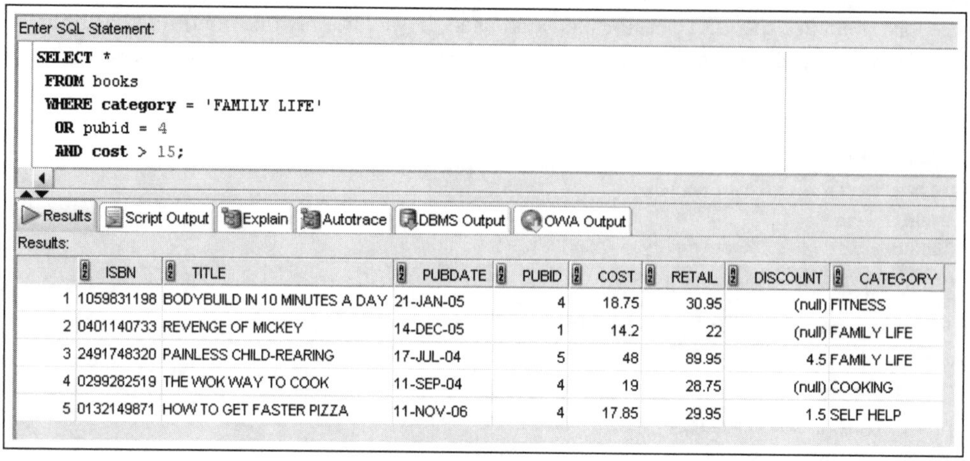

```
SELECT *
FROM books
WHERE category = 'FAMILY LIFE'
 OR pubid = 4
 AND cost > 15;
```

	ISBN	TITLE	PUBDATE	PUBID	COST	RETAIL	DISCOUNT	CATEGORY
1	1059831198	BODYBUILD IN 10 MINUTES A DAY	21-JAN-05	4	18.75	30.95	(null)	FITNESS
2	0401140733	REVENGE OF MICKEY	14-DEC-05	1	14.2	22	(null)	FAMILY LIFE
3	2491748320	PAINLESS CHILD-REARING	17-JUL-04	5	48	89.95	4.5	FAMILY LIFE
4	0299282519	THE WOK WAY TO COOK	11-SEP-04	4	19	28.75	(null)	COOKING
5	0132149871	HOW TO GET FASTER PIZZA	11-NOV-06	4	17.85	29.95	1.5	SELF HELP

FIGURE 8-26 Searching with both AND and OR operators

After examining the previous query's results, you realize that the order in which logical operators were evaluated didn't yield the output you want—to find any book costing more than $15.00 that's published by Publisher 4 *or* is in the Family Life category. To have Oracle 11*g* evaluate these conditions in the right order, you must use parentheses to identify any book that's published by Publisher 4 or categorized as Family Life first. After books meeting the Category or Publisher condition are found, the cost condition is then evaluated, and only those records with a cost higher than $15.00 are displayed. Notice that the query in Figure 8-27 returns results different from those in Figure 8-26.

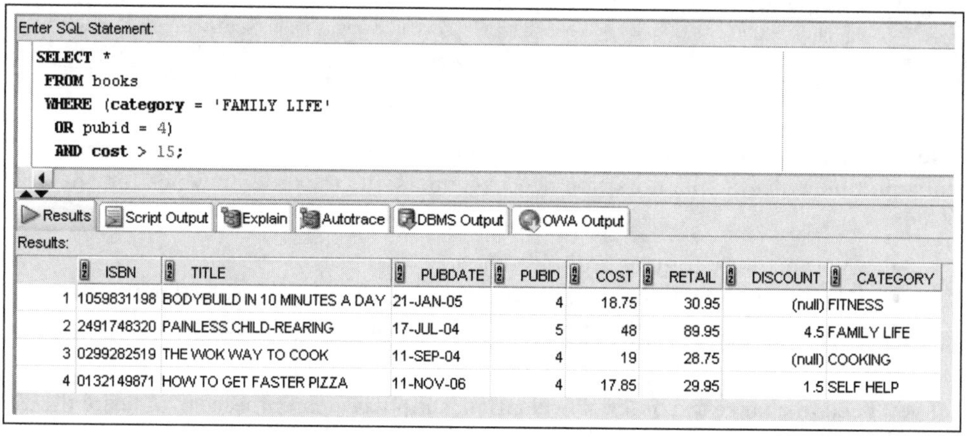

```
SELECT *
FROM books
WHERE (category = 'FAMILY LIFE'
 OR pubid = 4)
 AND cost > 15;
```

	ISBN	TITLE	PUBDATE	PUBID	COST	RETAIL	DISCOUNT	CATEGORY
1	1059831198	BODYBUILD IN 10 MINUTES A DAY	21-JAN-05	4	18.75	30.95	(null)	FITNESS
2	2491748320	PAINLESS CHILD-REARING	17-JUL-04	5	48	89.95	4.5	FAMILY LIFE
3	0299282519	THE WOK WAY TO COOK	11-SEP-04	4	19	28.75	(null)	COOKING
4	0132149871	HOW TO GET FASTER PIZZA	11-NOV-06	4	17.85	29.95	1.5	SELF HELP

FIGURE 8-27 Using parentheses to control the evaluation order for logical operators

TREATMENT OF NULL VALUES

When you're performing arithmetic operations or search conditions, NULL values can cause unexpected results. A **NULL value** means no value has been stored in that field. Don't confuse a NULL value with a blank space. A NULL is *the absence of data in a field*; a field containing a blank space *does* contain a value—a blank space—and is, therefore, *not* a NULL value. When searching for NULL values, you can't use the equal sign (=) because there's no value to use for comparison in the search condition. When checking for a NULL value, you're actually checking the status of the column: Does data exist or not? If you need to identify records that have a NULL value, you must use the **IS NULL** comparison operator.

For example, when an order is shipped to a customer, the shipping date is entered in the ORDERS table. If a date doesn't appear in the Shipdate field, the order hasn't been shipped yet. To find any order that hasn't been shipped, use the query shown in Figure 8-28.

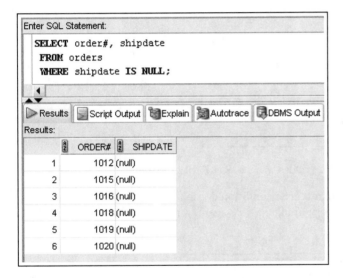

FIGURE 8-28 Searching for NULL values with the IS NULL operator

As shown in Figure 8-28, currently six orders are outstanding. Notice that when you're searching for a NULL value, you simply state the field to be searched followed by the words IS NULL in the WHERE clause. If you want a list of all orders that have shipped (that is, the Shipdate column contains an entry), simply add the logical operator NOT. When searching for a field with the **IS NOT NULL** operator, you instruct Oracle 11*g* to return any records with data available in the named field, as shown in Figure 8-29.

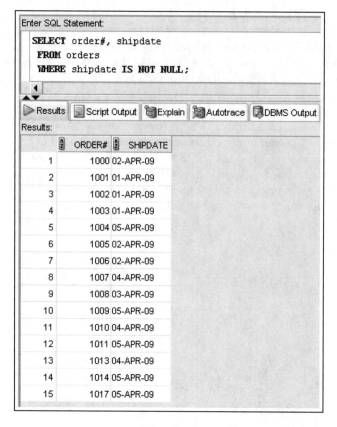

FIGURE 8-29   Searching for non-NULL values with the IS NOT NULL operator

Be aware that using = NULL in a search condition doesn't raise an error; however, it always returns no rows. Try the query in Figure 8-28 again, replacing IS NULL with = NULL to see the results, shown in Figure 8-30.

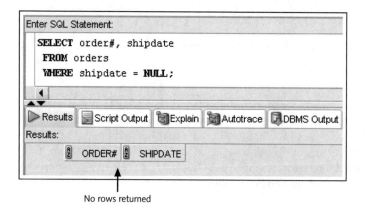

No rows returned

FIGURE 8-30   Using the = NULL operator by mistake

ORDER BY CLAUSE SYNTAX

The **ORDER BY** clause, used to display query results in a sorted order, is listed at the end of the SELECT statement, as shown in Figure 8-31 (which repeats the SELECT syntax shown earlier in the chapter).

```
SELECT   [DISTINCT | UNIQUE] (*, columnname [ AS alias], …)
         FROM      tablename
         [WHERE    condition]
         [GROUP BY group_by_expression]
         [HAVING   group_condition]
         [ORDER BY columnname];
```

FIGURE 8-31   Syntax of the SELECT statement

The columns used to sort the results are listed in the ORDER BY clause. For example, to see a list of all publishers sorted by the Name field, enter the SQL statement shown in Figure 8-32.

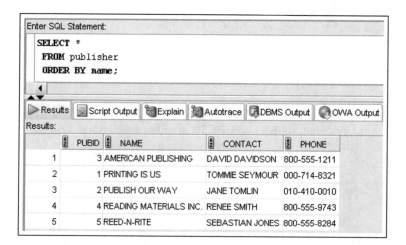

FIGURE 8-32   Sorting results by publisher name in ascending order

In the query results, the second column (Name) is listed in ascending alphabetical order. Note these important points:

- When sorting in ascending order, values are listed in this order:
  1. Blank and special characters
  2. Numeric values
  3. Character values (uppercase first)
  4. NULL values
- Unless you specify "DESC" for descending, the ORDER BY clause sorts in ascending order by default.

View the data stored in the TESTING table again. To see the default sorting treatment of character strings containing special characters, uppercase and lowercase characters, numbers, blank spaces, and NULL values, use a sorting operation on the Tvalue column, as shown in Figure 8-33. Review the order of rows in the results; the Descrip column explains how different characters control the sort order.

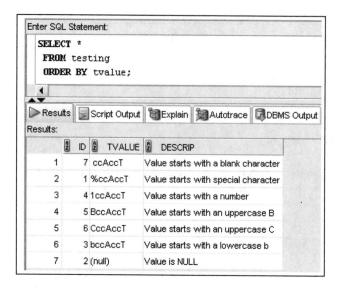

FIGURE 8-33   Sorting on the Tvalue column

Figure 8-32 showed publishers sorted in the default name order. To view publishers in descending alphabetical order by name, simply enter **DESC** after the column name. After changing the sort order to descending, you get the results shown in Figure 8-34.

```
Enter SQL Statement:
    SELECT *
    FROM publisher
    ORDER BY name DESC;
```

	PUBID	NAME	CONTACT	PHONE
1	5	REED-N-RITE	SEBASTIAN JONES	800-555-8284
2	4	READING MATERIALS INC.	RENEE SMITH	800-555-9743
3	2	PUBLISH OUR WAY	JANE TOMLIN	010-410-0010
4	1	PRINTING IS US	TOMMIE SEYMOUR	000-714-8321
5	3	AMERICAN PUBLISHING	DAVID DAVIDSON	800-555-1211

FIGURE 8-34   Sorting results by publisher name in descending order

If a column alias is given to a field in the SELECT clause, you can reference the field in the ORDER BY clause with the column alias—although doing so isn't required. Take a look at the example in Figure 8-35. Notice that "Publisher Year" is enclosed in double quotation marks because this column alias contains a space.

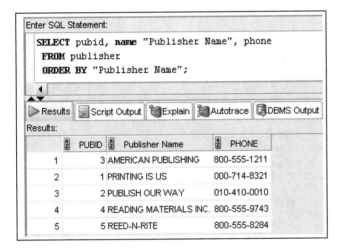

FIGURE 8-35   Referencing a column alias in the ORDER BY clause

You can also use the ORDER BY clause with the optional **NULLS FIRST** or **NULLS LAST** keywords to change the order for listing NULL values. By default, NULL values are listed *last* when results are sorted in ascending order and *first* when they're sorted in descending order. The query in Figure 8-36 lists the last and first name of each customer from California and the customer number of the person who referred the customer to JustLee Books. The results are sorted in ascending order by the Referred column.

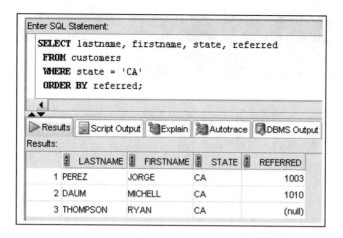

FIGURE 8-36 The default sort order for NULL values

Suppose, however, you want the results sorted in ascending order, but you need the NULL values listed first. To override the default order, you add NULLS FIRST in the ORDER BY clause, which instructs Oracle 11g to place NULL values at the beginning of the list and sort the remaining records in ascending order, as shown in Figure 8-37. (If you want a descending order with NULL values listed last, you use NULLS LAST to override the default order.)

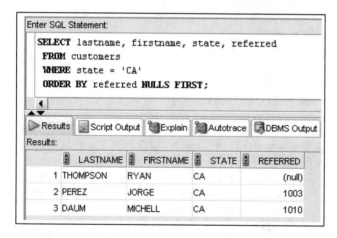

FIGURE 8-37 Using the NULLS FIRST option in the ORDER BY clause

Secondary Sort

In the previous examples, only one column was specified in the ORDER BY clause, which is called a **primary sort**. In some cases, you might want to include a **secondary sort**, which specifies a second field to sort by if an exact match occurs between two or more rows in the primary sort. For example, telephone books list residential customers alphabetically by last name. However, when two or more customers have the same last name, they're listed in alphabetical order by their first names. In other words, a primary sort is performed on the last name and then, when necessary, a secondary sort is performed on the first name.

N O T E

The limit on the number of columns that can be used in the ORDER BY clause is 255.

To illustrate, the query in Figure 8-38 specifies listing customers in descending order by state. When more than one customer lives in a state, customers are to be sorted by city—in ascending order. When looking at the query results, you can see that several states have multiple residents. In these states, customers are sorted in ascending order, according to the city in which they live. The descending sort order applies only to the column after which it's listed—State, in this example. Because City didn't reference a sort order, the default value of ascending is assumed.

271

Enter SQL Statement:

```
SELECT lastname, firstname, state, city
FROM customers
ORDER BY state DESC, city;
```

▷ Results | 📄 Script Output | 🗐 Explain | 🗐 Autotrace | 🗐 DBMS Output

Results:

	LASTNAME	FIRSTNAME	STATE	CITY
1	JONES	KENNETH	WY	CHEYENNE
2	LEE	JASMINE	WY	CODY
3	GIRARD	CINDY	WA	SEATTLE
4	GIANA	TAMMY	TX	AUSTIN
5	CRUZ	MESHIA	NY	ALBANY
6	SMITH	JENNIFER	NJ	MORRISTOWN
7	FALAH	KENNETH	NJ	TRENTON
8	NELSON	BECCA	MI	KALMAZOO
9	MCKENZIE	WILLIAM	MA	BOSTON
10	MCGOVERN	REESE	IL	CHICAGO
11	PIERSON	THOMAS	ID	BOISE
12	LUCAS	JAKE	GA	ATLANTA
13	MONTIASA	GREG	GA	MACON
14	NGUYEN	NICHOLAS	FL	CLERMONT
15	MORALES	BONITA	FL	EASTPOINT
16	SCHELL	STEVE	FL	MIAMI
17	SMITH	LEILA	FL	TALLAHASSEE
18	DAUM	MICHELL	CA	BURBANK
19	PEREZ	JORGE	CA	BURBANK
20	THOMPSON	RYAN	CA	SANTA MONICA

FIGURE 8-38   Using primary and secondary sort columns

Sorting by SELECT Order

The query statement in Figure 8-39 requests a list of customers who live in the states CA and FL. The statement also specifies listing the query results with a primary descending sort on State and a secondary sort on City.

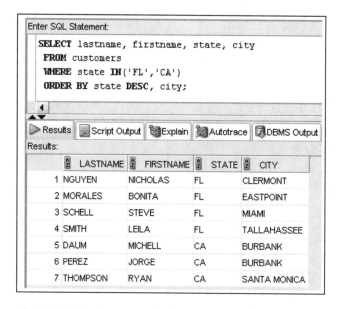

FIGURE 8-39   Sorting on the State and City columns

Oracle 11g also provides an abbreviated method for referencing the sort column if the name is used in the SELECT clause. In the previous example, State and City are used in both the SELECT and ORDER BY clauses. Instead of listing these column names again in the ORDER BY clause, you can reference them by their positions in the SELECT clause's column list. Because State is listed third and City is fourth in the SELECT clause, you can modify the SQL statement as shown in Figure 8-40 and get the same results.

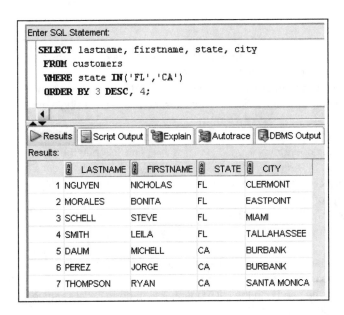

FIGURE 8-40   Referencing positions of sort columns in the ORDER BY clause

Chapter Summary

- The WHERE clause can be included in a SELECT statement to restrict the rows a query returns to only those meeting a specified condition.
- A column alias can't be used to reference columns in a WHERE clause condition.
- When searching a nonnumeric field, search values must be enclosed in single quotation marks.
- Comparison operators indicate how the record should relate to the search value.
- Mathematical comparison operators include $=$, $>$, $<$, $>=$, and $<=$ and the "not equal to" operators $<>$, $!=$, and $^=$.
- The BETWEEN ... AND comparison operator is used to search for records falling within a specified range of values.
- The IN comparison operator identifies a list of values to use for the search condition. A record must contain one of the values in the list to be included in the query results.
- The LIKE comparison operator is used with the percent and underscore symbols (% and _) to establish search patterns.
- The ESCAPE option can be used with the LIKE operator to perform literal searches for characters representing wildcards.
- Logical operators, such as AND and OR, can be used to combine several search conditions.
- The NOT logical operator can be used to reverse the comparison operation.
- Logical operators are always evaluated in the order NOT, AND, and OR. Parentheses can be used to override the evaluation order.
- When using the AND operator, all conditions must be TRUE for a record to be returned in the results. However, with the OR operator, only one condition must be TRUE.
- A NULL value is the absence of data, not a field with a blank space entered.
- Use the IS NULL comparison operator to match NULL values. The IS NOT NULL comparison operator finds records that don't contain NULL values in the indicated column.
- You can sort query results by using an ORDER BY clause. When used, the ORDER BY clause should be listed last in the SELECT statement.
- By default, records are sorted in ascending order. Entering DESC immediately after the column name sorts records in descending order.
- Multiple columns can be used for sorting by listing each column name in a single ORDER BY clause, separated by commas. The column you want used for the primary sort should be listed first. If an exact match occurs between two or more records, the next column listed (the secondary sort column) determines the correct order, and so on.
- A column doesn't have to be listed in the SELECT clause to serve as a basis for sorting.
- An ascending sort order lists blanks and special characters first.
- A column in the ORDER BY clause can be referenced with its position number in the SELECT clause instead of the column name.

Chapter 8 Syntax Summary

The following table summarizes the syntax you have learned in this chapter. You can use the table as a study guide and reference.

Syntax Guide

Element	Description	Example
Optional SELECT clauses		
WHERE clause	Specifies a search condition	`SELECT * FROM customers` `WHERE state = 'GA';`
ORDER BY clause	Specifies the display order of query results	`SELECT * FROM publisher` `ORDER BY name;`
Mathematical Comparison Operators		
=	"Equality" operator—requires an exact match of the record data and the search value	`WHERE cost = 55.95`
>	"Greater than" operator—requires a record to be greater than the search value	`WHERE cost > 55.95`
<	"Less than" operator—requires a record to be less than the search value	`WHERE cost < 55.95`

<>, !=, ^=	"Not equal to" operator—requires a record to not match the search value	`WHERE cost <> 55.95`
		or
		`WHERE cost != 55.95`
		or
		`WHERE cost ^= 55.95`

<=	"Less than or equal to" operator—requires a record to be less than or an exact match with the search value	`WHERE cost <= 55.95`
>=	"Greater than or equal to" operator—requires a record to be greater than or an exact match with the search value	`WHERE cost >= 55.95`

Element	Description	Example
Other Comparison Operators		
`[NOT] BETWEEN` x AND y	Searches for records in a specified range of values	**WHERE cost BETWEEN 40 AND 65**
`[NOT] IN` (x,y,...)	Searches for records matching one of the items in the list	**WHERE cost IN (22, 55.95, 13.50)**
`[NOT] LIKE`	Searches for records matching a search pattern—used with wildcard characters	**WHERE lastname LIKE '_A%'**
`IS[NOT] NULL`	Searches for records with a NULL value in the indicated column	**WHERE referred IS NULL**
Wildcard Characters		
`%`	Percent sign wildcard represents any number of characters	**WHERE lastname LIKE '%R%'**
`_`	Underscore symbol wildcard represents exactly one character in the indicated position	**WHERE lastname LIKE '_A'**
Logical Operators		
`AND`	Combines two conditions together—record must match both conditions	**WHERE cost > 20 AND retail < 50**
`OR`	Requires a record to match only one of the search conditions	**WHERE cost > 20 OR retail < 50**

Review Questions

1. Which clause of an SQL query is used to restrict the number of rows returned?
2. Which clause of an SQL query displays the results in a specific sequence?
3. Which operator can you use to find any books with a retail price of at least $24.00?
4. Which operator should you use to find NULL values?
5. The IN comparison operator is similar to which logical operator?

6. When should single quotation marks be used in a WHERE clause?

7. What's the effect of using the NOT operator in a WHERE clause?

8. When should a percent sign (%) be used with the LIKE operator?

9. When should an underscore symbol (_) be used with the LIKE operator?

10. Because % is a wildcard character, how can the LIKE operator search for a literal percent sign (%) in a character string?

Multiple Choice

To answer the following questions, refer to the tables in the JustLee Books database.

1. Which of the following SQL statements isn't valid?

   a. `SELECT address || city || state || zip "Address" FROM customers WHERE lastname = 'SMITH';`

   b. `SELECT * FROM publisher ORDER BY contact;`

   c. `SELECT address, city, state, zip FROM customers WHERE lastname = "SMITH";`

   d. All the above statements are valid and return the expected results.

2. Which clause is used to restrict rows or perform selection?

   a. SELECT

   b. FROM

   c. WHERE

   d. ORDER BY

3. Which of the following SQL statements is valid?

   a. `SELECT order# FROM orders WHERE shipdate = NULL;`

   b. `SELECT order# FROM orders WHERE shipdate = 'NULL';`

   c. `SELECT order# FROM orders WHERE shipdate = "NULL";`

   d. None of the statements are valid.

4. Which of the following returns a list of all customers' names sorted in descending order by city within state?

   a. `SELECT name FROM customers`
      `    ORDER BY desc state, city;`

   b. `SELECT firstname, lastname FROM customers`
      `    SORT BY desc state, city;`

   c. `SELECT firstname, lastname FROM customers`
      `    ORDER BY state desc, city;`

   d. `SELECT firstname, lastname FROM customers`
      `    ORDER BY state desc, city desc;`

   e. `SELECT firstname, lastname FROM customers`
      `    ORDER BY 5 desc, 6 desc;`

Restricting Rows and Sorting Data

5. Which of the following doesn't return a customer with the last name THOMPSON in the query results?

   a. `SELECT lastname FROM customers WHERE lastname = "THOMPSON";`

   b. `SELECT * FROM customers;`

   c. `SELECT lastname FROM customers WHERE lastname > 'R';`

   d. `SELECT * FROM customers WHERE lastname < 'V';`

6. Which of the following displays all books published by Publisher 1 with a retail price of at least $25.00?

   a. `SELECT * FROM books WHERE pubid = 1 AND retail >= 25;`

   b. `SELECT * FROM books WHERE pubid = 1 OR retail >= 25;`

   c. `SELECT * FROM books WHERE pubid = 1 AND WHERE retail > 25;`

   d. `SELECT * FROM books WHERE pubid = 1, retail >= 25;`

   e. `SELECT * FROM books WHERE pubid = 1, retail >= $25.00;`

7. What's the default sort sequence for the ORDER BY clause?

   a. ascending

   b. descending

   c. the order in which records are stored in the table

   d. There's no default sort sequence.

8. Which of the following doesn't include the display of books published by Publisher 2 and having a retail price of at least $35.00?

   a. `SELECT * FROM books WHERE pubid = 2, retail >= $35.00;`

   b. `SELECT * FROM books WHERE pubid = 2 AND NOT retail < 35;`

   c. `SELECT * FROM books WHERE pubid IN (1, 2, 5)`
      `    AND retail NOT BETWEEN 1 AND 29.99;`

   d. All the above statements display the specified books.

   e. None of the above statements display the specified books.

9. Which of the following includes a customer with the first name BONITA in the results?

   a. `SELECT * FROM customers WHERE firstname = 'B%';`

   b. `SELECT * FROM customers WHERE firstname LIKE '%N%';`

   c. `SELECT * FROM customers WHERE firstname = '%N%';`

   d. `SELECT * FROM customers WHERE firstname LIKE '_B%';`

10. Which of the following represents exactly one character in a pattern search?

    a. ESCAPE

    b. ?

    c. _

    d. %

    e. none of the above

11. Which of the following returns the book HANDCRANKED COMPUTERS in the results?

   a. `SELECT * FROM books WHERE title = 'H_N_%';`

   b. `SELECT * FROM books WHERE title LIKE "H_N_C%";`

   c. `SELECT * FROM books WHERE title LIKE 'H_N_C%';`

   d. `SELECT * FROM books WHERE title LIKE '_H%';`

12. Which of the following clauses is used to display query results in a sorted order?

   a. WHERE

   b. SELECT

   c. SORT

   d. ORDER

   e. none of the above

13. Which of the following SQL statements returns all books published after March 20, 2005?

   a. `SELECT * FROM books WHERE pubdate > 03-20-2005;`

   b. `SELECT * FROM books WHERE pubdate > '03-20-2005';`

   c. `SELECT * FROM books WHERE pubdate > '20-MAR-05';`

   d. `SELECT * FROM books WHERE pubdate > 'MAR-20-05';`

14. Which of the following lists all books published before June 2, 2004 *and* all books published by Publisher 4 or in the Fitness category?

   a. `SELECT * FROM books WHERE category = 'FITNESS' OR pubid = 4`
       `    AND pubdate < '06-02-2004';`

   b. `SELECT * FROM books WHERE category = 'FITNESS' AND pubid = 4`
       `    OR pubdate < '06-02-2004';`

   c. `SELECT * FROM books WHERE category = 'FITNESS' OR (pubid = 4`
       `    AND pubdate < '06-02-2004');`

   d. `SELECT * FROM books WHERE category = 'FITNESS'`
       `    OR pubid = 4, pubdate < '06-02-04';`

   e. none of the above

15. Which of the following finds all orders placed before April 5, 2009 that haven't yet shipped?

   a. `SELECT * FROM orders WHERE orderdate < '04-05-09'`
       `    AND shipdate = NULL;`

   b. `SELECT * FROM orders WHERE orderdate < '05-04-09'`
       `    AND shipdate IS NULL;`

   c. `SELECT * FROM orders WHERE orderdate < 'APR-05-09'`
       `    AND shipdate IS NULL;`

   d. `SELECT * FROM orders WHERE orderdate < '05-APR-09'`
       `    AND shipdate IS NULL;`

   e. none of the above

16. Which of the following symbols represents any number of characters in a pattern search?

    a.   *

    b.   ?

    c.   %

    d.   _

17. Which of the following lists books generating at least $12.00 in profit?

    a.   `SELECT * FROM books WHERE retail-cost > 12;`

    b.   `SELECT * FROM books WHERE retail-cost <= 12;`

    c.   `SELECT * FROM books WHERE profit >= 12;`

    d.   `SELECT * FROM books WHERE retail-cost => 12.00;`

    e.   none of the above

18. Which of the following lists each book having a profit of at least $10.00 in descending order by profit?

    a.   `SELECT * FROM books WHERE profit => 10.00`
        `ORDER BY "Profit" desc;`

    b.   `SELECT title, retail-cost "Profit" FROM books`
        `WHERE profit => 10.00`
        `ORDER BY "Profit" desc;`

    c.   `SELECT title, retail-cost "Profit" FROM books`
        `WHERE "Profit" => 10.00`
        `ORDER BY "Profit" desc;`

    d.   `SELECT title, retail-cost profit FROM books`
        `WHERE retail-cost >= 10.00`
        `ORDER BY "PROFIT" desc;`

    e.   `SELECT title, retail-cost "Profit" FROM books`
        `WHERE profit => 10.00`
        `ORDER BY 3 desc;`

19. Which of the following includes the book HOW TO GET FASTER PIZZA in the query results?

    a.   `SELECT * FROM books WHERE title LIKE '%AS_E%';`

    b.   `SELECT * FROM books WHERE title LIKE 'AS_E%';`

    c.   `SELECT * FROM books WHERE title = '%AS_E%'`

    d.   `SELECT * FROM books WHERE title = 'AS_E%';`

20. Which of the following returns all books published after March 20, 2005?

    a.   `SELECT * FROM books WHERE pubdate > 03-20-2005;`

    b.   `SELECT * FROM books WHERE pubdate > '03-20-2005';`

    c.   `SELECT * FROM books WHERE pubdate NOT < '20-MAR-05';`

    d.   `SELECT * FROM books WHERE pubdate NOT < 'MAR-20-05';`

    e.   none of the above

Hands-On Assignments

To perform the following assignments, refer to the tables created in the JLDB_Build_8.sql script at the beginning of the chapter. Give the SQL statements and output for the following data requests:

1. Which customers live in New Jersey? List each customer's last name, first name, and state.
2. Which orders shipped after April 1, 2009? List each order number and the date it shipped.
3. Which books aren't in the Fitness category? List each book title and category.
4. Which customers live in Georgia or New Jersey? Put the results in ascending order by last name. List each customer's customer number, last name, and state. Write this query two different ways.
5. Which orders were placed on or before April 1, 2009? List each order number and order date. Write this query two different ways.
6. List all authors whose last name contains the letter pattern "IN." Put the results in order of last name, then first name. List each author's last name and first name.
7. List all customers who were referred to the bookstore by another customer. List each customer's last name and the number of the customer who made the referral.
8. Display the book title and category for all books in the Children and Cooking categories. Create three different queries to accomplish this task: a) a search pattern operation, b) a logical operator, and c) another operator not used in a or b.
9. Use a search pattern to find any book title with "A" for the second letter and "N" for the fourth letter. List each book's ISBN and title. Sort the list by title in descending order.
10. List the title and publish date of any computer book published in 2005. Perform the task of searching for the publish date by using three different methods: a) a range operator, b) a logical operator, and c) a search pattern operation.

Advanced Challenge

To perform these activities, refer to the JustLee database tables.

During an afternoon at work, you receive various requests for data stored in the database. As you fulfill each request, you decide to document the SQL statements you used to find the data to assist with future requests. The following are two of the requests that were made:

1. A manager at JustLee Books requests a list of the titles of all books generating a profit of at least $10.00. The manager wants the results listed in descending order, based on each book's profit.
2. A customer service representative is trying to identify all books in the Computer or Family Life category *and* published by Publisher 1 or Publisher 3. However, the results shouldn't include any book selling for less than $45.00.

For each request, create a document showing the SQL statement and the query results.

Case Study: *City Jail*

Note: Run the CityJail_8.sql file provided by your instructor to ensure that all necessary tables and constraints are available for this case study. This script isn't included in student data files because case study assignments in previous chapters include table creation challenges. This script rebuilds the City Jail database. Don't be concerned with errors from the DROP TABLE commands, which delete any existing tables of the same names.

The following list reflects common data requests from city managers. Write the SQL statements to satisfy the requests. If the query can be accomplished by using different operators, supply alternative solutions so that the performance-tuning group can test them and identify the more efficient statements. Test the statements and show execution results.

1. List all criminal aliases beginning with the letter B.
2. List all crimes that occurred (were charged) during the month October 2008. List the crime ID, criminal ID, date charged, and classification.
3. List all crimes with a status of CA (can appeal) or IA (in appeal). List the crime ID, criminal ID, date charged, and status.
4. List all crimes classified as a felony. List the crime ID, criminal ID, date charged, and classification.
5. List all crimes with a hearing date more than 14 days after the date charged. List the crime ID, criminal ID, date charged, and hearing date.
6. List all criminals with the zip code 23510. List the criminal ID, last name, and zip code. Sort the list by criminal ID.
7. List all crimes that don't have a hearing date scheduled. List the crime ID, criminal ID, date charged, and hearing date.
8. List all sentences with a probation officer assigned. List the sentence ID, criminal ID, and probation officer ID. Sort the list by probation officer ID and then criminal ID.
9. List all crimes that are classified as misdemeanors and are currently in appeal. List the crime ID, criminal ID, classification, and status.
10. List all crime charges with a balance owed. List the charge ID, crime ID, fine amount, court fee, amount paid, and amount owed.
11. List all police officers who are assigned to the precinct OCVW or GHNT and have a status of active. List the officer ID, last name, precinct, and status. Sort the list by precinct and then by officer last name.

JOINING DATA FROM MULTIPLE TABLES

LEARNING OBJECTIVES

After completing this chapter, you should be able to do the following:

- Identify a Cartesian join
- Create an equality join with the WHERE clause
- Create an equality join with the JOIN keyword
- Create a non-equality join with the WHERE clause
- Create a non-equality join with the JOIN ... ON approach
- Create a self-join with the WHERE clause
- Create a self-join with the JOIN keyword
- Distinguish an inner join from an outer join
- Create an outer join with the WHERE clause
- Create an outer join with the OUTER keyword
- Use set operators to combine the results of multiple queries

INTRODUCTION

The main advantage of using a relational database is that you can eliminate data redundancy by structuring data in multiple tables. However, this structure requires combining or joining data rows from multiple tables before you can perform many kinds of queries. This chapter focuses on adding **join conditions**, which are instructions in queries that combine data from more than one table.

Traditionally, Oracle database users had to include join conditions in the WHERE clause to specify how data rows of different tables are related. Beginning with Oracle 9*i*, support for ANSI-compliant joins was introduced; these joins use the JOIN keyword in the FROM clause. (American National Standards Institute [ANSI] was introduced in Chapter 1.) The ANSI JOIN method has several advantages over a traditional WHERE clause join. It increases portability of SQL code between different DBMS platforms, as most relational databases are ANSI compliant. Also, because the WHERE clause is reserved for including only conditions that restrict rows from being returned, the ANSI JOIN method has improved statement clarity.

In this chapter, you examine several kinds of joins as well as the syntax for creating each join, first using the traditional WHERE clause approach and then using the ANSI JOIN method. You need to understand both approaches to creating joins to support existing code, to prepare new systems with increased portability goals, and to pass the Oracle 11*g* SQL exam. Because no performance advantages are associated with either join method, both methods are widely used. You also explore combining data rows from multiple tables with set operators, which allow you to combine rows from multiple queries. Table 9-1 gives you an overview of this chapter's topics.

TABLE 9-1   Types of Joins and Set Operators

Element	Description
Cartesian join (also known as a Cartesian product or cross join)	Replicates each row from the first table with every row from the second table. Creates a join between tables by displaying every possible record combination. Can be created by two methods: • Not including a joining condition in a WHERE clause • Using the JOIN method with the CROSS JOIN keywords
Equality join (also known as an equijoin, an inner join, or a simple join)	Creates a join by using a commonly named and defined column. Can be created by two methods: • Using the WHERE clause • Using the JOIN method with the NATURAL JOIN, JOIN ... ON, or JOIN ... USING keywords
Non-equality join	Joins tables when there are no equivalent rows in the tables to be joined—for example, to match values in one column of a table with a range of values in another table. Can be created by two methods: • Using the WHERE clause • Using the JOIN method with the JOIN ... ON keywords

TABLE 9-1   Types of Joins and Set Operators (continued)

Element	Description
Self-join	Joins a table to itself. Can be created by two methods: • Using the WHERE clause • Using the JOIN method with the JOIN … ON keywords
Outer join	Includes records of a table in output when there's no matching record in the other table. Can be created by two methods: • Using the WHERE clause with a (+) operator • Using the JOIN method with the OUTER JOIN keywords and the assigned type of LEFT, RIGHT, or FULL
Set operators	Combines results of multiple SELECT statements. Includes the keywords UNION, UNION ALL, INTERSECT, and MINUS.

DATABASE PREPARATION

Before attempting to work through the examples in this chapter, make sure you have completed the following two tasks. First, if you haven't already run the JLDB_Build_8.sql script from Chapter 8, execute this script to rebuild the JustLee Books database. Second, run the JLDB_Build_9.sql file in the Chapter9 folder of your data files to create the necessary additions for the JustLee Books database.

CARTESIAN JOINS

In a **Cartesian join**, also called a **Cartesian product** or **cross join**, each record in the first table is matched with each record in the second table. This type of join is useful when you're performing certain statistical procedures for data analysis. Therefore, if you have three records in the first table and four in the second table, the first record from the first table is matched with each of the four records in the second table. Then the second record of the first table is matched with each of the four records from the second table, and so on.

As shown in Figure 9-1, a Cartesian join of Table 1 and Table 2 results in 12 records being displayed. You can always identify a Cartesian join because the resulting number of rows is (# rows in Table 1) * (# rows in Table 2). In Figure 9-1, it's 3 rows * 4 rows, resulting in 12 rows.

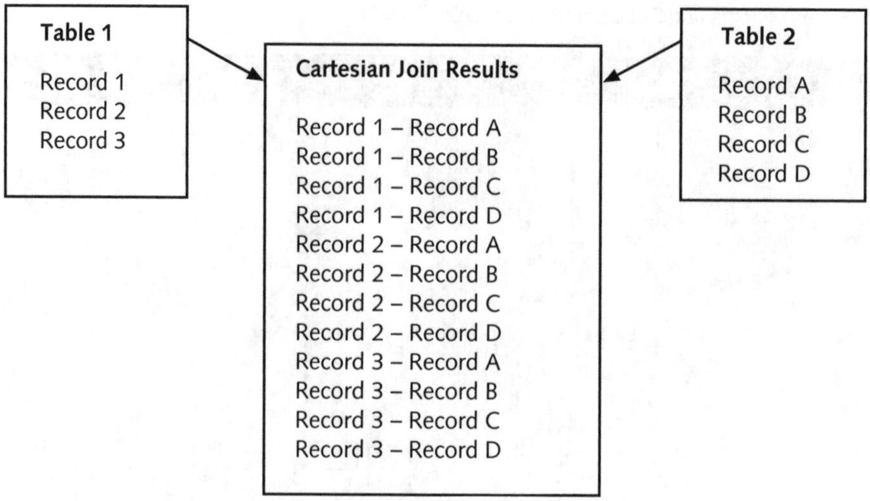

FIGURE 9-1   Results of a Cartesian join

Sometimes a user intends to generate a Cartesian product, but usually this intention is the exception rather than the rule. Be aware that selecting data from multiple tables in a query and accidentally omitting a correct join also produces a Cartesian product, which, in this case, is an incorrect result. Most Cartesian products are generated in error.

Cartesian Join: Traditional Method

JustLee Books needs to perform a manual book inventory in all three of its book warehouses. The manager has requested an inventory sheet listing each book for each warehouse along with a column to record the physical count. In this case, the WAREHOUSES table can be joined with the BOOKS table to produce a Cartesian product. Figure 9-2 shows the query including the two tables and the results. The results include 42 rows: 3 rows (WAREHOUSES table) * 14 rows (BOOKS table) = 42 rows. In this case, a Cartesian join produces the intended results. Note that a blank literal is used to create an empty Count column in the output.

> ### NOTE
>
> In Figure 9-2, only the first 15 rows of output are shown to conserve space. Several figures in this chapter show only partial output for the same reason.

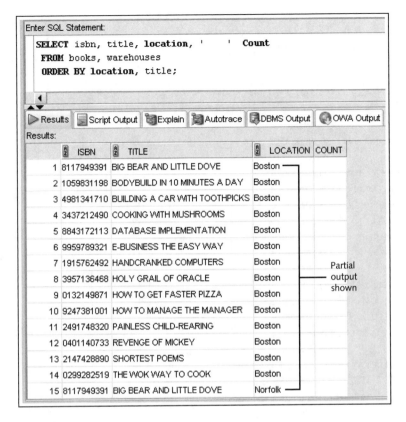

FIGURE 9-2 Producing an intended Cartesian join

Now suppose you need to find the publisher's name for each book in inventory. The SELECT statement in Figure 9-3 instructs Oracle 11*g* to list the Title column, which is stored in the BOOKS table, and the Name column, which is stored in the PUBLISHER table. (Again, only the first 15 rows of output are shown to conserve space.)

Although there are only 14 book titles in the database, 70 records are returned! These results should lead you to be suspicious of the output. The problem with the SQL statement in Figure 9-3 is that the columns to be retrieved from the two tables are specified, but no instructions on how to join the table rows correctly are included. Because a join condition isn't included, Oracle automatically replicates every possible combination of records, producing a Cartesian join. In "Equality Joins" later in this chapter, you learn how to instruct Oracle to match or join rows of multiple tables correctly, based on common column values.

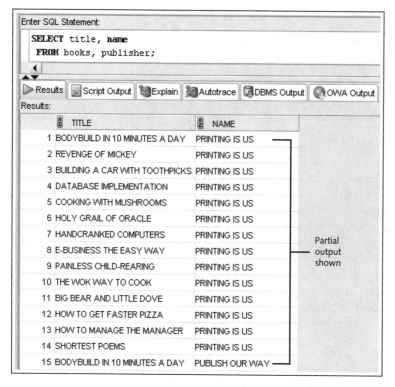

FIGURE 9-3   Producing an unintentional Cartesian join

288

NOTE

FOREIGN KEY constraints define relationships between table columns; however, these definitions aren't used by queries. A FOREIGN KEY constraint is activated only for DML actions.

Cartesian Join: JOIN Method

Beginning with Oracle 10g, the **CROSS** keyword, combined with the JOIN keyword, can be used in the FROM clause to explicitly instruct Oracle to create a Cartesian (cross) join. The CROSS JOIN keywords instruct the database system to create cross-products, using all records of the tables listed in the query. Figure 9-4 shows the same book inventory listing from the earlier example produced with the JOIN method.

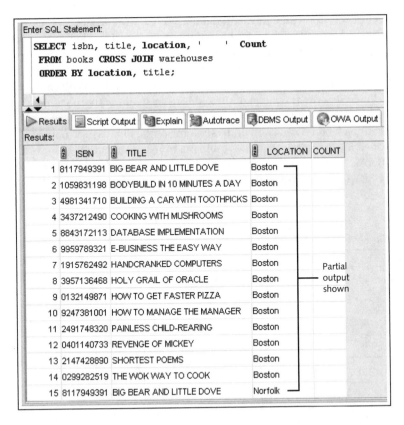

```
Enter SQL Statement:

SELECT isbn, title, location, '    ' Count
FROM books CROSS JOIN warehouses
ORDER BY location, title;
```

▶ Results 📄 Script Output 📄 Explain 📄 Autotrace 📄 DBMS Output 🌐 OWA Output

Results:

	ISBN	TITLE	LOCATION	COUNT
1	8117949391	BIG BEAR AND LITTLE DOVE	Boston	
2	1059831198	BODYBUILD IN 10 MINUTES A DAY	Boston	
3	4981341710	BUILDING A CAR WITH TOOTHPICKS	Boston	
4	3437212490	COOKING WITH MUSHROOMS	Boston	
5	8843172113	DATABASE IMPLEMENTATION	Boston	
6	9959789321	E-BUSINESS THE EASY WAY	Boston	
7	1915762492	HANDCRANKED COMPUTERS	Boston	
8	3957136468	HOLY GRAIL OF ORACLE	Boston	
9	0132149871	HOW TO GET FASTER PIZZA	Boston	
10	9247381001	HOW TO MANAGE THE MANAGER	Boston	
11	2491748320	PAINLESS CHILD-REARING	Boston	
12	0401140733	REVENGE OF MICKEY	Boston	
13	2147428890	SHORTEST POEMS	Boston	
14	0299282519	THE WOK WAY TO COOK	Boston	
15	8117949391	BIG BEAR AND LITTLE DOVE	Norfolk	

Partial output shown

FIGURE 9-4   Using the CROSS JOIN keywords

Notice the syntax of the SQL statement in Figure 9-4. In the FROM clause, the names of the tables to be used in the Cartesian join are separated by the CROSS JOIN keywords. Don't use commas to separate any parts of the FROM clause, as you would with the traditional method.

TIP

If you get an error message, make certain a comma is *not* entered after the BOOKS table name in the FROM clause.

EQUALITY JOINS

The query in Figure 9-3 returned an unintentional Cartesian join because Oracle didn't know what data the two tables had in common. The most common type of join used in the workplace is based on two (or more) tables having equivalent data stored in a common column. These joins are called **equality joins** but are also referred to as **equijoins**, **inner joins**, or **simple joins**.

A **common column** is a column with equivalent data existing in two or more tables. For example, both the BOOKS and PUBLISHER tables have a common column called Pubid containing an identification code assigned to each publisher. Therefore, when you want a list of publishers for each book in the BOOKS table, you need to match the publisher ID stored for each book in the BOOKS table with the corresponding publisher ID in the PUBLISHER table. The results *should* include only the name of the publisher whenever there's a match between the Pubid columns in each table.

To master join operations, you need a solid understanding of the database structure. An E-R model, as developed in Chapter 1, identifies table relationships. Figure 9-5 shows the E-R model for JustLee Books. Each of the one-to-many relationship lines should be supported with a FOREIGN KEY (FK) constraint to ensure consistency of data for common columns.

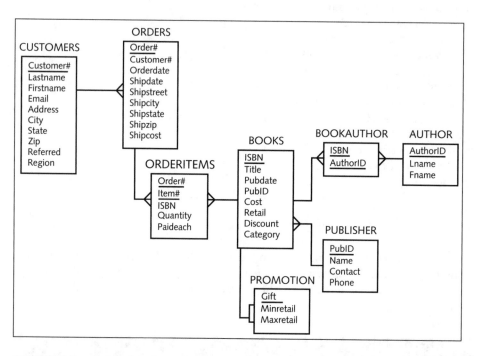

FIGURE 9-5　The JustLee Books table structure

A review of referential integrity or FK constraints can help you identify the common columns between tables. Keep in mind that even though common columns typically have the same name, this isn't a requirement; you can't depend on column names alone. Review the FK constraints in the JustLee Books table creation script from Chapter 8 (JLDB_Build_8.sql) to identify the common column for each relationship line in the E-R model. These common columns are used in query join conditions to instruct the database system how to relate rows of multiple tables logically.

The following section explains how to create joins based on a common column containing equivalent data stored in multiple tables.

Equality Joins: Traditional Method

The traditional way to include join conditions and avoid an unintended Cartesian result is to use the WHERE clause to instruct Oracle 11g how to join tables correctly. You have used the WHERE clause in previous chapters to provide conditions that restrict the rows affected by the SQL statement. A traditional join adds join conditions with other conditions in the WHERE clause. In other words, the WHERE clause can perform two different activities: joining tables and providing conditions to limit or filter the rows that are affected.

Using the same example from the section on Cartesian joins (Figure 9-3), include the WHERE clause to retrieve a list of books that includes the publisher name. Figure 9-6 shows the correct query with a join condition in the WHERE clause.

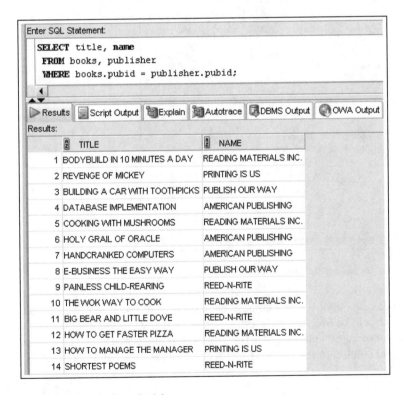

FIGURE 9-6 An equality join

The WHERE clause tells Oracle 11*g* that the BOOKS table and the PUBLISHER table are related by the Pubid column. The equal sign specifies that the contents of the Pubid column in each table must be exactly equal for the rows to be joined and returned in the results. Also, notice that the Pubid column names are prefixed with their corresponding table names. Any time Oracle 11*g* references multiple tables having the same column name, the column name *must* be prefixed with the table name. If your query is ambiguous and doesn't specify exactly which column is the common column, you get an error message. By entering `publisher.pubid`, for example, you're specifying the Pubid column in the PUBLISHER table, which is known as "qualifying" the column name. A **column qualifier** indicates the table containing the column being referenced.

Suppose you want to include additional information—the publisher ID—in the output. If the publisher ID is also listed in the SELECT clause to be included in the query output, the table name prefix is required. If this prefix is omitted, you get a column ambiguity error message, as shown in Figure 9-7.

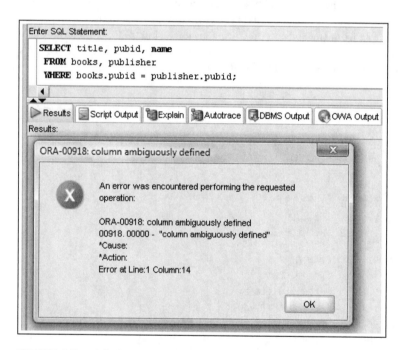

FIGURE 9-7   A "column ambiguously defined" error

The error occurs because Oracle finds a column named Pubid in both tables and is confused as to which table should be used to retrieve the values. In this case, the Pubid column is the common column of these two tables and is the same value in a proper join. Therefore, you can qualify the Pubid column in the SELECT clause with either table, and the query will work.

Search conditions can be added to the WHERE clause along with join conditions, as shown in Figure 9-8. Notice the AND logical operator in the WHERE clause. Including this operator limits query results to only those from Publisher 4. Any of the search conditions used in Chapter 8 can be issued in the WHERE clause when you're joining a table.

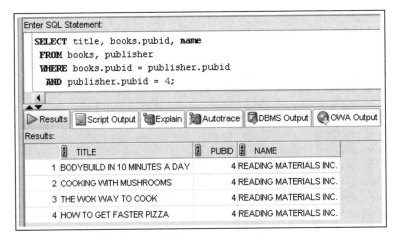

FIGURE 9-8   Including search and join conditions in a WHERE clause

You can also use table aliases to simplify the process of qualifying columns with the table name. In Figure 9-9, the SELECT statement requests the title, publisher ID, and publisher name for any book costing less than $15.00 or any book from Publisher 1.

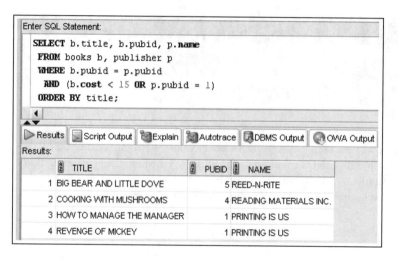

```
Enter SQL Statement:
SELECT b.title, b.pubid, p.name
 FROM books b, publisher p
 WHERE b.pubid = p.pubid
   AND (b.cost < 15 OR p.pubid = 1)
 ORDER BY title;
```

| Results | Script Output | Explain | Autotrace | DBMS Output | OWA Output |

Results:

	TITLE	PUBID	NAME
1	BIG BEAR AND LITTLE DOVE	5	REED-N-RITE
2	COOKING WITH MUSHROOMS	4	READING MATERIALS INC.
3	HOW TO MANAGE THE MANAGER	1	PRINTING IS US
4	REVENGE OF MICKEY	1	PRINTING IS US

FIGURE 9-9   Equality join with table aliases

Take a closer look at some elements in Figure 9-9:

- The SELECT clause not only lists the columns to be displayed, but also includes **table aliases** for the PUBLISHER table (p) and BOOKS table (b). A period is used to separate a table alias from a column name, as when you qualify a column with the full table name. These aliases are assigned in the FROM clause (discussed next).
- The table aliases in the FROM clause work like a column alias by temporarily giving a table a different name. Table aliases offer a couple of advantages. First, they improve processing efficiency, as the system no longer needs to identify which table a specified column is in. Second, coding is simplified when the full table name doesn't have to be indicated (although a table alias can have as many as 30 characters). There's one important rule you must remember when using a table alias: If a table alias is assigned in the FROM clause, *it must be used any time the table is referenced in that SQL statement.*
- The WHERE clause includes the join condition for the BOOKS and PUBLISHER tables as well as other search conditions using the AND and OR logical operators.
- The statement concludes with an ORDER BY clause to display the results in a sorted order.

TIP

Make sure you use the letter "p" (the alias for the PUBLISHER table) or the letter "b" (the alias for the BOOKS table) before the Pubid column name, or you'll get an error message.

Up to this point, the join examples have only included two tables, but suppose you need a list of all customer names along with all books each customer has purchased. The book titles are in the BOOKS table and the customer names are in the CUSTOMERS table. Would you join the BOOKS and CUSTOMERS tables to produce the needed list? No! Joins need to follow logical relationships between tables. In this case, the BOOKS and CUSTOMERS tables aren't directly related. Review the E-R models for the JustLee Books database in Figure 9-5. These models are helpful in planning join operations because they show table relationship lines.

In this example, the join operation needs to include four tables: CUSTOMERS, ORDERS, ORDERITEMS, and BOOKS. This task requires three join operations, as follows:

- Join CUSTOMERS to ORDERS based on Customer#
- Join ORDERS to ORDERITEMS based on Order#
- Join ORDERITEMS to BOOKS based on ISBN

TIP

The number of join operations needed is the number of tables in the query minus one.

Multiple join operations are included in a WHERE clause by using the AND logical operator, as shown in Figure 9-10. Although not all rows are shown in this figure, this query produces 32 rows of output.

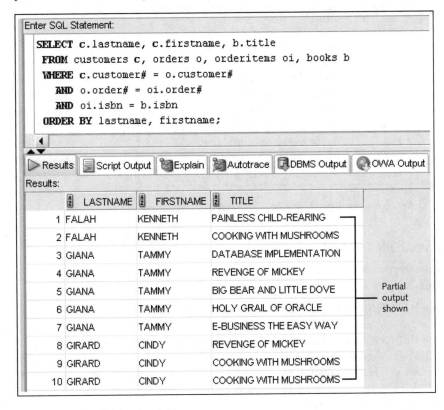

FIGURE 9-10 Joining four tables

Keep in mind that even though no columns from the ORDERS and ORDERITEMS tables are specifically selected for output, these tables are needed in the query to join tables together logically.

Regardless of how many join operations are required, other conditions can still be added in the WHERE clause. For example, Figure 9-11 shows the previous query modified to select books only in the Computer category. The results have been reduced from 32 rows to 10 rows of output.

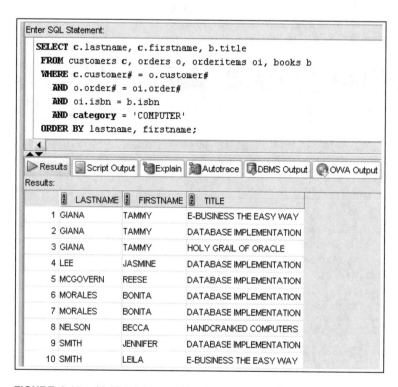

FIGURE 9-11   Multiple joins combined with a search condition

Equality Joins: JOIN Method

You can use three approaches to create an equality join that uses the JOIN keyword: NATU-RAL JOIN, JOIN ... USING, and JOIN ... ON:

- The **NATURAL JOIN** keywords create a join automatically between two tables, based on columns with matching names.
- The **USING** clause allows you to create joins based on a column that has the same name and definition in both tables.
- When the tables to be joined in a USING clause don't have a commonly named and defined field, you must add the **ON** clause to the JOIN keyword to specify how the tables are related.

The query in Figure 9-12 uses the NATURAL JOIN keywords to instruct Oracle 11g to list the title of each book in the BOOKS table—and the corresponding publisher ID number and publisher name.

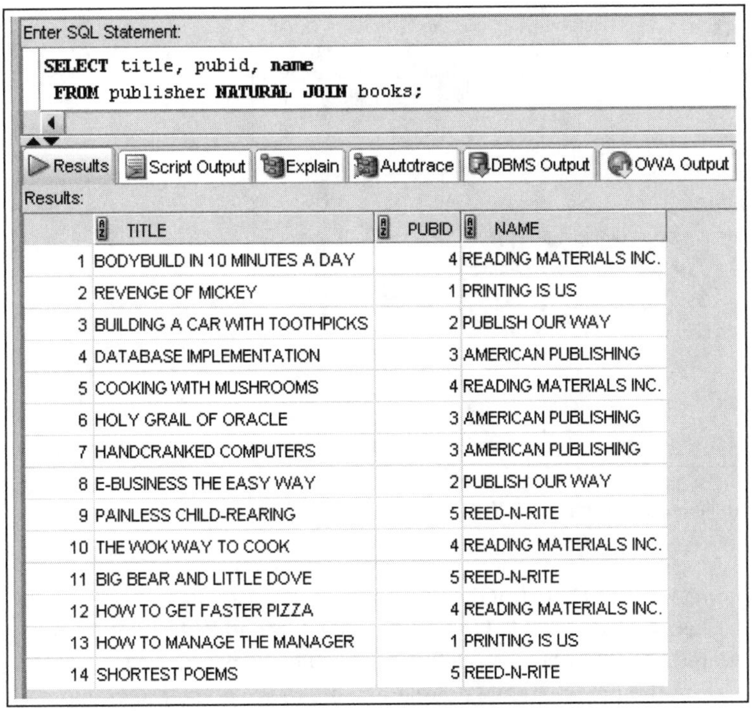

```
Enter SQL Statement:
  SELECT title, pubid, name
  FROM publisher NATURAL JOIN books;
```

	TITLE		PUBID		NAME
1	BODYBUILD IN 10 MINUTES A DAY		4		READING MATERIALS INC.
2	REVENGE OF MICKEY		1		PRINTING IS US
3	BUILDING A CAR WITH TOOTHPICKS		2		PUBLISH OUR WAY
4	DATABASE IMPLEMENTATION		3		AMERICAN PUBLISHING
5	COOKING WITH MUSHROOMS		4		READING MATERIALS INC.
6	HOLY GRAIL OF ORACLE		3		AMERICAN PUBLISHING
7	HANDCRANKED COMPUTERS		3		AMERICAN PUBLISHING
8	E-BUSINESS THE EASY WAY		2		PUBLISH OUR WAY
9	PAINLESS CHILD-REARING		5		REED-N-RITE
10	THE WOK WAY TO COOK		4		READING MATERIALS INC.
11	BIG BEAR AND LITTLE DOVE		5		REED-N-RITE
12	HOW TO GET FASTER PIZZA		4		READING MATERIALS INC.
13	HOW TO MANAGE THE MANAGER		1		PRINTING IS US
14	SHORTEST POEMS		5		REED-N-RITE

FIGURE 9-12   Using the NATURAL JOIN keywords

Because both the BOOKS and the PUBLISHER tables contain the Pubid column, this column is a common column and should be used to relate the two tables. When using the NATURAL JOIN keywords, you aren't required to specify columns the two tables have in common. The NATURAL keyword implies that the two specified tables have at least one column in common with the same name and contain the same datatype. Oracle 11g compares the two tables and uses the common columns to join the table.

Unlike the traditional method, you aren't allowed to use a qualifier for the column used to create the join. In essence, because the data value in a column is equivalent in both tables when the records match, it doesn't make sense to identify the column from only one of the tables. Therefore, Oracle 11g returns an error message if a qualifier is used on the join column anywhere in a SELECT statement including the NATURAL JOIN keywords, as shown in Figure 9-13.

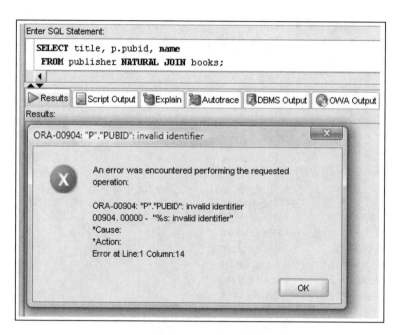

FIGURE 9-13 Column qualifier error with a NATURAL JOIN

Most developers avoid using a NATURAL JOIN because it can cause unexpected results. What if a column named "Description" is added to both the BOOKS and PUBLISHER tables but the columns aren't related to each other? A NATURAL JOIN attempts to use these columns in a join operation, even though they have no relationship. Developers prefer a join that explicitly specifies what columns should be used to join the rows. This join is accomplished by including a USING clause immediately after the FROM clause. Figure 9-14 reissues the previous query with the JOIN ... USING keywords. Notice that the query returns the same results as the NATURAL JOIN keywords.

N O T E

Because the statement in Figure 9-14 doesn't include an ORDER BY clause, your results might be listed in a different order for this query and other queries in the chapter.

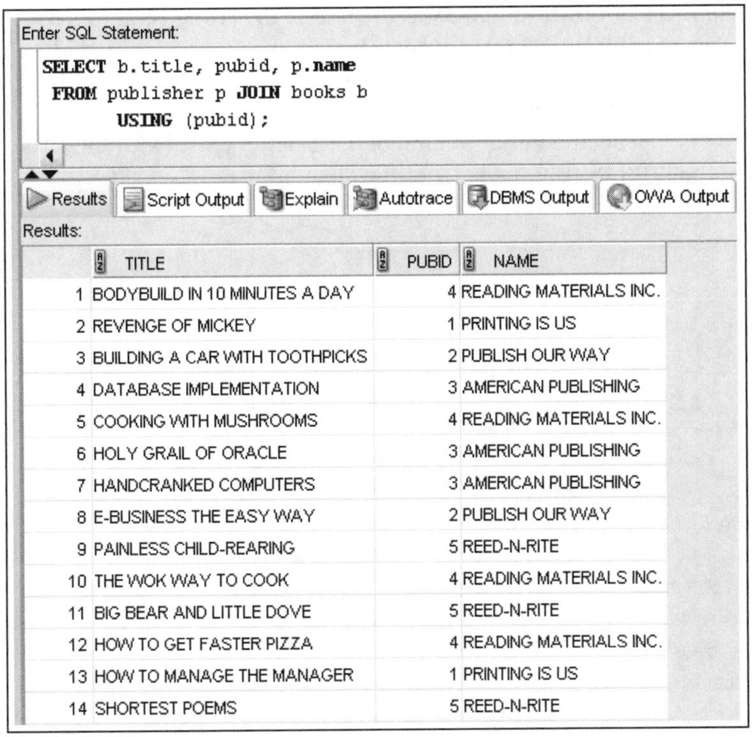

```
Enter SQL Statement:
SELECT b.title, pubid, p.name
  FROM publisher p JOIN books b
       USING (pubid);
```

Results | Script Output | Explain | Autotrace | DBMS Output | OWA Output

Results:

	TITLE	PUBID	NAME
1	BODYBUILD IN 10 MINUTES A DAY	4	READING MATERIALS INC.
2	REVENGE OF MICKEY	1	PRINTING IS US
3	BUILDING A CAR WITH TOOTHPICKS	2	PUBLISH OUR WAY
4	DATABASE IMPLEMENTATION	3	AMERICAN PUBLISHING
5	COOKING WITH MUSHROOMS	4	READING MATERIALS INC.
6	HOLY GRAIL OF ORACLE	3	AMERICAN PUBLISHING
7	HANDCRANKED COMPUTERS	3	AMERICAN PUBLISHING
8	E-BUSINESS THE EASY WAY	2	PUBLISH OUR WAY
9	PAINLESS CHILD-REARING	5	REED-N-RITE
10	THE WOK WAY TO COOK	4	READING MATERIALS INC.
11	BIG BEAR AND LITTLE DOVE	5	REED-N-RITE
12	HOW TO GET FASTER PIZZA	4	READING MATERIALS INC.
13	HOW TO MANAGE THE MANAGER	1	PRINTING IS US
14	SHORTEST POEMS	5	REED-N-RITE

FIGURE 9-14   Performing a join with the JOIN … USING keywords

As with the NATURAL JOIN keywords, a column referenced by a USING clause can't contain a column qualifier anywhere in the SELECT statement. In addition, the column referenced in the USING clause must be enclosed in parentheses. Qualifying all other columns in the SELECT clause, as shown in Figure 9-14, is still a good practice.

TIP

You might come across equijoin statements using the INNER JOIN keywords instead of just the JOIN keyword. If a join type isn't specified (CROSS, INNER, and so forth), the INNER type is used implicitly (by default). All equijoins in this textbook use just the JOIN keyword for brevity. As an example of using the INNER keyword, the statement shown in Figure 9-14 could be written as follows:

```
SELECT b.title, pubid, p.name
  FROM publisher p INNER JOIN books b
       USING (pubid);
```

The USING clause technique requires a commonly named column to perform the join. However, perhaps you have created tables with common columns but without a common name. When there are no commonly named columns to use in a join, you need to use the JOIN keyword in the FROM clause and add an ON clause immediately after the FROM clause to specify which fields are related.

To demonstrate this join, a PUBLISHER2 table was created in the JLDB_Build_9.sql script, which is the same as the PUBLISHER table except the Pubid column is named Id. Figure 9-15 uses the same example as in previous queries, listing the Title, Pubid (or Id), and Name columns from the BOOKS and PUBLISHER2 tables. The ON clause in this case instructs Oracle to join rows by using the Pubid column of the BOOKS table (with the alias "b") and the Id column of the PUBLISHER2 table (with the alias "p").

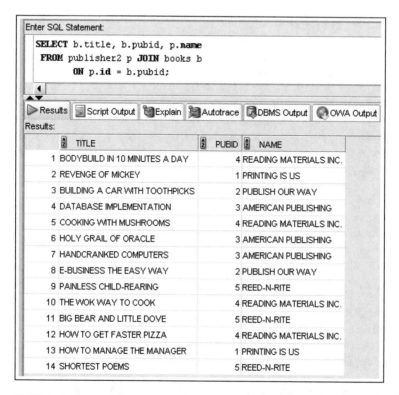

FIGURE 9-15   Performing a join with the JOIN … ON keywords

Column qualifiers are used in this example, but they aren't required because all field names referenced in the query are unique to a particular table. Even though the ON clause is required to join columns with different names, this technique can also be used instead of the USING clause when common column names exist. Again, with columns having the same name, there might be ambiguity when referencing columns with the JOIN … ON keywords, so Oracle 11g requires column qualifiers to avoid possible ambiguity. Using the ON clause or USING clause in a SELECT statement gives you the freedom to use the WHERE clause only for restricting rows to be included in the results, which can improve the readability of complex SELECT statements. Figure 9-16 shows the JOIN … ON approach to return the title, publisher ID, and publisher name for all books published by Publisher 4.

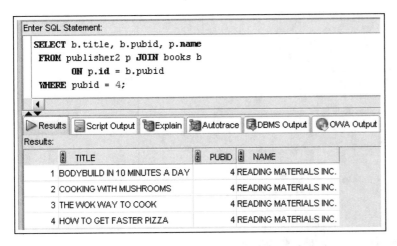

```
Enter SQL Statement:
  SELECT b.title, b.pubid, p.name
  FROM publisher2 p JOIN books b
       ON p.id = b.pubid
  WHERE pubid = 4;
```

	TITLE	PUBID	NAME
1	BODYBUILD IN 10 MINUTES A DAY	4	READING MATERIALS INC.
2	COOKING WITH MUSHROOMS	4	READING MATERIALS INC.
3	THE WOK WAY TO COOK	4	READING MATERIALS INC.
4	HOW TO GET FASTER PIZZA	4	READING MATERIALS INC.

FIGURE 9-16   The JOIN method reserves the WHERE clause for search conditions

There are two main differences between using the USING and ON clauses with the JOIN keyword:

- The USING clause can be used *only* if the tables being joined have a common column with the same name. This rule isn't a requirement for the ON clause.
- A condition is specified in the ON clause; this isn't allowed in the USING clause. The USING clause can contain only the name of the common column.

To see how to perform ANSI-compliant joins involving more than two tables, return to the request for a list of all customer names with all book titles each customer has purchased from the Computer category. As mentioned, to perform this join operation, you need to include four tables: CUSTOMERS, ORDERS, ORDERITEMS, and BOOKS. The query in Figure 9-17 shows the required join operation with the USING clause.

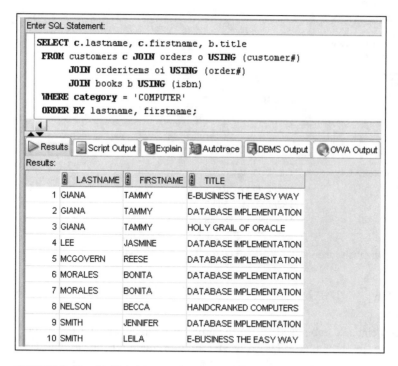

```
Enter SQL Statement:
SELECT c.lastname, c.firstname, b.title
 FROM customers c JOIN orders o USING (customer#)
      JOIN orderitems oi USING (order#)
      JOIN books b USING (isbn)
 WHERE category = 'COMPUTER'
 ORDER BY lastname, firstname;
```

Results | Script Output | Explain | Autotrace | DBMS Output | OWA Output

Results:

	LASTNAME	FIRSTNAME	TITLE
1	GIANA	TAMMY	E-BUSINESS THE EASY WAY
2	GIANA	TAMMY	DATABASE IMPLEMENTATION
3	GIANA	TAMMY	HOLY GRAIL OF ORACLE
4	LEE	JASMINE	DATABASE IMPLEMENTATION
5	MCGOVERN	REESE	DATABASE IMPLEMENTATION
6	MORALES	BONITA	DATABASE IMPLEMENTATION
7	MORALES	BONITA	DATABASE IMPLEMENTATION
8	NELSON	BECCA	HANDCRANKED COMPUTERS
9	SMITH	JENNIFER	DATABASE IMPLEMENTATION
10	SMITH	LEILA	E-BUSINESS THE EASY WAY

FIGURE 9-17　Multiple joins combined with a search condition

NON-EQUALITY JOINS

With an equality join, the data value of a record stored in the common column for the first table must match the data value in the second table. However, in many cases, there's no exact match. A **non-equality join** is used when the related columns can't be joined with an equal sign—meaning there are no equivalent rows in the tables to be joined.

For example, the shipping fee charged by many freight companies is based on the weight of the item being shipped. To use a database table to determine shipping fees, you could store every possible weight and its corresponding fee in a table. When an item is shipped, you could use an equality join to match the item's weight to the equivalent weight stored in the table and find the correct fee. However, most shipping fees are based on a scale, or range, of weights. For example, an item weighing between 3 and 5 pounds might have one fee, and an item weighing between 5 and 8 pounds might have another fee.

A non-equality join enables you to store a range's minimum value in one column of a record and the maximum value in another column. So instead of finding a column-to-column match, you can use a non-equality join to determine whether the item being shipped falls between minimum and maximum ranges in the columns. If the join does find a matching range for the item, the corresponding shipping fee can be returned in the results.

As with the traditional method of equality joins, a non-equality join can be performed in a WHERE clause. In addition, the JOIN keyword can be used with the ON clause to specify relevant columns for the join. First, take a look at creating a non-equality join with the WHERE clause, and then see how to use the JOIN ... ON approach.

Non-Equality Joins: Traditional Method

Once a year, JustLee Books offers a weeklong promotion in which customers receive a gift based on the value of each book purchased. If a customer purchases a book with a retail price of $12 or less, the customer receives a bookmark. If the retail price is more than $12 but less than or equal to $25, the customer receives a box of book-owner labels. For books retailing for more than $25 and less than or equal to $56, the customer is entitled to a free book cover. For books retailing for more than $56, the customer receives free shipping. Figure 9-18 shows the PROMOTION table.

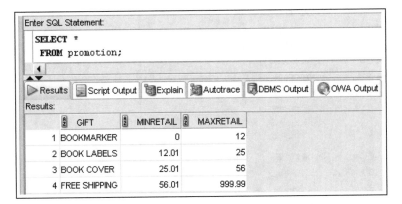

FIGURE 9-18   Contents of the PROMOTION table

Because the rows in the BOOKS and PROMOTION tables don't contain equivalent values, you must use a non-equality join to determine which gift a customer receives during the promotion, as shown in Figure 9-19. The BETWEEN operator is used in the WHERE clause to determine the range in which a book's retail price falls. Based on the range specified by the Minretail and Maxretail columns, the query determines which gift is appropriate for each purchase.

TIP

You might want to include additional columns in your queries to verify the results of join operations. For example, including the Retail, Minretail, and Maxretail columns in the statement in Figure 9-19 enables you to verify that the join operation correctly matches the book retail price with an associated gift value range.

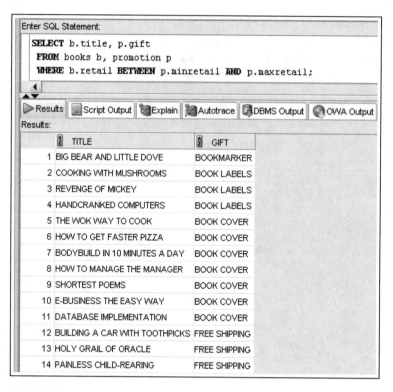

FIGURE 9-19 A traditional non-equality join

> **NOTE**
>
> If JustLee Books wanted to use the actual price a customer paid for a book, the Paideach column of the ORDERITEMS table would be used. The Retail column in the BOOKS table represents the base retail price of a book.

When you use a non-equality join to determine where a value falls within a range, you must make sure *no range values overlap*. If you select all records from the PROMOTION table to see the values stored in each field, you'll notice that the Minretail value in one row of the PROMOTION table doesn't equal the Maxretail value in another row. If any of the values did overlap, a customer could be returned twice in the results (and receive two gifts for a book rather than one!). You should check output from a non-equality join to make sure rows appearing more than once aren't the result of overlapping values in ranges.

Non-Equality Joins: JOIN Method

A non-equality join using the JOIN keyword has the same syntax as an equality join with the JOIN keyword. The only difference is that an equal sign isn't used to establish the relationship in the ON clause. Figure 9-20 uses the JOIN keyword to list the gift for each book.

Notice that the joining condition in the ON clause is the same as the one in the WHERE clause in Figure 9-19.

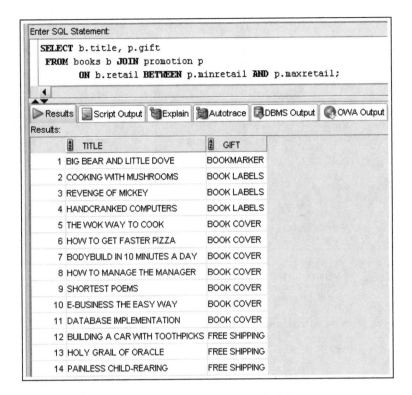

FIGURE 9-20 The JOIN method for a non-equality join

SELF-JOINS

Sometimes data in one column of a table has a relationship with another column in the same table. For example, customers who refer a new customer to JustLee Books receive a discount certificate for a future purchase. The Referred column of the CUSTOMERS table stores the customer number of the person who referred the new customer. If you need to determine the name of the customer who referred another customer, you face a challenge: The CUSTOMERS table serves as the master table for all customer information. Therefore, the Referred column in the CUSTOMERS table relates to other rows in the same table. To retrieve all the information you need, you must join a table to itself. In this case, you need to join the Referred and Customer# columns of the CUSTOMERS table, as shown in Figure 9-21. This type of join is known as a **self-join**.

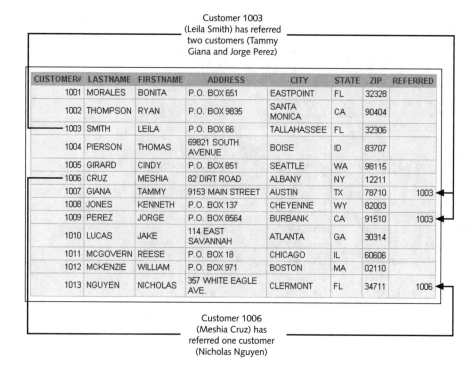

FIGURE 9-21 Two columns of the same table are related

Self-Joins: Traditional Method

To perform a self-join, you list the CUSTOMERS table twice in the FROM clause. However, you must make it appear as though the query is referencing two different tables. To do this, you assign each listing of the CUSTOMERS table a different table alias. In Figure 9-22, the table alias "c" identifies the table containing information for the new customer, and the table alias "r" identifies the table storing the person who referred the new customer. Because the table aliases are different, Oracle 11g operates as though two copies of the CUS-TOMERS table exist and can examine and match up different records in the same table while executing the query.

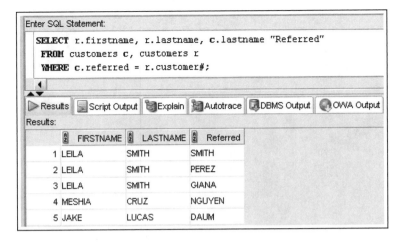

FIGURE 9-22 A self-join with the traditional method

In Figure 9-22's query, the Referred column alias is used to more easily distinguish between the customer making the referral and the customer who was referred. This output shows that customer Lela Smith referred three customers to JustLee Books.

TIP

If an error message is returned, make sure the CUSTOMERS table is listed twice in the FROM clause, each with a different table alias. Also, remember to precede each column in the WHERE clause with the table alias so that there are no ambiguity errors.

Self-Joins: JOIN Method

Regardless of the method used, the concept behind a self-join is the same—to use table aliases to make it appear that you're joining two different tables. To see how a self-join with the JOIN keyword works, use the same example as in the previous section. Figure 9-23 shows a self-join query with the JOIN keyword used to accomplish this task.

Again, the CUSTOMERS table is listed twice in the FROM clause, but each listing has a different table alias to mimic using two different tables. The columns used to relate the table's two occurrences are identified in the ON clause. Keep in mind that a USING clause can't be used with a self-join because two different column names are used in the join.

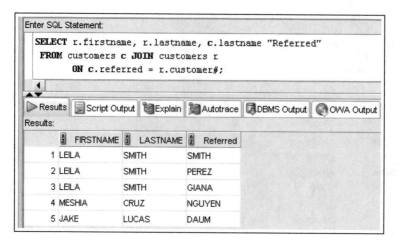

Enter SQL Statement:

```
SELECT r.firstname, r.lastname, c.lastname "Referred"
 FROM customers c JOIN customers r
      ON c.referred = r.customer#;
```

Results | Script Output | Explain | Autotrace | DBMS Output | OWA Output

Results:

	FIRSTNAME	LASTNAME	Referred
1	LEILA	SMITH	SMITH
2	LEILA	SMITH	PEREZ
3	LEILA	SMITH	GIANA
4	MESHIA	CRUZ	NGUYEN
5	JAKE	LUCAS	DAUM

FIGURE 9-23   A self-join with the JOIN ... ON keywords

OUTER JOINS

With the equality, non-equality, and self-joins you've used so far, a row is returned only if a corresponding record in each table is queried. These types of joins can be categorized as **inner joins** because records are listed in the results only if a match is found in each table. In fact, the default **INNER** keyword can be included with the JOIN keyword to specify that only records having a matching row in the corresponding table should be returned in the results.

However, suppose you want a list of *all* customers (not just ones who've placed an order) and order numbers for orders the customers have recently placed. (Recall that the CUSTOMERS table lists all customers who have ever placed an order, but the ORDERS table lists just the current month's orders and unfilled orders from previous months.) An inner join might not give you the exact results you want because some customers might not have placed a recent order. The query in Figure 9-24 produces an equality join that returns all order numbers stored in the ORDERS table and the name of the customer placing the order.

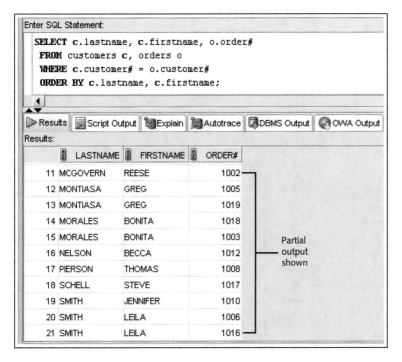

```
Enter SQL Statement:

SELECT c.lastname, c.firstname, o.order#
  FROM customers c, orders o
  WHERE c.customer# = o.customer#
  ORDER BY c.lastname, c.firstname;
```

	LASTNAME	FIRSTNAME	ORDER#
11	MCGOVERN	REESE	1002
12	MONTIASA	GREG	1005
13	MONTIASA	GREG	1019
14	MORALES	BONITA	1018
15	MORALES	BONITA	1003
16	NELSON	BECCA	1012
17	PIERSON	THOMAS	1008
18	SCHELL	STEVE	1017
19	SMITH	JENNIFER	1010
20	SMITH	LEILA	1006
21	SMITH	LEILA	1016

Partial output shown

FIGURE 9-24   An inner join omits nonmatching rows

Although this query identifies any customer who has placed an order stored in the ORDERS table, it does not list customers who haven't placed an order recently. For example, Ryan Thompson is listed in the CUSTOMERS tables but not in Figure 9-24's output. Because Ryan Thompson has no matching records in the ORDERS table, he's omitted from the query results. You want a list of all customers, however, so you need to change this query. To include records in the join results that exist in one table but don't have a matching row in the other table, you use an **outer join**. In essence, Oracle 11g joins the unmatched record to a NULL record in the other table. An outer join can be created by using the WHERE clause with an outer join operator or the OUTER JOIN keywords.

Outer Joins: Traditional Method

To tell Oracle 11g to create NULL rows for records that don't have a matching row, use an **outer join operator**, which looks like this: (+). It's placed in the WHERE clause immediately *after* the column name from the table that's missing the matching row and tells Oracle to create a NULL row in that table to join with the row in the other table.

Figure 9-25 corrects the problem in the customer query from Figure 9-24. It shows a partial list of all customers, and for those who have placed orders, it shows the corresponding order number. Note that Ryan Thompson is now included in the results, even though he hasn't placed a recent order.

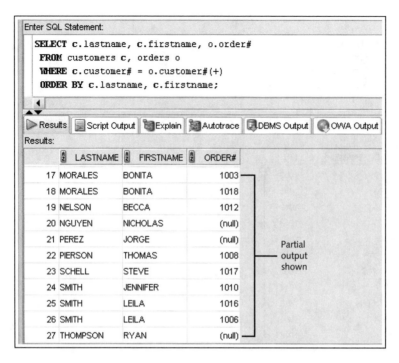

FIGURE 9-25 A traditional outer join with the (+) operator

If a customer in the CUSTOMERS table hasn't placed a recent order, the customer isn't listed in the ORDERS table; therefore, the ORDERS table is referred to as the "deficient table" (that is, the table with missing data). So for this query, the outer join operator is placed immediately after the reference to the deficient ORDERS table in the WHERE clause.

You need to be aware of some limitations when using the traditional approach to outer joins:

- The outer join operator can be used for only *one* table in the joining condition. In other words, you can't create NULL rows in both tables at the same time.
- A condition that includes the outer join operator can't use the IN or OR operator.

Outer Joins: JOIN Method

When creating a traditional outer join with the outer join operator, the join can be applied to only one table—not both. However, with the JOIN keyword, you can specify which table the join should be applied to by using a left, right, or full outer join. Left and right outer joins specify which table the outer join should be applied to, based on the table's location in the join condition. For example, a left outer join instructs Oracle to keep any rows in the table listed on the left side of the join condition, even if no matches are found with the table listed on the right. A full outer join keeps all rows from both tables in the results, no matter which table is deficient when matching rows. (That is, it performs a combination of left and right outer joins.)

By default, the JOIN keyword creates an inner join. To use it to create an outer join, you include the keyword **LEFT**, **RIGHT**, or **FULL** to identify the join type. You can also include the OUTER keyword; however, using it isn't necessary to specify an outer join. Figure 9-26, which shows partial results, re-creates the query in Figure 9-25 by using a left outer join. With a left outer join, if the table listed on the left side of the join has an unmatched record, it's matched with a NULL record and displayed in the results.

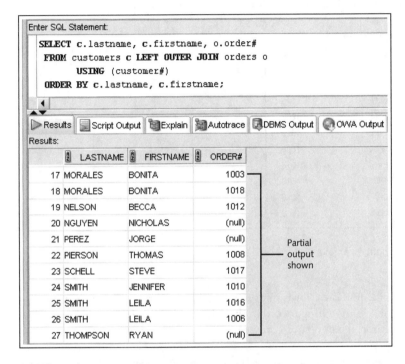

FIGURE 9-26   Using a left outer join

As shown in Figure 9-27's partial results, Oracle 11*g* interprets a right outer join to mean the results should include any order that doesn't have a match in the CUSTOMERS table. (This happens only if the customer record has been deleted from the CUSTOMERS table, but the order still exists in the ORDERS table.) Because a customer exists for every order that has been placed recently, no NULL rows are created. However, notice that customers who aren't listed in the ORDERS table are no longer displayed (Thompson, Perez, and so on) because the CUSTOMERS table is referenced in the *left* side of the joining condition, and this statement is executing a *right* outer join.

> **N O T E**
>
> Recall that the FOREIGN KEY constraint on the Customer# column of the CUSTOMERS and ORDERS table prevents deleting a customer who has an order record in the database.

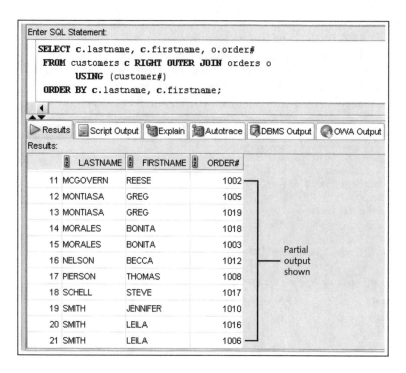

FIGURE 9-27 Using a right outer join

Substituting the FULL JOIN keywords instructs Oracle 11g to return records from either table that don't have a matching record in the other table. A full join can't be used with the outer join operator; it can be used only with the JOIN keyword.

NOTE

When joining three or more tables, keep in mind that each join operation, by default, omits nonmatching rows. To retain nonmatching rows, you might need an outer join operation for each join in the query.

SET OPERATORS

Set operators are used to combine the results of two (or more) SELECT statements. Valid set operators in Oracle 11g are UNION, UNION ALL, INTERSECT, and MINUS. When used with two SELECT statements, the **UNION** set operator returns the results of both queries. However, if there are any duplicates, they are removed, and the duplicated record is listed only once. To include duplicates in the results, use the **UNION ALL** set operator. **INTERSECT** lists only records that are returned by both queries; the **MINUS** set operator removes the second query's results from the output if they are also found in the first query's results. INTERSECT and MINUS set operations produce unduplicated results. Table 9-2 summarizes the set operators.

TABLE 9-2   List of Set Operators

Set Operator	Description
UNION	Returns the results of both queries and removes duplicates
UNION ALL	Returns the results of both queries but includes duplicates
INTERSECT	Returns only the rows included in the results of both queries
MINUS	Subtracts the second query's results if they're also returned in the first query's results

N O T E

Another set operator, EXISTS, is available and is discussed in Chapter 12.

Suppose you want a list of all author IDs with books in the Children or Family Life categories. As mentioned, the UNION set operator displays all rows returned by both queries. In Figure 9-28, authors with books in the Family Life category are retrieved in the first SELECT, and authors with books in the Children category are retrieved in the second SELECT. Because the UNION set operator combines the two SELECT statements, each book is listed only once, even if an author ID appears several times in the ORDERS table. For example, author R100 has a book in each category but is listed only once in the results.

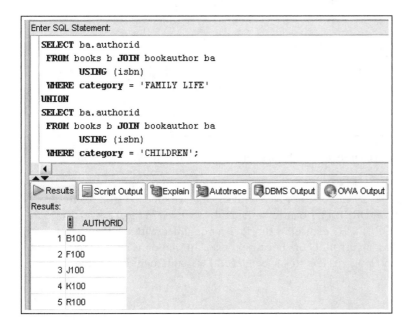

FIGURE 9-28   Producing an unduplicated combined list with the UNION set operator

NOTE

The UNION set operator was used to create full outer joins before the ANSI full outer join capability was available.

Unlike the UNION set operator, the UNION ALL set operator displays every row returned by the combined SELECT statements. In Figure 9-29's partial results, the query in Figure 9-28 is modified to use UNION ALL instead of UNION. As you can see, the author ID R100 is now listed twice, indicating that two rows total have been returned for this author. The UNION ALL set operator doesn't suppress duplicate rows, so author IDs are displayed more than once for authors who have published multiple books.

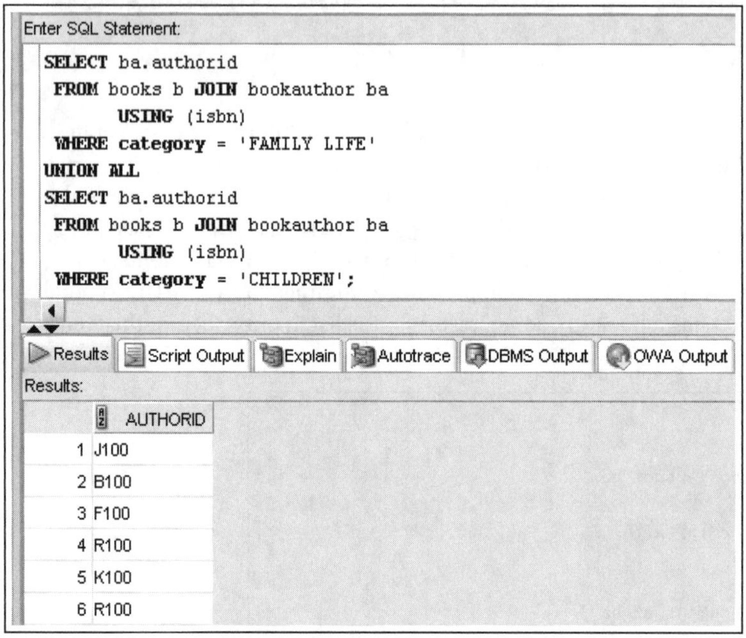
Enter SQL Statement:
SELECT ba.authorid
FROM books b JOIN bookauthor ba
USING (isbn)
WHERE category = 'FAMILY LIFE'
UNION ALL
SELECT ba.authorid
FROM books b JOIN bookauthor ba
USING (isbn)
WHERE category = 'CHILDREN';
Results
Script Output
Explain
Autotrace
DBMS Output
OWA Output
Results:
AUTHORID
1 J100
2 B100
3 F100
4 R100
5 K100
6 R100

FIGURE 9-29 Producing a combined list with duplication by using the UNION ALL set operator

The output in Figure 9-29 can be improved with sorting, which can be added to set operations by including an ORDER BY clause at the end, as shown in Figure 9-30. Notice that the column in the ORDER BY clause, Authorid, isn't qualified with a table reference. If this column *is* qualified, an error occurs because the sorting action is applied to the set operation's results and no longer refers to a specific SELECT statement in the query.

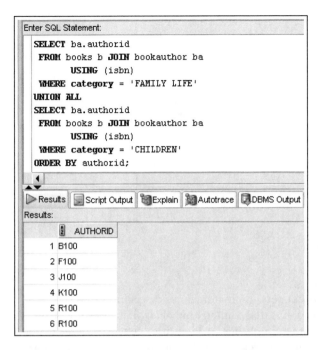

FIGURE 9-30 Add sorting to a set operation

So far, the UNION examples have included only queries involving a single column in the SELECT clause, but set operations can involve multiple-column queries. To see how this works, compare the data in the PUBLISHER and PUBLISHER3 tables, listed in Figures 9-31 and 9-32. First, the two tables use different column names for the publisher ID values: Pubid and Id. Second, the data rows in the two tables differ. For example, the PUBLISHER3 table contains two rows that don't exist in the PUBLISHER table (Id 6 and 7), one row that's the same in the PUBLISHER table (Id 2), and one row that exists in each table but contains a different name value to represent American Publishing (Id 3).

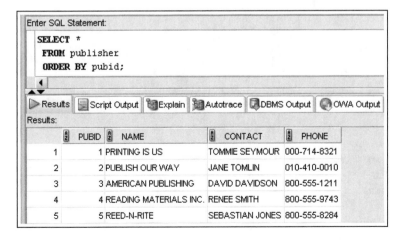

FIGURE 9-31 The PUBLISHER table

```
Enter SQL Statement:

SELECT *
FROM publisher3
ORDER BY id;
```

◄▲▼

▷ Results | Script Output | Explain | Autotrace | DBMS Output | OWA Output

Results:

	ID	NAME	CONTACT	PHONE
1	2	PUBLISH OUR WAY	JANE TOMLIN	010-410-0010
2	3	AMERICAN PUB	DAVID DAVIDSON	800-555-1211
3	6	PRINTING HERE	SAM HUNT	000-714-8321
4	7	PRINT THERE	CINDY TIKE	010-410-0010

FIGURE 9-32   The PUBLISHER3 table

Using these tables, test a UNION set operation that includes two columns, as shown in Figure 9-33. The results include an unduplicated combination of all publishers in both tables. However, two rows are listed for the publisher ID 3 because the two columns included in the UNION don't match. The ID number is the same in both rows; however, a different name is recorded in each row, so they aren't considered matching rows that should count as duplicates.

```
Enter SQL Statement:

SELECT pubid, name
 FROM publisher
UNION
SELECT id, name
 FROM publisher3;
```

◄▲▼

▷ Results | Script Output | Explain | Autotrace | DBMS Output | OWA Output

Results:

	PUBID	NAME
1	1	PRINTING IS US
2	2	PUBLISH OUR WAY
3	3	AMERICAN PUB
4	3	AMERICAN PUBLISHING
5	4	READING MATERIALS INC.
6	5	REED-N-RITE
7	6	PRINTING HERE
8	7	PRINT THERE

FIGURE 9-33   Multiple-column UNION set operation

Keep in mind some guidelines for multiple-column set operations:

- All columns are included to perform the set comparison.
- Each query must contain the same number of columns, which are compared positionally.
- Column names can be different in the queries.

The UNION set operation in Figure 9-33 ran successfully with different column names in position 1 of the SELECT clauses; this position references the publisher ID information. The results display the first column name provided; therefore, you see the column title Pubid in the output. If you want to sort on the first column, you could use Pubid or 1 in the ORDER BY clause. (Recall from Chapter 8 that using numbers in the ORDER BY clause simply references column position.)

You might not always need to combine data rows in queries. The query in Figure 9-34 contains two SELECT statements joined with the INTERSECT set operator. The first SELECT statement asks for all customer numbers in the CUSTOMERS table—basically, a list of all customers. The second SELECT statement lists all customer numbers for customers who have placed an order recently. By using the INTERSECT set operator to combine the two SELECT statements, you instruct Oracle 11g to list all customers who have placed an order recently *and* who exist in the CUSTOMERS table. In other words, only customers who are retrieved in both SELECTs should be in the results.

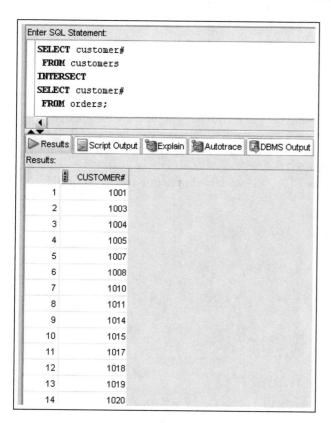

FIGURE 9-34 Identify overlapping values with the INTERSECT set operator

The query shown in Figure 9-35 requests a list of customer numbers for customers who are stored in the CUSTOMERS table but haven't placed an order recently. To do this, you use the MINUS set operator to remove customer numbers returned by the second SELECT statement (customers in the ORDERS table) from the results of the first SELECT statement (customers in the CUSTOMERS table).

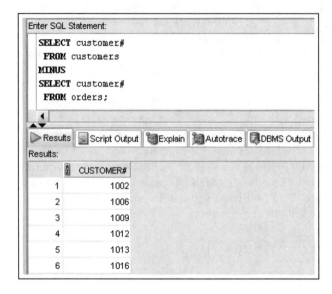

FIGURE 9-35   Subtract result sets with the MINUS set operator

In response to the first SELECT statement, 20 customer numbers are returned; however, only 14 of these customers have placed orders. Because you used the MINUS set operator, the 14 customers who placed orders are deleted from the results. As you can see in the output, six customers remain. They are the ones who haven't placed an order recently, meaning they're listed in the CUSTOMERS table but not in the ORDERS table.

Chapter Summary

- Data stored in multiple tables for a single entity can be linked by using joins.
- A Cartesian join between two tables returns every possible combination of rows from the tables. A Cartesian join can be produced by not including a join operation in the query or by using a CROSS JOIN.
- Broadly speaking, a join can be an inner join, in which the only records returned have a matching record in all tables, or an outer join, in which records can be returned regardless of whether there's a matching record in the join.
- Inner joins are categorized as equality, non-equality, or self-joins.
- An equality join is created when data joining records from two different tables is an exact match (that is, an equality condition creates the relationship). The traditional approach uses an equal sign as the comparison operator in the WHERE clause. The JOIN approach can use the NATURAL JOIN, JOIN ... USING, or JOIN ... ON keywords.
- Columns that exist in more than one table in a query must be qualified by the table name when a traditional join is used.
- Columns used to perform an ANSI JOIN can't be qualified with a table name.
- The NATURAL JOIN keywords don't require a condition to establish the relationship between two tables. However, a common column must exist. Column qualifiers can't be used with the NATURAL JOIN keywords.
- The JOIN ... USING approach is similar to the NATURAL JOIN approach, except the common column is specified in the USING clause. A condition can't be included in the USING clause to indicate how the tables are related. In addition, column qualifiers can't be used for the common column specified in the USING clause.
- The JOIN ... ON approach joins tables based on a specified condition. The JOIN keyword in the FROM clause indicates the tables to be joined, and the ON clause indicates how the two tables are related. This approach must be used if the tables being joined don't have a common column with the same name in each table.
- A non-equality join establishes a relationship based on anything other than an equal condition. Range values used with non-equality joins must be mutually exclusive.
- Self-joins are used when a table must be joined to itself to retrieve the data you need. Table aliases are required in the FROM clause to perform a self-join.
- An outer join is created when records need to be included in the results without having corresponding records in the join tables. These records are matched with NULL records so that they're included in the output.
- When using the WHERE clause to create an outer join, records from only one table can be matched to a NULL record. The outer join operator (+) is placed after the "deficient" table (the one that doesn't contain rows matching existing rows in the other table).
- With the JOIN method for outer joins, you can add the LEFT, RIGHT, or FULL keywords. A left outer join includes all records from the table listed on the left side of the join, even if no match is found with the other table in the join operation. A full outer join includes all records from both tables, even if no corresponding record in the other table is found.
- Set operators, such as UNION, UNION ALL, INTERSECT, and MINUS, can be used to combine the results of multiple queries.

Chapter 9 Syntax Summary

The following table summarizes the syntax and information you have learned in this chapter. You can use the table as a study guide and reference.

Syntax Guide

Element	Description	Example
WHERE clause	In the traditional approach, the WHERE clause indicates which columns should be used to join tables.	SELECT *columnname* [, . . .] FROM *tablename1,* *tablename2* WHERE *tablename1.* *columnname* *<comparison operator>* *tablename2.columnname;*
NATURAL JOIN keywords	These keywords are used in the FROM clause to join tables containing a common column with the same name and definition.	SELECT *columnname* [, . . .] FROM *tablename1* NATURAL JOIN *tablename2;*
JOIN ... USING keywords	The JOIN keyword is used in the FROM clause; combined with the USING clause, it identifies the common column used to join the tables. Normally, it's used if the tables have more than one commonly named column and only one is being used for the join.	SELECT *columnname* [, . . .] FROM *tablename1* JOIN *tablename2* USING (*columnname*) ;
JOIN ... ON keywords	The JOIN keyword is used in the FROM clause. The ON clause identifies the column used to join the tables.	SELECT *columnname* [, . . .] FROM *tablename1* JOIN *tablename2* ON *tablename1.* *columnname* *<comparison operator>* *tablename2.* *columnname;*
OUTER JOIN keywords Can be a RIGHT, LEFT, or FULL outer join	Indicates that at least one of the tables doesn't have a matching row in the other table.	SELECT *columnname* [, . . .] FROM *tablename1* [RIGHT|LEFT|FULL] OUTER JOIN *tablename2* ON *tablename1.* *columnname =* *tablename2.columnname;*

Type of Join	Traditional Method	JOIN Method

Cartesian Join
 	
Also known as a Cartesian product or cross join; matches each record in one table with each record in another table
	Example
	
	`SELECT title, name`
	`    FROM books, publisher;`
		Use keywords CROSS JOIN.
		
		Example
		`SELECT title, name`
		`    FROM books CROSS JOIN`
		`        publisher;`

Equality Join
 	
Also known as an equijoin, an inner join, or a simple join. Joins data in tables having equivalent data in a common column. With the traditional approach, you must qualify common columns with a table name.
	Use the keyword WHERE.
	
	Example
	`SELECT title,`
	`    books.pubid, name`
	`FROM publisher, books`
	`WHERE publisher.pubid =`
	`    books.pubid AND`
	`    publisher.pubid = 4;`
		Use keywords NATURAL JOIN or JOIN … USING to join tables having a commonly defined field.
		
		Use keywords JOIN … ON when tables don't have a commonly defined field. The column qualifier ON tells Oracle 11*g* how tables are related.
		
		Examples
		`SELECT title, pubid, name`
		`FROM publisher`
		`    NATURAL JOIN books;`
		
		`SELECT title, pubid, name`
		`    FROM publisher JOIN books`
		`        USING (pubid);`
		
		`SELECT title, name`
		`    FROM books b`
		`        JOIN publisher p`
		`        ON b.pubid = p.pubid;`

Non-Equality Join
 	
Joins tables when there are no equivalent rows in the tables to be joined. Often used to match values in one column with a range of values in another column. Can use any comparison operator *except* the equal sign (=).
	Use the keyword WHERE.
	
	Example
	`SELECT title, gift`
	`    FROM books, promotion`
	`    WHERE retail`
	`    BETWEEN minretail AND`
	`    maxretail;`
		Use keywords JOIN … ON and the same syntax as the JOIN method's equality join.
		
		Example
		`SELECT title, gift`
		`    FROM books JOIN promotion`
		`        ON retail BETWEEN`
		`        minretail AND`
		`        maxretail;`

321

Syntax Guide (continued)

Type of Join	Traditional Method	JOIN Method

Self-Join
 	
Joins a table to itself so that columns in the table can be joined. Must create table aliases.
	Use the keyword WHERE.
	
	Example
	`SELECT r.firstname,`
	`   r.lastname, c.lastname`
	`      "Referred"`
	`FROM customers c,`
	`   customers r`
	`WHERE c.referred =`
	`   r.customer#;`
		Use the keywords JOIN ... ON.
		
		Example
		`SELECT r.firstname,`
		`   r.lastname, c.lastname`
		`      "Referred"`
		`FROM customers c`
		`   JOIN customers r`
		`   ON c.referred =`
		`   r.customer#;`

Outer Join
 	
Includes a table's records in the output when there's no matching record in the other table.
	Use the keyword WHERE; use the outer join operator (+) to create NULL rows in the deficient table for records that don't have a matching row.
	
	Example
	`SELECT lastname,`
	`   firstname, order#`
	`FROM customers c,`
	`   orders o`
	`WHERE c.customer# =`
	`   o.customer#(+)`
	`ORDER BY c.customer#;`
		Include the keyword LEFT, RIGHT, or FULL with the OUTER JOIN keywords.
		
		Example
		`SELECT lastname, firstname,`
		`   order#`
		`FROM customers c LEFT OUTER`
		`   JOIN orders o`
		`   USING (customer#) ORDERBY`
		`      c.customer#;`

Set Operators
 	
These operators include UNION, UNION ALL, INTERSECT, and MINUS.
	Combine results of multiple SELECT statements.	`SELECT customer#`
		`   FROM customers`
		`UNION`
		`SELECT customer#`
		`   FROM orders;`

Review Questions

1. Explain the difference between an inner join and an outer join.
2. How many rows are returned in a Cartesian join between one table having 5 records and a second table having 10 records?
3. Describe problems you might encounter when using the NATURAL JOIN keywords to perform join operations.
4. Why are the NATURAL JOIN keywords not an option for producing a self-join? (*Hint*: Think about what happens if you use a table alias with the NATURAL JOIN keywords.)
5. What's the purpose of a column qualifier? When are you required to use one?

6. In an OUTER JOIN query, the outer join operator (+) is placed after which table?
7. What's the difference between the UNION and UNION ALL set operators?
8. How many join conditions are needed for a query that joins five tables?
9. What's the difference between an equality and a non-equality join?
10. What are the differences between the JOIN ... USING and JOIN ... ON approaches for joining tables?

Multiple Choice

To answer the following questions, refer to the tables in the JustLee Books database.

1. Which of the following queries creates a Cartesian join?

   a. `SELECT title, authorid`
      `FROM books, bookauthor;`

   b. `SELECT title, name`
      `FROM books CROSS JOIN publisher;`

   c. `SELECT title, gift`
      `FROM books NATURAL JOIN promotion;`

   d. all of the above

2. Which of the following operators is not allowed in an outer join?

   a. AND

   b. =

   c. OR

   d. >

3. Which of the following queries contains an equality join?

   a. `SELECT title, authorid`
      `FROM books, bookauthor`
      `WHERE books.isbn = bookauthor.isbn`
      `AND retail > 20;`

   b. `SELECT title, name`
      `FROM books CROSS JOIN publisher;`

   c. `SELECT title, gift`
      `FROM books, promotion`
      `WHERE retail >= minretail`
      `AND retail <= maxretail;`

   d. none of the above

4. Which of the following queries contains a non-equality join?

   a. `SELECT title, authorid`
      `FROM books, bookauthor`
      `WHERE books.isbn = bookauthor.isbn`
      `AND retail > 20;`

b. SELECT title, name
FROM books JOIN publisher
USING (pubid);

c. SELECT title, gift
FROM books, promotion
WHERE retail >= minretail
AND retail <= maxretail;

d. none of the above

5. The following SQL statement contains which type of join?

SELECT title, order#, quantity
FROM books FULL JOIN orderitems
ON books.isbn = orderitems.isbn;

a. equality

b. self-join

c. non-equality

d. outer join

6. Which of the following queries is valid?

a. SELECT b.title, b.retail, o.quantity
FROM books b NATURAL JOIN orders od
NATURAL JOIN orderitems o
WHERE od.order# = 1005;

b. SELECT b.title, b.retail, o.quantity
FROM books b, orders od, orderitems o
WHERE orders.order# = orderitems.order#
AND orderitems.isbn=books.isbn
AND od.order#=1005;

c. SELECT b.title, b.retail, o.quantity
FROM books b, orderitems o
WHERE o.isbn = b.isbn
AND o.order#=1005;

d. none of the above

7. Given the following query:

SELECT zip, order#
FROM customers NATURAL JOIN orders;

Which of the following queries is equivalent?

a. SELECT zip, order#
FROM customers JOIN orders
WHERE customers.customer# = orders.customer#;

b. SELECT zip, order#
FROM customers, orders
WHERE customers.customer# = orders.customer#;

c. SELECT zip, order#
   FROM customers, orders
     WHERE customers.customer# = orders.customer# (+);

d. none of the above

8. Which line in the following SQL statement raises an error?

  1. SELECT name, title
  2. FROM books NATURAL JOIN publisher
  3. WHERE category = 'FITNESS'
  4. OR
  5. books.pubid = 4;

a. line 1

b. line 2

c. line 3

d. line 4

e. line 5

9. Given the following query:

```
SELECT lastname, firstname, order#
  FROM customers LEFT OUTER JOIN orders
    USING (customer#)
    ORDER BY customer#;
```

Which of the following queries returns the same results?

a. SELECT lastname, firstname, order#
   FROM customers c OUTER JOIN orders o
    ON c.customer# = o.customer#
    ORDER BY c.customer#;

b. SELECT lastname, firstname, order#
   FROM orders o RIGHT OUTER JOIN customers c
    ON c.customer# = o.customer#
    ORDER BY c.customer#;

c. SELECT lastname, firstname, order#
   FROM customers c, orders o
    WHERE c.customer# = o.customer# (+)
    ORDER BY c.customer#;

d. none of the above

10. Given the following query:

```
SELECT DISTINCT zip, category
  FROM customers NATURAL JOIN orders NATURAL JOIN orderitems
    NATURAL JOIN books;
```

Which of the following queries is equivalent?

a.
```
SELECT zip FROM customers
   UNION
      SELECT category FROM books;
```

b.
```
SELECT DISTINCT zip, category
   FROM customers c, orders o, orderitems oi, books b
     WHERE c.customer# = o.customer# AND o.order# =
       oi.order#
     AND oi.isbn = b.isbn;
```

c.
```
SELECT DISTINCT zip, category
   FROM customers c JOIN orders o
     JOIN orderitems oi JOIN books b
     ON c.customer# = o.customer#
     AND o.order# = oi.order#
     AND oi.isbn = b.isbn;
```

d. all of the above

e. none of the above

11. Which line in the following SQL statement raises an error?

```
1.  SELECT name, title
2.  FROM books JOIN publisher
3.  WHERE books.pubid = publisher.pubid
4.  AND
5.  cost <45.95
```

a. line 1

b. line 2

c. line 3

d. line 4

e. line 5

12. Given the following query:

```
SELECT title, gift
  FROM books CROSS JOIN promotion;
```

Which of the following queries is equivalent?

a.
```
SELECT title, gift
   FROM books NATURAL JOIN promotion;
```

b.
```
SELECT title
   FROM books INTERSECT
     SELECT gift
     FROM promotion;
```

c. ```
    SELECT title
      FROM books UNION ALL
        SELECT gift
        FROM promotion;
    ```

d.  all of the above

13. If the CUSTOMERS table contains seven records and the ORDERS table has eight records, how many records does the following query produce?
    ```
    SELECT *
      FROM customers CROSS JOIN orders;
    ```

    a.  0

    b.  8

    c.  7

    d.  15

    e.  56

14. Which of the following SQL statements is not valid?

    a.  ```
        SELECT b.isbn, p.name
          FROM books b NATURAL JOIN publisher p;
        ```

    b. ```
        SELECT isbn, name
          FROM books b, publisher p
            WHERE b.pubid = p.pubid;
        ```

    c.  ```
        SELECT isbn, name
          FROM books b JOIN publisher p
            ON b.pubid = p.pubid;
        ```

    d. ```
        SELECT isbn, name
          FROM books JOIN publisher
            USING (pubid);
        ```

    e.  None—all the above are valid SQL statements.

15. Which of the following lists all books published by the publisher named Printing Is Us?

    a.  ```
        SELECT title
          FROM books NATURAL JOIN publisher
            WHERE name = 'PRINTING IS US';
        ```

    b. ```
        SELECT title
          FROM books, publisher
            WHERE pubname = 1;
        ```

    c.  ```
        SELECT *
          FROM books b, publisher p
            JOIN tables ON b.pubid = p.pubid
            WHERE name = 'PRINTING IS US';
        ```

    d. none of the above

16. Which of the following SQL statements is *not* valid?

    a.   `SELECT isbn`
           `FROM books`
            `MINUS`
            `SELECT isbn`
            `FROM orderitems;`

    b.   `SELECT isbn, name`
            `FROM books, publisher`
            `WHERE books.pubid (+) = publisher.pubid (+);`

    c.   `SELECT title, name`
            `FROM books NATURAL JOIN publisher`

    d.   All the above SQL statements are valid.

17. Which of the following statements about an outer join between two tables is true?

    a.   If the relationship between the tables is established with a WHERE clause, both tables can include the outer join operator.

    b.   To include unmatched records in the results, the record is paired with a NULL record in the deficient table.

    c.   The RIGHT, LEFT, and FULL keywords are equivalent.

    d.   all of the above

    e.   none of the above

18. Which line in the following SQL statement raises an error?

    1.   `SELECT name, title`
    2.   `FROM books b, publisher p`
    3.   `WHERE books.pubid = publisher.pubid`
    4.   `AND`
    5.   `(retail > 25 OR retail-cost > 18.95);`

    a.   line 1

    b.   line 3

    c.   line 4

    d.   line 5

19. What is the maximum number of characters allowed in a table alias?

    a.   10

    b.   30

    c.   255

    d.   256

20. Which of the following SQL statements is valid?

    a.   `SELECT books.title, orderitems.quantity`
            `FROM books b, orderitems o`
            `WHERE b.isbn= o.ibsn;`

b. `SELECT title, quantity`
    `FROM books b JOIN orderitems o;`

c. `SELECT books.title, orderitems.quantity`
    `FROM books JOIN orderitems`
    `ON books.isbn = orderitems.isbn;`

d. none of the above

Hands-On Assignments

To perform these assignments, refer to the tables in the JustLee Books database.

Generate and test two SQL queries for each of the following tasks: a) the SQL statement needed to perform the stated task with the traditional approach, and b) the SQL statement needed to perform the stated task with the JOIN keyword. Apply table aliases in all queries.

1. Create a list that displays the title of each book and the name and phone number of the contact at the publisher's office for reordering each book.

2. Determine which orders haven't yet shipped and the name of the customer who placed the order. Sort the results by the date on which the order was placed.

3. Produce a list of all customers who live in the state of Florida and have ordered books about computers.

4. Determine which books customer Jake Lucas has purchased. Perform the search using the customer name, not the customer number. If he has purchased multiple copies of the same book, unduplicate the results.

5. Determine the profit of each book sold to Jake Lucas, using the actual price the customer paid (not the book's regular retail price). Sort the results by order date. If more than one book was ordered, sort the results by profit amount in descending order. Perform the search using the customer name, not the customer number.

6. Which books were written by an author with the last name Adams? Perform the search using the author name.

7. What gift will a customer who orders the book *Shortest Poems* receive? Use the actual book retail value to determine the gift.

8. Identify the authors of the books Becca Nelson ordered. Perform the search using the customer name.

9. Display a list of *all* books in the BOOKS table. If a book has been ordered by a customer, also list the corresponding order number and the state in which the customer resides.

10. An EMPLOYEES table was added to the JustLee Books database to track employee information. Display a list of each employee's name, job title, and manager's name. Use column aliases to clearly identify employee and manager name values. Include all employees in the list and sort by manager name.

Advanced Challenge

To perform this activity, refer to the tables in the JustLee Books database.

The Marketing Department of JustLee Books is preparing for its annual sales promotion. Each customer who places an order during the promotion will receive a free gift with each book purchased. Each gift will be based on the book's retail price. JustLee Books also participates in co-op advertising programs with certain publishers. If the publisher's name is included in advertisements, JustLee Books is reimbursed a certain percentage of the advertisement costs. To determine the projected costs of this year's sales promotion, the Marketing Department needs the publisher's name, profit amount, and free gift description for each book in the JustLee Books inventory.

Also, the Marketing Department is analyzing books that don't sell. A list of ISBNs for all books with no sales recorded is needed. Use a set operation to complete this task.

Create a document that includes a synopsis of these requests, the necessary SQL statements, and the output requested by the Marketing Department.

Case Study: *City Jail*

Note: Use the City Jail database created with the CityJail_8.sql script that you ran for the Chapter 8 case study.

The following list reflects the current data requests from city managers. Provide the SQL statements that satisfy the requests. For each request, include one solution using the traditional method and one using an ANSI JOIN statement. Test the statements and show execution results.

1. List all criminals along with the crime charges filed. The report needs to include the criminal ID, name, crime code, and fine amount.

2. List all criminals along with crime status and appeal status (if applicable). The reports need to include the criminal ID, name, crime classification, date charged, appeal filing date, and appeal status. Show all criminals, regardless of whether they have filed an appeal.

3. List all criminals along with crime information. The report needs to include the criminal ID, name, crime classification, date charged, crime code, and fine amount. Include only crimes classified as "Other." Sort the list by criminal ID and date charged.

4. Create an alphabetical list of all criminals, including criminal ID, name, violent offender status, parole status, and any known aliases.

5. A table named Prob_Contact contains the required frequency of contact with a probation officer, based on the length of the probation period (the number of days assigned to probation). Review the data in this table, which indicates ranges for the number of days and applicable contact frequencies. Create a list containing each criminal who has been assigned a probation period, which is indicated by the sentence type. The list should contain the criminal name, probation start date, probation end date, and required frequency of contact. Sort the list by criminal name and probation start date.

6. A column named Mgr_ID has been added to the Prob_Officers table and contains the ID number of the probation supervisor for each officer. Produce a list showing each probation officer's name and his or her supervisor's name. Sort the list alphabetically by probation officer name.

SELECTED SINGLE-ROW FUNCTIONS

LEARNING OBJECTIVES

After completing this chapter, you should be able to do the following:

- Use the UPPER, LOWER, and INITCAP functions to change the letter case of field values and character strings
- Manipulate character substrings with the SUBSTR and INSTR function
- Nest functions inside other functions
- Determine the length of a character string with the LENGTH function
- Use the LPAD and RPAD functions to ensure that a string is a certain width
- Use the LTRIM and RTRIM functions to remove specific character strings
- Substitute character string values with the REPLACE and TRANSLATE functions
- Combine character strings with the CONCAT function
- Round and truncate numeric and date data with the ROUND and TRUNC functions
- Return only the remainder of a division operation with the MOD function
- Use the ABS function to set numeric values as positive
- Use the POWER function to raise a number to a specified power
- Calculate the number of months between two dates with the MONTHS_BETWEEN function
- Manipulate date data with the ADD_MONTHS, NEXT_DAY, LAST_DAY, and TO_DATE functions
- Differentiate between CURRENT_DATE and SYSDATE values
- Extend pattern-matching capabilities with regular expressions
- Identify and correct problems in calculations involving NULL values by using the NVL function
- Manipulate NULL values with the NVL2 and NULLIF functions
- Display dates and numbers in a specific format with the TO_CHAR function
- Perform condition processing similar to an IF statement with the DECODE function and CASE expression
- Use the SOUNDEX function to identify character phonetics
- Convert string values to numeric with the TO_NUMBER function
- Use the DUAL table to test functions

INTRODUCTION

In this chapter, you learn about single-row SQL functions. A **function** is a predefined block of code that accepts one or more **arguments**—values listed inside parentheses—and then returns a single value as output. The nature of an argument depends on the syntax of the function being executed. **Single-row functions** return one row of results for each record processed. By contrast, **multiple-row functions** return only one result per group or category of rows processed, such as counting the number of books published by each publisher. (Multiple-row functions are discussed in Chapter 11.)

Single-row functions range from converting letter case to calculating the number of months between two dates. The functions in this chapter have been grouped into character functions (case conversion and character manipulation functions), number functions, date functions, regular expressions, and other functions. Table 10-1 gives you an overview of these functions.

TABLE 10-1   Functions Covered in This Chapter

Type of Function	Functions
Case conversion functions	UPPER, LOWER, INITCAP
Character manipulation functions	SUBSTR, INSTR, LENGTH, LPAD/RPAD, LTRIM/RTRIM, REPLACE, TRANSLATE, CONCAT
Numeric functions	ROUND, TRUNC, MOD, ABS, POWER
Date functions	MONTHS_BETWEEN, ADD_MONTHS, NEXT_DAY, LAST_DAY, TO_DATE, ROUND, TRUNC, CURRENT_DATE
Regular expressions	REGEXP_LIKE, REGEXP_SUBSTR
Other functions	NVL, NVL2, NULLIF, TO_CHAR, DECODE, CASE expression, SOUNDEX, TO_NUMBER

Oracle 11g supports a wide variety of single-row functions. This chapter covers the most commonly used ones. However, you should review the Functions chapter in Oracle's SQL Reference to become familiar with the available functions. You can find all reference books in the documentation area of the Oracle Technology Network Web site.

DATABASE PREPARATION

Before attempting to work through the examples in this chapter, make sure you have completed the following tasks. First, if you haven't already run the JLDB_Build_8.sql script from Chapter 8, execute it to rebuild the JustLee Books database. Second, run the JLDB_Build_10.sql file in the Chapter 10 folder of your student data files to make the necessary additions to the JustLee Books database.

CASE CONVERSION FUNCTIONS

You can use **character functions** to change the case of characters (for example, to convert uppercase letters to lowercase letters) or to manipulate characters (such as substituting one character for another). Although most database administrators rarely need to use character functions, application developers often use them to create user-friendly database interfaces.

Case conversion functions alter the case of a character string. Used in a query, the case conversion is only temporary—it changes how Oracle 11g views data while executing a query, but it doesn't affect how data is stored. A case conversion function could be used in an INSERT statement to set the case of a stored value, for example. The case conversion functions Oracle 11g supports are LOWER, UPPER, and INITCAP.

The LOWER Function

The data in the JustLee Books database is stored in uppercase letters. However, when users perform data searches, they might enter character strings for search conditions in lowercase letters if they don't know the data is in uppercase letters. If this happens, Oracle doesn't return any rows because the data is in uppercase letters. You can solve this problem in several ways. One simple method is using the **LOWER** function, which converts character strings to lowercase letters. You can use the LOWER function to convert a table's data to lowercase letters temporarily during query execution, as shown in Figure 10-1.

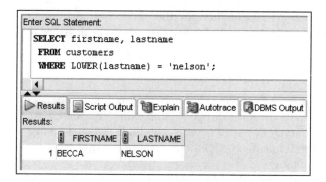

FIGURE 10-1   LOWER function in the WHERE clause

In Figure 10-1, you're searching for a customer with the last name Nelson. The syntax of the LOWER function is LOWER(c), where c is the field or character string to convert. In this case, you need data in the Lastname field converted to lowercase letters during the search. Therefore, the Lastname field is inserted between the parentheses after the function name.

Notice that the query results in Figure 10-1 are still in uppercase letters. Because the LOWER function isn't used in the SELECT clause, the first and last name are displayed in the same case in which they're stored. If you want data displayed in lowercase letters, simply include the LOWER function in the SELECT clause for each field to convert. Figure 10-2 shows the LOWER function in the SELECT and WHERE clauses.

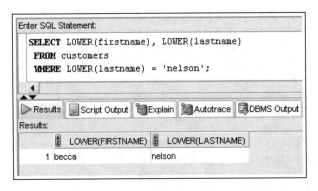

FIGURE 10-2 LOWER function in the SELECT and WHERE clauses

In Figure 10-2, the LOWER function in the SELECT clause makes Oracle display the query results in lowercase letters. However, you must still include the LOWER function in the WHERE clause because the character string `'nelson'` is, for comparison purposes, still entered in its lowercase form. In other words, when a function is used in a SELECT clause, it affects *only how data is displayed in the results*. By contrast, when a function is used in a WHERE clause, it's used only during the specified comparison operation.

Because the columns contain the results of the LOWER function, the column headings show the actual function used in the SELECT clause. Oracle 11g includes the function in the column heading to indicate that the data was manipulated or altered before being listed in the output. If you don't want the function name included in the column heading, simply add a column alias after the function in the SELECT clause, and the alias is displayed as the column header in the results.

The UPPER Function

The **UPPER** function converts characters into uppercase letters. This function can be used in the same way as the LOWER function to affect the display of characters (when used in a SELECT clause) and to modify the character case for a search condition (when used in a WHERE clause). The syntax of the UPPER function is UPPER(`c`), where `c` is the character string or field to convert into uppercase letters.

As discussed in Chapter 5, substitution variables allow using user input to complete an SQL statement. If the user input provides a value to use in a search condition, the UPPER function could be used to make sure this value is converted to uppercase letters before performing the comparison in the search condition. For example, the query shown in Figure 10-3 requests providing a last name value at execution. The UPPER function is included in the condition of the WHERE clause to instruct Oracle 11g to convert the user-entered character string `nelson` to uppercase letters while executing the SELECT query.

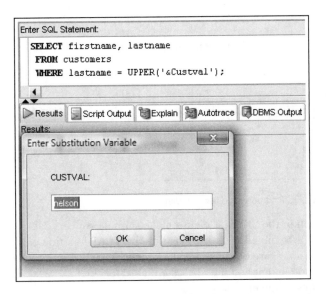

FIGURE 10-3   Using the UPPER function to manage user input

Converting the single search condition to the same case as data stored in the table is more efficient for the following reasons:

- You don't need to be concerned with whether the user knows which case to use when entering search strings.
- Oracle 11g doesn't need to convert the case of each row value in the field during query execution, which reduces the processing burden on the Oracle server.

The INITCAP Function

Although having table data and search criteria in the same case enables you to find records, the output might not look appealing. Generally, it's easier for people to read data in mixed-case letters rather than all uppercase or lowercase letters. Oracle 11g includes the **INITCAP** function to convert character strings to mixed case, with each word beginning with a capital letter—for example, *Great Mushroom Recipes*. The syntax of the INITCAP function is INITCAP(c), where c represents the field or character string to convert. The INITCAP function converts the first (INITial) letter of each word in the character string to uppercase (CAPital letter) and the remaining characters into lowercase, as shown in Figure 10-4.

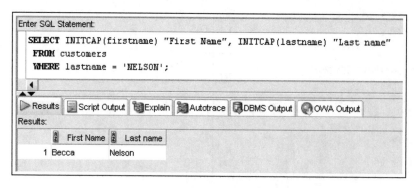

```
Enter SQL Statement:
    SELECT INITCAP(firstname) "First Name", INITCAP(lastname) "Last name"
    FROM customers
    WHERE lastname = 'NELSON';
```

> Results | Script Output | Explain | Autotrace | DBMS Output | OWA Output
Results:

First Name	Last name
1 Becca	Nelson

FIGURE 10-4 INITCAP function in a SELECT clause modifies the display

In this query, the INITCAP function in the SELECT clause makes Oracle convert data in the Firstname and Lastname columns to mixed case. Although the INITCAP function can also be used in a WHERE clause, doing so is rare because not all users are consistent in the case they use for entering data. Also, notice that column aliases are used for the headings of the Firstname and Lastname columns in the output.

CHARACTER MANIPULATION FUNCTIONS

Although most data that JustLee Books management needs is already stored in the correct form in the database, data might need to be manipulated to yield a different query output. At times you might need to determine a string's length, extract portions of a string, or reposition a string by using **character manipulation functions**. For example, to identify the state distribution center for an address, you need to extract the first three digits from the zip code. The following sections explain some commonly used manipulation functions.

The SUBSTR Function

You can use the **SUBSTR** function to return a **substring** (a portion of a string). Many organizations code data values such as inventory, benefits information, and general ledger accounts according to some type of coding scheme. For example, a benefits code might contain three characters, with the second character indicating the health plan the employee has selected. Similarly, the area code of a customer's telephone number indicates a region in a state.

One way to determine where a customer resides is to look at the first three numbers of the zip code. The United States Postal Service assigns the first three digits of the zip code to a geographical distribution area in each state. The Marketing Department can use this data to determine where to target certain promotional campaigns. For this purpose, the SUBSTR function can be used to extract the first three digits of the zip code stored for each customer.

The syntax of this function is SUBSTR(c, p, l), where c represents the character string, p represents the beginning character position for the extraction, and l represents the length of the string to return in the query results. As shown in Figure 10-5, the SELECT clause includes the DISTINCT keyword to eliminate duplication in the results. The arguments (zip, 1, 3) instruct Oracle to read the value in the Zip field, starting at the first character position, and extract three characters as the substring. Because the function is used in the SELECT clause, the extracted value is simply displayed.

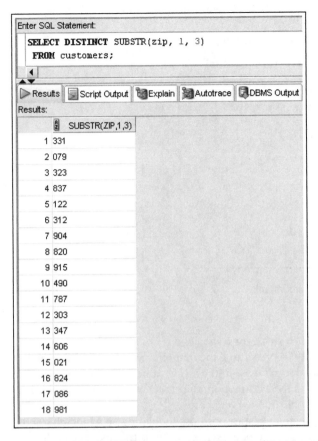

Enter SQL Statement:

```
SELECT DISTINCT SUBSTR(zip, 1, 3)
  FROM customers;
```

Results | Script Output | Explain | Autotrace | DBMS Output

Results:

	SUBSTR(ZIP,1,3)
1	331
2	079
3	323
4	837
5	122
6	312
7	904
8	820
9	915
10	490
11	787
12	303
13	347
14	606
15	021
16	824
17	086
18	981

FIGURE 10-5   SUBSTR function extracting part of the zip code

NOTE

If you're using SQL*Plus, the column header in Figure 10-5 might be truncated to show only a portion of the SUBSTR function. The header is truncated to the default width setting for displaying data, but you can modify this setting to display the entire header. Chapter 14 (in the online materials) covers formatting column widths in SQL*Plus.

The SUBSTR function can also extract substrings from the end of data stored in the field if negative position numbers are used. For example, if you enter -3 to indicate the beginning position, Oracle counts backward three positions from the end of the field and starts there to extract the substring. If you enter (zip, -3, 2) as arguments, the third and fourth digits of the zip code are returned.

Notice that in Figure 10-6, the first column of the results displays the entire zip code for each unique zip code stored in the CUSTOMERS table. The second column contains just the first three digits of the zip code, and the third column contains just the third and fourth digits of the zip code, as requested by the SUBSTR(zip, -3, 2) portion of the SELECT clause. The WHERE clause uses the SUBSTR function to filter the results, based on the value of the zip code's third and fourth digits.

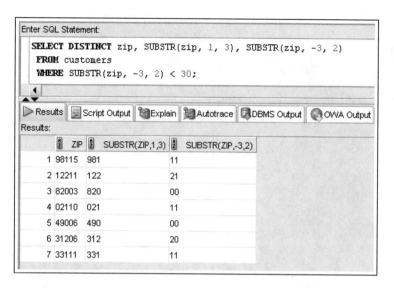

FIGURE 10-6 Comparison of SUBSTR arguments

The INSTR Function

The **INSTR** (instring) function searches a string for a specified set of characters or a substring, and then returns a numeric value representing the first character position in which the substring is found. If the substring doesn't exist in the string value, a 0 (zero) is returned. Two arguments must be provided to the INSTR function: the string value to search and the characters or substring (enclosed in single quotes) to locate.

Two optional arguments are also available: start position, indicating on which character of the string value the search should begin, and occurrence, which is the instance of the search value to locate (that is, first occurrence, second occurrence, and so on). By default, the search begins at the beginning of the string value, and the position of the first occurrence is located. Figure 10-7 shows using different arguments in the INSTR function.

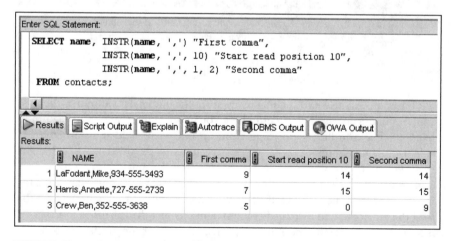

FIGURE 10-7 Comparison of INSTR arguments

The INSTR function is often used with the SUBSTR function. For example, in Figure 10-7, the INSTR function identifies the location of commas in the Name field. However, what if you want to extract the first and last name from the Name field? In this case, you can nest INSTR and SUBSTR to perform this task. **Nesting** simply means using one function as an argument inside another function. Figure 10-8 shows nesting these functions in a SELECT clause to extract the first name from the Name field.

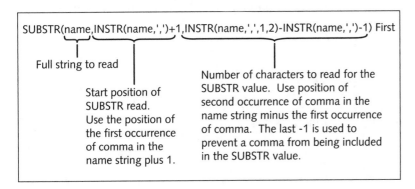

FIGURE 10-8 INSTR nested inside SUBSTR

Any single-row function can be nested inside other single-row functions. When nesting functions, you should remember the following important rules:

- All arguments required for each function must be provided.
- Every opening parenthesis must have a corresponding closing parenthesis.
- The nested, or inner, function is evaluated first. The inner function's result is then passed to the outer function, and the outer function is executed.

Figure 10-9 shows a SELECT statement that uses nested functions to extract the first and last name from the Name field.

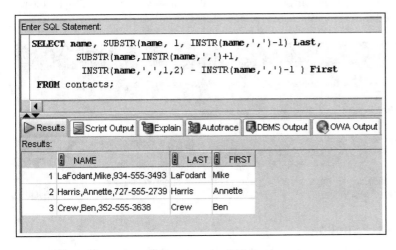

FIGURE 10-9 Using nested functions in a query

The LENGTH Function

When you plan the width of table columns, design text areas for forms, or determine the size of mailing labels, you might ask "What's the maximum number of characters that will be entered on this line?" For example, you're creating mailing labels and need labels wide enough to accommodate the longest mailing address. To determine the number of characters in a string, you can use the **LENGTH** function, shown in Figure 10-10. The syntax of the LENGTH function is LENGTH(c), where c represents the character string to analyze.

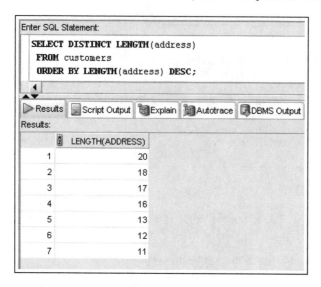

FIGURE 10-10   Checking data width with the LENGTH function

The (address) argument in Figure 10-10 determines the number of characters, or the length of data, in the Address field for each customer. The DISTINCT keyword eliminates duplicate values in the results. The sort in the ORDER BY clause places the highest length value at the top of the results. As the output shows, a mailing label accommodating at least 20 characters is required to send mail to current customers.

N O T E

Using the LENGTH function on a column with a CHAR datatype always returns the column's total width or size.

T I P

The LENGTH function is also used to determine whether a column might need resizing.

The LPAD and RPAD Functions

Have you ever received a check with the dollar amount preceded by a series of asterisks? Many companies fill in blank spaces on checks and forms with symbols to make it difficult for someone to alter the numbers. The **LPAD** function can be used to pad, or fill in, the area to the left of a character string with a specific character—or even a blank space.

The syntax of the LPAD function is LPAD(`c`, `l`, `s`), where `c` represents the character string to pad, `l` represents the length of the character string *after* padding, and `s` represents the symbol or character (enclosed in single quotes) to use as padding, as shown in Figure 10-11.

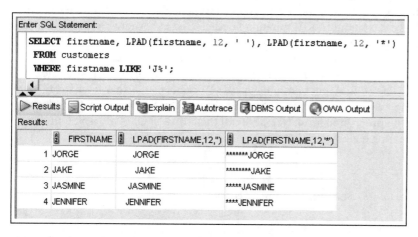

FIGURE 10-11   Using the LPAD function

N O T E

If you're using SQL*Plus, you need to set output to text format rather than the default HTML format. Click Preferences, System Configuration, Script Formatting, and set the Preformatted Output option to On.

In Figure 10-11, the LPAD function is used twice on the Firstname column, each time with different padding characters. The first LPAD function contains the arguments (firstname, 12, ' '), explained in the following list:

- The first argument, firstname, specifies padding the Firstname field.
- The second argument, 12, means that data in the Firstname column should be padded to a total length of 12 characters—the total length includes both data and the padding symbol.
- The third argument, ' ', is an instruction to use a blank space as the padding symbol. Because the LPAD function is used, blank spaces are added to the left of the customer's first name until the data's total length is 12. Notice that because blank spaces are used to left-pad the data, the customers' first names are right-aligned in the second column of output.

The only difference in the second LPAD function is that the padding symbol is set to an asterisk (*) rather than a blank.

Oracle 11g also provides the **RPAD** function, which uses a symbol to pad the right side of a character string to a specific width. The syntax of the RPAD function is

RPAD(*c*, *l*, *s*), where *c* represents the character string to pad, *l* represents the total length of the character string *after* padding, and *s* represents the symbol or character to use as padding.

The LTRIM and RTRIM Functions

You can use the **LTRIM** function to remove a specific string of characters from the left side of data values. The syntax of the LTRIM function is LTRIM(*c*, *s*), where *c* represents the field to modify, and *s* represents the string to remove from the left side of data.

For example, suppose the preprinted forms JustLee Books uses contain the string "P.O. BOX." However, some customers have "P.O. BOX" as part of their address in the CUSTOMERS table. The LTRIM function in Figure 10-12 removes the character string "P.O. BOX" from each customer's address before displaying it in the output so that it's not displayed twice on the preprinted form. Notice that if the string "P.O. BOX" isn't found in the address, no trimming occurs.

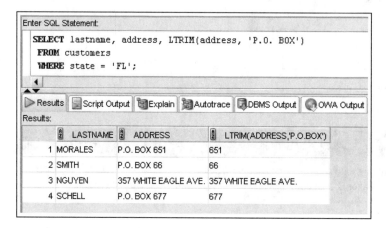

FIGURE 10-12 Using the LTRIM function

Oracle 11*g* also supports the **RTRIM** function to remove specific characters from the right side of data values. The syntax of the RTRIM function is RTRIM(*c*, *s*). The *c* represents the field to modify, and *s* represents the string to remove from the right side of data.

The REPLACE Function

The **REPLACE** function is similar to the "search and replace" function used in many programs. It looks for the occurrence of a specified string of characters and, if found, substitutes it with another set of characters. The syntax of the REPLACE function is REPLACE(*c*, *s*, *r*), where *c* represents the field to search, *s* represents the string of characters to find, and *r* represents the string of characters to substitute for *s*. In Figure 10-13, REPLACE(address, 'P.O.', 'POST OFFICE') indicates substituting the string POST OFFICE in the display every time Oracle 11*g* encounters the string P.O. in the Address field of a customer.

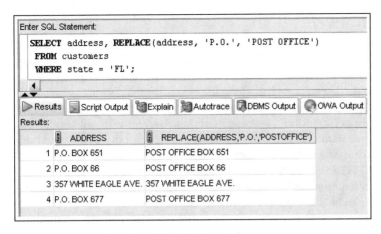

FIGURE 10-13 Using the REPLACE function

The TRANSLATE Function

The **TRANSLATE** function is used to replace a character in a string with a new value. It's different from the REPLACE function, in that it modifies single characters rather than a character string. In addition, you can make more than one substitution operation with a single use of the TRANSLATE function. This function has three arguments: the field to modify, the character to search for, and the character to substitute. Figure 10-14 shows two examples of using the TRANSLATE function.

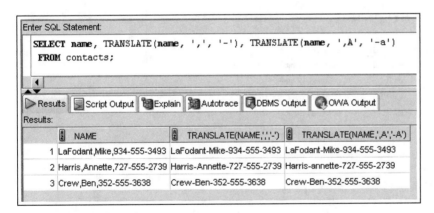

FIGURE 10-14 Using TRANSLATE to substitute character values

In the first use of TRANSLATE in Figure 10-14, there's only one character substitution. This operation searches the Name field for every comma and changes it to a hyphen. So if the string contains four commas, for example, each one is changed to a hyphen.

The second use of TRANSLATE includes two substitution operations. Notice that the character substitution arguments are listed in positional order: (',A', '-a'). The first search character is a comma, which should be replaced with a hyphen, as indicated by the first substitution character. The second search character, an uppercase "A," should be replaced with a lowercase "a," as specified by the second substitution character.

Selected Single-Row Functions

The CONCAT Function

You have learned how to use the concatenation operator (| |) to concatenate, or combine, data from columns with string literals. The **CONCAT** function can also be used to concatenate data from two columns. The main difference between the concatenation operator and the CONCAT function is that you can combine a long list of columns and string literals with the concatenation operator; by contrast, you can combine only *two items* (columns or string literals) with the CONCAT function. The concatenation operator is usually preferred because it's not limited to two items. If you need to combine more than two items with the CONCAT function, you must nest a CONCAT function inside another CONCAT function.

The syntax of the CONCAT function is CONCAT(*c1*, *c2*), where *c1* represents the first item to include in the concatenation, and *c2* represents the second item to include in the operation. Both *c1* and *c2* can be a column name or a string literal.

In Figure 10-15, the label "Customer number:" has been added before each customer number, so someone reading the output knows that the column is a customer number without having to look at the column heading. A column alias has also been added to identify the column's contents.

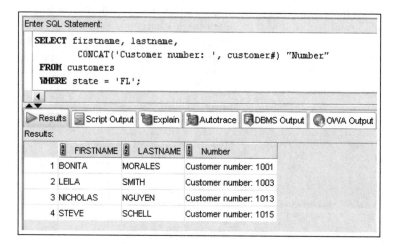

FIGURE 10-15 Using the CONCAT function

NUMBER FUNCTIONS

Oracle 11g provides a set of functions specifically designed for manipulating numeric data. In many organizations' daily operations, some of the most needed number functions are ROUND, TRUNC, MOD, ABS, and POWER.

The ROUND Function

The **ROUND** function is used to round numeric fields to the stated precision. The syntax of the ROUND function is ROUND(*n*, *p*), where *n* represents the numeric data, or field, to round, and *p* represents the position of the digit to which data should be rounded. If the value of *p* is a positive number, the function refers to the right side of the decimal. However, if a negative value is entered, Oracle 11g rounds to the left side of the decimal position.

In the third column of the results in Figure 10-16, each book's retail price has been rounded to the nearest tenth (or a single digit to the right of the decimal). As shown in the second column, the actual retail price of most books ends with .95. In Oracle 11*g* (as in most programs), values of 5 or more are rounded up, and values of less than 5 are rounded down. In Figure 10-16, the retail price for most books appears to be rounded to the nearest dollar. However, notice the book *The Wok Way to Cook*. Its retail price is $28.75, and when rounded to the nearest tenth, the price is $28.80 (28.8 in the results).

```
Enter SQL Statement:
  SELECT title, retail, ROUND(retail,1), ROUND(retail,0), ROUND(retail,-1)
  FROM books;
```

Results | Script Output | Explain | Autotrace | DBMS Output | OWA Output

Results:

	TITLE	RETAIL	ROUND(RETAIL,1)	ROUND(RETAIL,0)	ROUND(RETAIL,-1)
1	BODYBUILD IN 10 MINUTES A DAY	30.95	31	31	30
2	REVENGE OF MICKEY	22	22	22	20
3	BUILDING A CAR WITH TOOTHPICKS	59.95	60	60	60
4	DATABASE IMPLEMENTATION	55.95	56	56	60
5	COOKING WITH MUSHROOMS	19.95	20	20	20
6	HOLY GRAIL OF ORACLE	75.95	76	76	80
7	HANDCRANKED COMPUTERS	25	25	25	30
8	E-BUSINESS THE EASY WAY	54.5	54.5	55	50
9	PAINLESS CHILD-REARING	89.95	90	90	90
10	THE WOK WAY TO COOK	28.75	28.8	29	30
11	BIG BEAR AND LITTLE DOVE	8.95	9	9	10
12	HOW TO GET FASTER PIZZA	29.95	30	30	30
13	HOW TO MANAGE THE MANAGER	31.95	32	32	30
14	SHORTEST POEMS	39.95	40	40	40

FIGURE 10-16   Using the ROUND function to round numbers to various places

The fourth column in Figure 10-16 shows the retail price of books rounded to the nearest dollar by using ROUND(retail, 0). The 0 indicates that the retail price should be rounded to no decimal places. The last column in Figure 10-16 shows the results of rounding the retail price to the nearest tens of dollars, using -1 to indicate that the amount to the *left* of the decimal position should be rounded.

The TRUNC Function

At times you need to truncate, rather than round, numeric data. You can use the **TRUNC** (truncate) function to truncate a numeric value to a specific position. Any numbers after that position are simply removed (truncated). The syntax of the TRUNC function is TRUNC(n, p), where n represents the numeric data or field to truncate, and p represents the position of the digit where data should be truncated. As with the ROUND function, entering a positive value for p indicates a position to the right of the decimal, and a negative number indicates a position to the left of the decimal.

Figure 10-17 shows the TRUNC function using the same arguments as for the ROUND function. The third column of the output displays the results of `TRUNC(retail, 1)`. Compare these results with the results of the ROUND function. Unlike the ROUND function, the results of the TRUNC function don't depend on what value comes after the tenths' position; the TRUNC function simply drops any value beyond the first number after the decimal, without changing the first number's value.

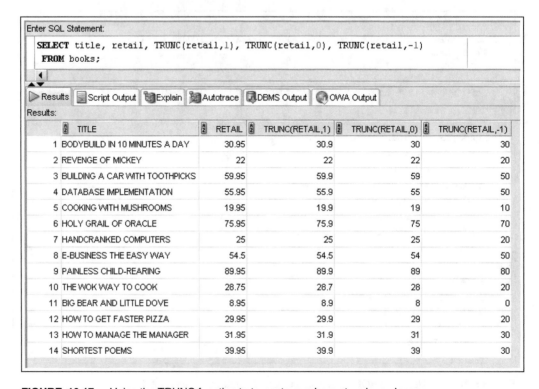

FIGURE 10-17   Using the TRUNC function to truncate numbers at various places

Again, refer to the book *The Wok Way to Cook*, with the retail price $28.75. After the retail price is truncated, the result is $28.7 instead of the $28.8 that resulted from rounding the price. The value after the tenths' position has no effect on the result—it's simply removed from the retail price.

When a zero is included as the second argument of the TRUNC function, no decimal positions are displayed for the retail price. However, remember that the dollar amount displayed isn't rounded; the decimals are simply dropped.

The MOD Function

The **MOD** (modulus) function returns only the remainder of a division operation. Two arguments are needed: the numerator and denominator of the division operation. Say you have an amount, 235, representing the total ounces of a liquid product, and you need to convert the amount to pounds and ounces. The first query in Figure 10-18 shows the result of a division operation (with 16 as the denominator because 16 ounces equal 1 pound), which is 14.6875 pounds. The remaining ounces over 14 pounds are in decimal form (.6875 of a pound). This decimal amount might be hard to figure out for someone who needs to know the weight in pounds and ounces to package the liquid for shipping. The second query uses the MOD function to capture the remainder 11, which represents the ounces over 14 pounds.

FIGURE 10-18   Using the MOD function to return the remainder

> **N O T E**
>
> Use the Run Script button to execute both queries in SQL Developer.

The ABS Function

The **ABS** (absolute) function returns the absolute, or positive, value of the numeric values supplied as the argument. For example, when subtracting two date values to determine the difference in number of days, you might have noticed that the result is negative if the more recent date is subtracted from the earlier date. Suppose you perform the operation 18-OCT-09 minus 20-OCT-09: The result is -2. If you're concerned only with the difference of two dates in number of days and don't want to be concerned with the order of dates in the calculation, use the ABS function to always return the absolute value of the result. The query in Figure 10-19 shows the effect of using the ABS function on a date calculation.

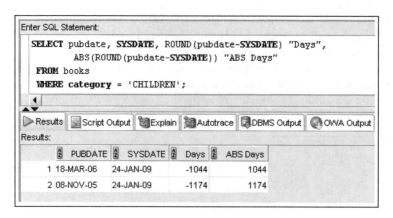

```
Enter SQL Statement:

SELECT pubdate, SYSDATE, ROUND(pubdate-SYSDATE) "Days",
       ABS(ROUND(pubdate-SYSDATE)) "ABS Days"
FROM books
WHERE category = 'CHILDREN';
```

	PUBDATE	SYSDATE	Days	ABS Days
1	18-MAR-06	24-JAN-09	-1044	1044
2	08-NOV-05	24-JAN-09	-1174	1174

FIGURE 10-19   The effect of using the ABS function

The POWER Function

The **POWER** function raises the number in first argument to the power indicated as the second argument. The syntax of the POWER function is POWER(x, y), where x represents the number you're raising, and y represents the power to which you're raising it. For example, POWER(2, 3) produces the result 8 (2 * 2 * 2). Many queries don't involve advanced mathematical operations; however, the POWER function is included in this chapter as a reminder that many mathematical functions are available. Scientific and financial applications are some examples in which data analysis might require advanced calculations.

DATE FUNCTIONS

By default, Oracle 11*g* displays date values in a DD-MON-YY format: a two-digit number for day of the month, a three-letter month abbreviation, and a two-digit number for the year. For example, the date February 2, 2009 is displayed as 02-FEB-09. Although users reference a date as a nonnumeric field (that is, a character string that must be enclosed in single quotation marks), it's actually stored in a numeric format that includes century, year, month, day, hours, minutes, and seconds. The valid range of dates that Oracle 11*g* can reference is January 1, 4712 B.C. to December 31, 9999 A.D.

NOTE

A default date format is set in the Oracle database configuration. The DBA can set a different date format in the configuration files, so some installations might not display dates in the default DD-MON-YY format.

Although dates appear as nonnumeric fields, users can perform calculations with dates because they're stored as numeric data. The numeric version of a date used by Oracle 11*g* is a **Julian date**, which represents the number of days that have passed between a specified date and January 1, 4712 B.C. For example, if you need to calculate the number of days between two dates, Oracle 11*g* first converts the dates to the Julian date numeric format, and then determines the difference between the two dates. If Oracle 11*g* didn't have a numeric equivalent for a date, trying to derive the solution for the arithmetic expression `'02-JAN-06' - '08-SEP-05'` would be a problem.

Look at the calculation with date columns in Figure 10-20. Order 1004 didn't ship the same day it was ordered. To determine how many days shipment was delayed, the Orderdate column is subtracted from the Shipdate column. In calculations between two date fields, the results are returned in number of days because dates are stored internally as numeric values.

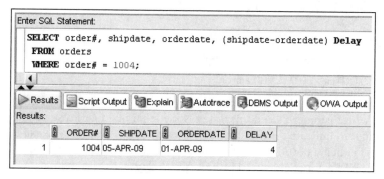

FIGURE 10-20   A calculation with date columns

By contrast, if you need the results returned in terms of weeks rather than days, simply divide the results by 7. For example, to have the results of the query in Figure 10-20 reported in terms of weeks (or portion of a week, in this case), the equation in the SELECT clause must be changed to `(shipdate-orderdate)/7`. However, converting date calculation results to a unit you need can be difficult sometimes. For example, suppose you need to know the number of months between two dates. Do you divide by 30 or by 31? What number should you use for February? As shown in the following sections, Oracle 11*g* provides several functions to assist with these calculations.

NOTE

The calculation shown in Figure 10-20 has an earlier date (Orderdate, 01-APR-09) subtracted from a more recent date (Shipdate, 05-APR-09), so the result is positive. If the order of the calculation is reversed (Orderdate-Shipdate), the result is the same number of days, except the resulting number is negative.

The MONTHS_BETWEEN Function

Suppose management wants to know whether customers' orders are higher for more recently released books or for books published several months (or even years) ago. To find the answer, you might simply subtract the publication date (Pubdate) from the order date (Orderdate) for a book to determine how many days a book was available to the public before it was ordered. However, as mentioned previously, what number should be used to convert days to months? Oracle 11*g* provides the **MONTHS_BETWEEN** function, shown in Figure 10-21, to determine the number of months between two dates. The syntax is `MONTHS_BETWEEN(d1, d2)`, where *d1* and *d2* are the two dates in question, and *d2* is subtracted from *d1*.

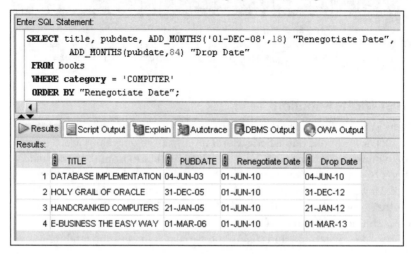

```
Enter SQL Statement:
SELECT title, MONTHS_BETWEEN(orderdate,pubdate) MTHS
  FROM books JOIN orderitems USING (isbn)
       JOIN orders USING (order#)
  WHERE order# = 1004;
```

Results | Script Output | Explain | Autotrace | DBMS Output | OWA Output
Results:

	TITLE	MTHS
1	PAINLESS CHILD-REARING	56.4838709677419354838709677419354838709

FIGURE 10-21　Using the MONTHS_BETWEEN function

In Figure 10-21, the user wants to determine how many months the book from order 1004 was available before the customer placed this order. Notice that the query results don't just provide a whole number to indicate the number of months that have elapsed. Instead, the digits after the decimal indicate portions of a month. To remove the portions of a month, include the MONTHS_BETWEEN function nested inside a TRUNC or ROUND function.

The ADD_MONTHS Function

The management of JustLee Books renegotiates contract pricing for books every 18 months and stocks books for up to 7 years after they're published. Management believes that after 7 years, sales for most books will decline to a level that makes keeping them in inventory no longer profitable. Therefore, management periodically requests a list of current books in inventory, along with the date each book contract should be renegotiated and the date the book should be dropped from inventory.

One method of calculating the "renegotiate date" is to use an average of 30 days for a month and add 540 (30 * 18) days to the most recent negotiation date of each book. However, an even better approach is using the **ADD_MONTHS** function, as shown in Figure 10-22. Assume that books in the Computer category were last negotiated on December 1, 2008.

```
Enter SQL Statement:
SELECT title, pubdate, ADD_MONTHS('01-DEC-08',18) "Renegotiate Date",
       ADD_MONTHS(pubdate,84) "Drop Date"
  FROM books
  WHERE category = 'COMPUTER'
  ORDER BY "Renegotiate Date";
```

Results | Script Output | Explain | Autotrace | DBMS Output | OWA Output
Results:

	TITLE	PUBDATE	Renegotiate Date	Drop Date
1	DATABASE IMPLEMENTATION	04-JUN-03	01-JUN-10	04-JUN-10
2	HOLY GRAIL OF ORACLE	31-DEC-05	01-JUN-10	31-DEC-12
3	HANDCRANKED COMPUTERS	21-JAN-05	01-JUN-10	21-JAN-12
4	E-BUSINESS THE EASY WAY	01-MAR-06	01-JUN-10	01-MAR-13

FIGURE 10-22　Using the ADD_MONTHS function

In Figure 10-22, the ADD_MONTHS function is applied to the most recent negotiation date and Pubdate column for each book to determine its "Renegotiate Date" and "Drop Date." The syntax of the ADD_MONTHS function is `ADD_MONTHS(d, m)`, where *d* represents the beginning date for the calculation, and *m* represents the number of months to add to the date. As shown, the result of the ADD_MONTHS function is a new date with the correct number of months added to the old date. The database system understands the calendar and can manipulate dates in terms of months accurately.

The NEXT_DAY and LAST_DAY Functions

JustLee Books has a policy that books must be shipped by the first Monday after receiving a customer's order. Whenever an order is received, the customer is informed of the latest date the order is expected to ship. The **NEXT_DAY** function, shown in Figure 10-23, can determine the next occurrence of a specific day of the week after a given date. The syntax of the NEXT_DAY function is `NEXT_DAY(d, DAY)`, where *d* represents the starting date, and *DAY* represents the day of the week to identify.

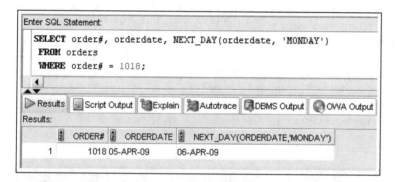

FIGURE 10-23 Using the NEXT_DAY function

In Figure 10-23, order 1018 was ordered on April 5, 2009. Because JustLee's policy is to inform a customer of the latest possible ship date, the query requests the date of the first Monday following the order date. As shown in the results, the customer can expect the order to be shipped by April 6—the first Monday after April 5.

NOTE

If the order date in this example happens to be on a Monday, the NEXT_DAY function returns the date for the following Monday, or one week away.

The **LAST_DAY** function is similar to the NEXT_DAY function, except it always determines the last day of the month for a given date. The example in Figure 10-24 retrieves the last day of the month for the month of two selected order dates.

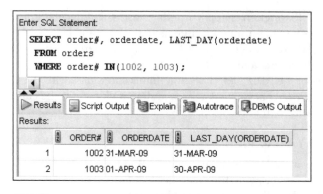

```
Enter SQL Statement:
  SELECT order#, orderdate, LAST_DAY(orderdate)
  FROM orders
  WHERE order# IN(1002, 1003);
```

	ORDER#	ORDERDATE	LAST_DAY(ORDERDATE)
1	1002	31-MAR-09	31-MAR-09
2	1003	01-APR-09	30-APR-09

FIGURE 10-24   Using the LAST_DAY function

The TO_DATE Function

The **TO_DATE** function is of particular interest to application developers. Many database users are uncomfortable entering a date in the default DD-MON-YY format because they're more accustomed to entering dates as MM/DD/YY or Month DD, YYYY. The TO_DATE function allows users to enter a date in any format, and then it converts the entry into the default format used by Oracle 11g. The syntax of the TO_DATE function is TO_DATE('d', 'f'), where d represents the date entered by the user, and f is the formatting instruction for the date. A **format argument** consists of a series of elements representing exactly what the data should look like and must be entered in single quotation marks. Table 10-2 shows valid formats for entering a date in the TO_DATE function.

TABLE 10-2   Format Arguments for Dates

Element	Description	Example
MONTH	Name of the month spelled out and padded with blank spaces to a total width of nine characters	APRIL
MON	Three-letter abbreviation for the name of the month	APR
MM	Two-digit numeric value for the month	04
RM	Roman numeral representing the month	IV
D	Numeric value for the day of the week	Wednesday = 4
DD	Numeric value for the day of the month	28
DDD	Numeric value for the day of the year	December 31 = 365
DAY	Name of the day of the week, padded with blank spaces to a length of nine characters	WEDNESDAY

TABLE 10-2   Format Arguments for Dates (continued)

Element	Description	Example
DY	Three-letter abbreviation for the day of the week	WED
YYYY	Displays the four-digit numeric value of the year	2009
YYY or YY or Y	The last three, two, or single digits of the year	2009 = 009; 2009 = 09; 2009 = 9
YEAR	Spelled-out version of the year	TWO THOUSAND NINE
B.C. or A.D.	Value indicating B.C. or A.D.	2009 A.D.

NOTE

For a list of additional format arguments, see the section "The TO_CHAR Function" later in this chapter.

When working with the TO_DATE function, you enter the specified date as the first argument. The second argument is a formatting instruction that allows Oracle 11g to distinguish different parts of the date. Because both arguments are character strings, each argument must be enclosed in single quotation marks.

Suppose you need a list of orders placed on March 31, 2009, and shipped in April, and you enter the order date in the format Month DD, YYYY. As shown in Figure 10-25, using the TO_DATE function in a WHERE clause enables you to enter the order date in the preferred format, and then include the format argument Oracle 11g needs to interpret the order date. Notice that literals, such as commas, are included in the format argument string.

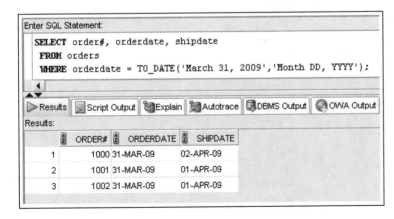

FIGURE 10-25   Using the TO_DATE function

Rounding Date Values

Even though the **ROUND** function is typically associated with numeric data, you can use it on date values as well. The syntax of the ROUND function is ROUND(*d*, *u*), where *d* represents the date data, or field, to round, and *u* represents the unit to use for rounding. A date can be rounded by the unit of month or year. Figure 10-26 shows each book's publish date rounded by month and year. For month rounding, if the day of the month is 16 or later, the date is rounded up to the first of the next month. If the day of the month is less than 16, the date is rounded down to the first of that month. Year rounding works similarly, with July 16th as the cutoff to determine rounding.

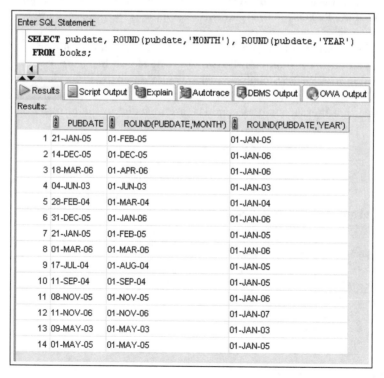

FIGURE 10-26   Rounding dates by month and year

Truncating Date Values

The **TRUNC** function, used for truncating numeric results, can also be quite useful in date calculations and is applied to dates by using the unit of month or year. The query shown previously in Figure 10-21 returned the number of months between the order date and the publication date of the book that was ordered. However, the output included portions of a month. To eliminate the decimal portion of the output and return only the number of whole months between the two dates, you can nest the MONTHS_BETWEEN function inside the TRUNC function, as shown in Figure 10-27.

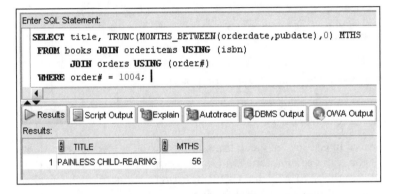

FIGURE 10-27   Using the TRUNC function on date calculation results

Remember that the TRUNC function has two arguments—the value to truncate and the position at which the truncation should occur. The value to truncate is the result of the MONTHS_BETWEEN function. Therefore, the MONTHS_BETWEEN function is entered as the first argument of the TRUNC function in Figure 10-27. The closing parenthesis after the Pubdate column name completes the MONTHS_BETWEEN function.

The nested MONTHS_BETWEEN function is followed by a comma, a zero, and a parenthesis, which serve to complete the TRUNC function. The zero indicates there should be no decimal positions after the truncation occurs, and the parenthesis closes the TRUNC function. The value calculated by the inner function (MONTHS_BETWEEN) is used to complete the outer function (TRUNC), and the result is the whole number of months between the book's publication date and order date.

CURRENT_DATE Versus SYSDATE

Many companies conduct transactions in different parts of the world, so time zone recognition is essential. Both the **CURRENT_DATE** and **SYSDATE** functions identify the current date and time, but you should know the differences between them. The SYSDATE function returns the current date and time set on the operating system where the database resides, but the CURRENT_DATE function returns the current date and time from the user session. A client software tool, such as SQL*Plus or SQL Developer, connecting to a database on a separate machine must be used in this example so that the session's time zone setting is different from the database's time zone setting. Because iSQL*Plus runs from the database server, it's not actually a client session and doesn't allow identifying a different client session time zone.

To demonstrate, the local computer is set to a different time zone from the database server (by resetting the time zone in Windows). Figure 10-28 shows an SQL Developer connection from the local computer (set to the Asia/Tokyo time zone) to the Oracle 11g server (set to the Eastern U.S./Canada time zone). The query verifies the time zone settings of the client (CURRENT_TIMESTAMP) and the database server (SYSTIMESTAMP).

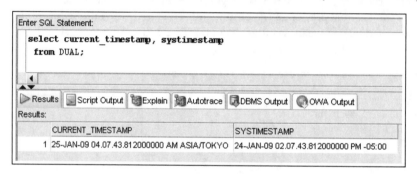

FIGURE 10-28   Identifying time zone information

The time zone is indicated in Greenwich mean time (also referred to as Universal Coordinated Time). Notice that the client session time is 14 hours ahead of the database time. The query in Figure 10-29 uses the SYSDATE and CURRENT_DATE values to highlight the difference in time for the client session and database server.

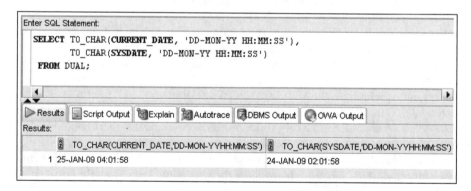

FIGURE 10-29   Retrieving session and database server times

NOTE

A TIMESTAMP datatype is available for storing time zone information as part of the date value in a column. The DATE datatype doesn't store time zone information.

REGULAR EXPRESSIONS

Oracle 11*g* recognizes **regular expressions**, which allow describing complex patterns in textual data. Although these tools have roots in the UNIX environment, they're used in many languages. Regular expressions adhere to Portable Operating System Interface for UNIX (POSIX) standards—a set of IEEE and ISO standards that define an interface between programs and operating systems.

You already know that simple pattern matching can be accomplished by using the LIKE operator. However, the LIKE operator is limited to the _ and % wildcard characters and must reference a whole string value instead of just a portion of it. Assume JustLee Books is researching the possibility of expanding operations into used book sales. A list of used book dealers has been purchased and loaded into a table named SUPPLIERS. Review the data in this table, shown in Figure 10-30.

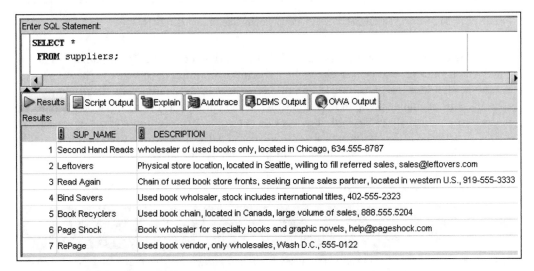

FIGURE 10-30 SUPPLIERS table data

Notice that the Description column might or might not contain a phone number. In addition, phone numbers are recorded in different formats, such as with hyphens or periods, and one phone number is missing an area code. What if JustLee Books wants to produce a list of only the vendors with a complete phone number recorded? The **REGEXP_LIKE** function handles this task, as it can reference an extended set of pattern operators available for regular expression operations. Figure 10-31 shows the REGEXP_LIKE function in a query to accomplish this task. Review the regular expression along with the pattern operators used as the second argument to the function.

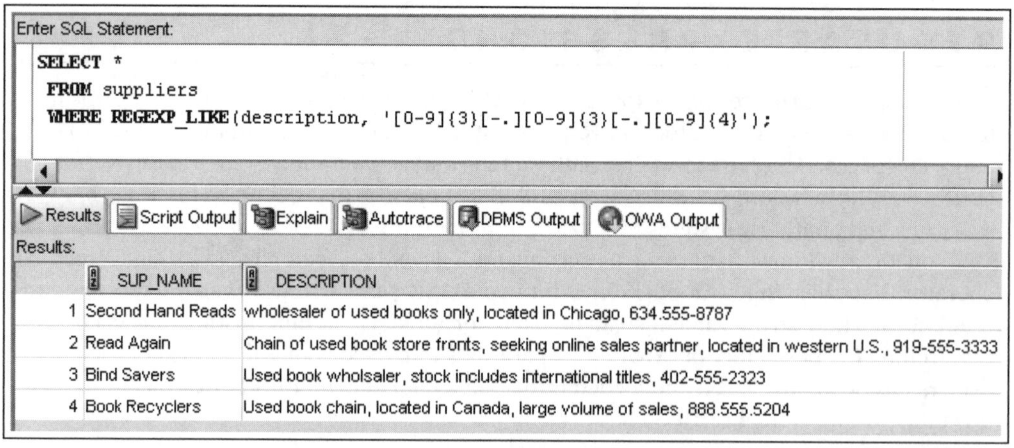

```
Enter SQL Statement:
    SELECT *
    FROM suppliers
    WHERE REGEXP_LIKE(description, '[0-9]{3}[-.][0-9]{3}[-.][0-9]{4}');
```

Results | Script Output | Explain | Autotrace | DBMS Output | OWA Output

Results:

	SUP_NAME	DESCRIPTION
1	Second Hand Reads	wholesaler of used books only, located in Chicago, 634.555-8787
2	Read Again	Chain of used book store fronts, seeking online sales partner, located in western U.S., 919-555-3333
3	Bind Savers	Used book wholsaler, stock includes international titles, 402-555-2323
4	Book Recyclers	Used book chain, located in Canada, large volume of sales, 888.555.5204

FIGURE 10-31   Using REGEXP_LIKE to identify complete phone numbers

Notice that the REGEXP_LIKE function has two arguments. The first argument indicates which value should be searched (the Description column, in this case). The second argument identifies the pattern it's attempting to locate in the search value (a 12-character string, in this example). Keep in mind that the search pattern needs to match only a part of the entire value being searched. It doesn't need to reflect the contents of the entire search string, as with the LIKE operator. Table 10-3 lists the pattern operators used in this example and describes what each one searches for in the 12-character string.

TABLE 10-3   Regular Expression Pattern Operators

Pattern Operator	Description
[0-9]{3}	The [0-9] operator indicates that it's looking for a single character that must be a digit. The {3} indicates repeating the previous operator three times. In this case, these two operators together search for a three-digit number.
[-.]	This operator indicates that the next character must be a hyphen (-) or a period.
[0-9]{3}	The next three places in the string must contain three digits.
[-.]	The next place in the string can be a hyphen (-) or period.
[0-9]{4}	The last four places in the string must contain four digits.

TIP

If you want a list of the suppliers who *don't* have a complete phone number recorded, you could use the NOT operator in the WHERE clause to identify these records.

What if JustLee Books wants a list of just supplier names and phone numbers instead of the entire description? You can use the **REGEXP_SUBSTR** function to extend the capabilities of the SUBSTR function, using the same pattern-matching operators. Figure 10-32 shows a SELECT clause modified to extract the phone number from the Description column by using REGEXP_SUBSTR.

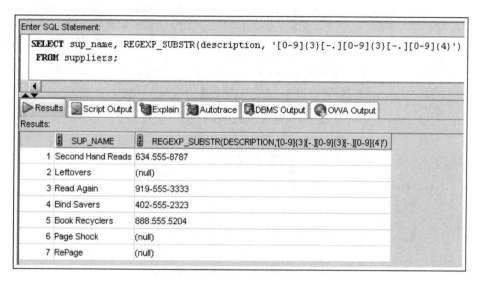

FIGURE 10-32   Using REGEXP_SUBSTR to extract phone numbers

This discussion simply introduces the concept of using regular expressions, which is an extensive topic. Many other pattern-matching operators are available and provide tremendously flexible tools for handling complex pattern searches. You can find more information on regular expressions in the SQL reference on the Oracle Technology Network Web site.

OTHER FUNCTIONS

Some functions provided by Oracle 11g don't fall neatly into a character, numeric, or date category. However, these functions are important and widely used in business. The following sections discuss these functions: NVL, NVL2, NULLIF, TO_CHAR, DECODE, CASE expression, SOUNDEX, and TO_NUMBER.

The NVL Function

You can use the **NVL** function to address problems caused when performing arithmetic operations with fields that might contain NULL values. (Recall that a NULL value is the absence of data, not a blank space or a zero.) When a NULL value is used in a calculation, the result is always a NULL value. The NVL function is used to substitute a value for the existing NULL so that the calculation can be completed. The syntax of the NVL function is NVL(x, y), where y represents the value to substitute if x is NULL. In many cases, the substitute for a NULL value in a calculation is zero (0).

In practice, the NVL function is commonly used to calculate an employee's gross pay as "salary + commission." What happens when an employee's sales aren't high enough to earn a commission, however? If you add the employee's salary to a NULL commission, the resulting gross pay is NULL—which means no paycheck! Unfortunately, you'll probably realize the error has occurred when the employee storms into your office, angry about the absence of a paycheck. To avoid this problem, instead of calculating gross pay as salary plus commission, use `salary+NVL(commission, 0)`. The NVL function simply substitutes a zero when the commission is NULL, and the person still gets paid because "salary + 0" still equals "salary."

Say you get a request to produce a list of all books along with the current sales price—the retail price less the discount amount. After reviewing the data, you discover some books don't have a discount amount, so this value is NULL. If you don't address the NULL value issue in the calculation, incorrect results are produced. Figure 10-33 shows a query using subtraction to calculate the current sales price.

Enter SQL Statement:

```
SELECT title, retail, discount, retail-discount "Sales price"
FROM books;
```

Results | Script Output | Explain | Autotrace | DBMS Output | OWA Output

Results:

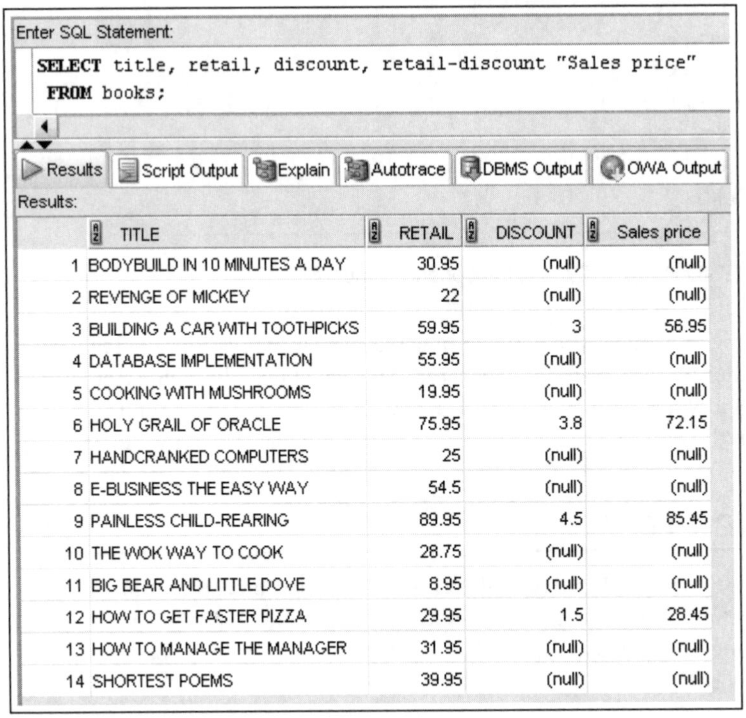

	TITLE	RETAIL	DISCOUNT	Sales price
1	BODYBUILD IN 10 MINUTES A DAY	30.95	(null)	(null)
2	REVENGE OF MICKEY	22	(null)	(null)
3	BUILDING A CAR WITH TOOTHPICKS	59.95	3	56.95
4	DATABASE IMPLEMENTATION	55.95	(null)	(null)
5	COOKING WITH MUSHROOMS	19.95	(null)	(null)
6	HOLY GRAIL OF ORACLE	75.95	3.8	72.15
7	HANDCRANKED COMPUTERS	25	(null)	(null)
8	E-BUSINESS THE EASY WAY	54.5	(null)	(null)
9	PAINLESS CHILD-REARING	89.95	4.5	85.45
10	THE WOK WAY TO COOK	28.75	(null)	(null)
11	BIG BEAR AND LITTLE DOVE	8.95	(null)	(null)
12	HOW TO GET FASTER PIZZA	29.95	1.5	28.45
13	HOW TO MANAGE THE MANAGER	31.95	(null)	(null)
14	SHORTEST POEMS	39.95	(null)	(null)

FIGURE 10-33 Calculation involving a NULL value

Notice that each book with a NULL discount value shows a NULL sales price as a result of the calculation. To solve this problem, use the NVL function, as shown in Figure 10-34, to modify the query and produce a sales price for every book.

```
Enter SQL Statement:

SELECT title, retail, discount, retail-NVL(discount,0) "Sales price"
 FROM books;
```

▶ Results   Script Output   Explain   Autotrace   DBMS Output   OWA Output

Results:

	TITLE		RETAIL		DISCOUNT		Sales price
1	BODYBUILD IN 10 MINUTES A DAY		30.95		(null)		30.95
2	REVENGE OF MICKEY		22		(null)		22
3	BUILDING A CAR WITH TOOTHPICKS		59.95		3		56.95
4	DATABASE IMPLEMENTATION		55.95		(null)		55.95
5	COOKING WITH MUSHROOMS		19.95		(null)		19.95
6	HOLY GRAIL OF ORACLE		75.95		3.8		72.15
7	HANDCRANKED COMPUTERS		25		(null)		25
8	E-BUSINESS THE EASY WAY		54.5		(null)		54.5
9	PAINLESS CHILD-REARING		89.95		4.5		85.45
10	THE WOK WAY TO COOK		28.75		(null)		28.75
11	BIG BEAR AND LITTLE DOVE		8.95		(null)		8.95
12	HOW TO GET FASTER PIZZA		29.95		1.5		28.45
13	HOW TO MANAGE THE MANAGER		31.95		(null)		31.95
14	SHORTEST POEMS		39.95		(null)		39.95

FIGURE 10-34   Handle NULL calculations with the NVL function

The NVL function isn't restricted to use with number values. Suppose management requests a report showing the number of days it took to ship each order placed on or after April 3, 2009. Some of these orders might not have shipped yet, but they're expected to ship on the following Monday. If the Shipdate column for an order is left blank (NULL) for pending shipments, the number of days to ship can't be calculated by simply subtracting Shipdate and Orderdate. However, an anticipated Shipdate could be included by using the NVL function for calculating the days. The query in Figure 10-35 returns the number of days to ship, using an anticipated shipping date of April 6, 2009 for any orders not yet shipped.

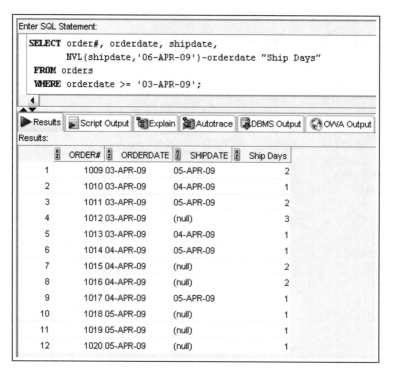

```
Enter SQL Statement:
SELECT order#, orderdate, shipdate,
       NVL(shipdate,'06-APR-09')-orderdate "Ship Days"
FROM orders
WHERE orderdate >= '03-APR-09';
```

Results | Script Output | Explain | Autotrace | DBMS Output | OWA Output

Results:

	ORDER#	ORDERDATE	SHIPDATE	Ship Days
1	1009	03-APR-09	05-APR-09	2
2	1010	03-APR-09	04-APR-09	1
3	1011	03-APR-09	05-APR-09	2
4	1012	03-APR-09	(null)	3
5	1013	03-APR-09	04-APR-09	1
6	1014	04-APR-09	05-APR-09	1
7	1015	04-APR-09	(null)	2
8	1016	04-APR-09	(null)	2
9	1017	04-APR-09	05-APR-09	1
10	1018	05-APR-09	(null)	1
11	1019	05-APR-09	(null)	1
12	1020	05-APR-09	(null)	1

FIGURE 10-35   Date calculations involving a NULL value

In Figure 10-35, the days between the Orderdate and Shipdate columns for each order are calculated. Because some orders haven't been shipped yet, the actual ship date is a NULL value, and the NVL function is used to substitute the anticipated shipping date of April 6, 2009 before completing the calculation. The NVL function instructs Oracle 11g that if the Shipdate column is NULL, substitute April 6, 2009 as the shipping date for the order, and then subtract the order date from the assigned shipment date. After the NVL function has made the substitution, the subtraction is performed and the number of days is calculated.

The NVL2 Function

The **NVL2** function is a variation of the NVL function with different options based on whether a NULL value exists. The syntax of the NVL2 function is NVL2(x, y, z), where y represents what should be substituted if x isn't NULL, and z represents what should be substituted if x is NULL. This variation gives you a little more flexibility when working with NULL values.

Returning to the gross pay calculation example, instead of using the equation salary+NVL(commission, 0) to substitute a zero when the commission is NULL, you could use NVL2(commission, salary, salary+commission). The NVL2 function is read as "If the commission IS NOT NULL, the gross pay is just salary. If the commission IS NULL, calculate gross pay as salary plus commission."

As another example, suppose management needs a report describing the shipment status of orders. The report should list an order as shipped or not shipped. You could use the NVL2 function to display each order's status, based on whether the Shipdate column contains a

NULL value. Because a date indicates that an order has been shipped and a NULL value indicates that an order hasn't shipped, it's the ideal situation for using the NVL2 function.

In Figure 10-36, the NVL2 function displays one of two character strings, depending on whether the Shipdate column contains a NULL value. Because character strings are used in the function, they must be enclosed in single quotation marks. If they aren't, Oracle 11g assumes an existing column is being referenced and returns an error because these columns don't exist in the table.

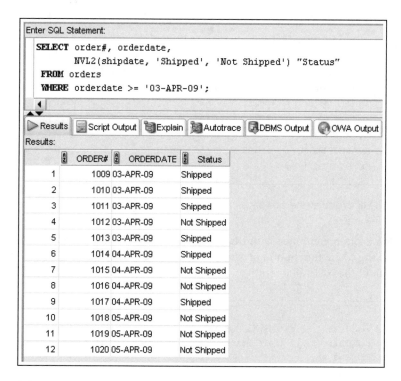

FIGURE 10-36   Using NVL2 to substitute values

The NULLIF Function

The **NULLIF** function is used to compare two values for equality. If the two values are equal, the function returns a NULL value. If the two values aren't equal, the function returns the first of the two values compared. For example, JustLee Books management might need to find out whether items in an order were purchased on sale or at the normal retail price. Figure 10-37 shows a query that compares the Paideach value of an order to the Retail value of the book. Notice that the NULLIF function value is NULL if the Paideach and Retail amounts are equal. Otherwise, the Paideach value is displayed. This query includes only orders 1001 and 1007.

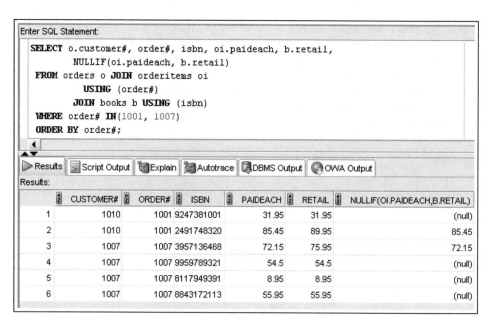

FIGURE 10-37 Using NULLIF to display the sale price

The NULLIF function is often combined with the NVL2 function to display a descriptive status. Figure 10-38 adds the NVL2 function to produce the value "Regular Price" or "Sale Price," based on the result of the NULLIF operation.

```
Enter SQL Statement:
    SELECT o.customer#, order#, isbn, oi.paideach, b.retail,
        NVL2(NULLIF(oi.paideach, b.retail), 'Sale Price', 'Regular Price') "Price"
    FROM orders o JOIN orderitems oi
            USING (order#)
            JOIN books b USING (isbn)
    WHERE order# IN(1001, 1007)
    ORDER BY order#;
```

Results | Script Output | Explain | Autotrace | DBMS Output | OWA Output
Results:

	CUSTOMER#	ORDER#	ISBN	PAIDEACH	RETAIL	Price
1	1010	1001	9247381001	31.95	31.95	Regular Price
2	1010	1001	2491748320	85.45	89.95	Sale Price
3	1007	1007	3957136468	72.15	75.95	Sale Price
4	1007	1007	9959789321	54.5	54.5	Regular Price
5	1007	1007	8117949391	8.95	8.95	Regular Price
6	1007	1007	8843172113	55.95	55.95	Regular Price

FIGURE 10-38 Using NVL2 with NULLIF

The TO_CHAR Function

The **TO_CHAR** function is often used to convert dates and numbers to a formatted character string. It's the opposite of the TO_DATE function for handling date data discussed previously. The TO_DATE function allows you to *enter* a date in any type of format and use the format argument to read the value as a date. The TO_CHAR function, on the other hand, is used to have Oracle 11g *display* dates in a particular format. The syntax of the TO_CHAR function is TO_CHAR(n, 'f'), where n is the date or number to format, and f is the formatting instruction to use. Figure 10-39 shows an example of formatting values with the TO_CHAR function.

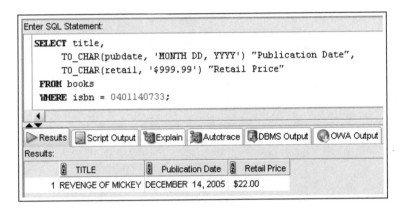

FIGURE 10-39 Formatting values for display with TO_CHAR

The query in Figure 10-39 contains two TO_CHAR functions. The first TO_CHAR is used to convert the publication date (Pubdate) to the date format MONTH DD, YYYY: month of the year spelled out, followed by the two-digit day of the month, a comma, and then the four-digit year. If you prefer displaying the month name in mixed case (that is, "December"), simply use this case in the format argument: ('Month DD, YYYY').

The second TO_CHAR function in Figure 10-39 is used to format the book's retail price (Retail column) to display a dollar sign and two decimal positions. If the format argument isn't used, the retail price is displayed as 22—without the dollar sign or any decimals.

Oracle 11g provides a wide variety of elements you can use to create format arguments for dates and numbers. Table 10-4 describes some commonly used format arguments. Notice that some of the same format arguments can be used with both the TO_DATE and TO_CHAR functions.

TABLE 10-4   Format Arguments

Element	Description	Example
Dates		
MONTH	Name of the month spelled out	APRIL
MON	Three-letter abbreviation for the name of the month	APR
MM	Two-digit numeric value for the month	04
RM	Roman numeral representing the month	IV
D	Numeric value for day of the week	4 (indicates Wednesday)
DD	Numeric value for day of the month	28
DDD	Numeric value for day of the year	365 (indicates December 31)
DAY	Name of day of the week	WEDNESDAY
DY	Three-letter abbreviation for day of the week	WED
YYYY	Four-digit numeric value for the year	2009
YYY or YY or Y	Numeric value for the last three, two, or single digit of the year	009, 09, or 9
YEAR	Spelled-out version of the year	TWO THOUSAND NINE
BC or AD	Value indicating B.C. or A.D.	2009 A.D.
Time		
SS	Seconds	Value between 0–59
SSSS	Seconds past midnight	Value between 0–86399
MI	Minutes	Value between 0–59
HH or HH12	Hours	Value between 1–12
HH24	Hours (for military time)	Value between 0–23
A.M. or P.M.	Value indicating morning or evening hours	A.M. (before noon) or P.M. (after noon)

TABLE 10-4   Format Arguments (continued)

Element	Description	Example
Numbers		
9	Indicates display width with a series of 9s but doesn't display insignificant leading zeros	99999
0	Displays insignificant leading zeros	00099999
$	Displays a floating dollar sign	$99999
.	Indicates the decimal position	999.99
,	Displays a comma in the indicated position	9,999
Other		
. , (punctuation symbols)	Displays the indicated punctuation	DD, YYYY = 24, 2009
"string"	Displays the exact characters inside the double quotation marks	"of the year" YYYY = of the year 2009
TH	Displays the ordinal number	DDTH = 8th
SP	Spells out the number	DDSP = EIGHT
SPTH	Spells out the ordinal number	DDSPTH = EIGHTH

NOTE

An RR format argument was used in the past to address potential problems raised by Y2K and storing only two-digit years. However, most people in the industry use the YYYY format argument to specify the exact century for a date.

The DECODE Function

The **DECODE** function takes a specified value and compares it to values in a list. If a match is found, the specified result is returned. If no match is found, a default result is returned. If no default result is defined, a NULL is returned. The DECODE function enables you to specify different actions to take, depending on the circumstances (for example, the exact value is or isn't contained in a column). It saves you from having to enter multiple statements for each possible situation.

The syntax of the DECODE function is DECODE(V, $L1$, $R1$, $L2$, $R2$, . . ., D), where V is the value you're searching for, $L1$ represents the first value in the list, $R1$ represents the result to return if *L1* and *V* are equivalent, and so on, and *D* is the default result to return if no match is found.

> **NOTE**
>
> The DECODE function is similar to the CASE or IF ... THEN ... ELSE structures used in many programming languages.

JustLee Books is required to collect sales tax from customers who live in Florida and California but not from customers in other states. Suppose Florida's sales tax is 7%, and California's is 8%. So if a customer resides in California, he or she must pay 8% of the total order price as sales tax; if the customer lives in Florida, he or she must pay 7% sales tax; if the customer lives in any other state, no sales tax is paid.

To determine the sales tax rate that applies to each customer, you can use the DECODE function to compare the state where each customer lives to a list of states. If a match occurs, the sales tax rate for that state is returned. However, if the customer lives in a state that isn't listed, a default sales tax rate of 0 is applied, as shown in Figure 10-40.

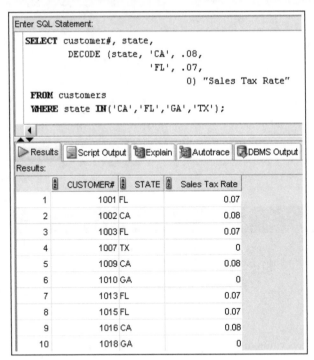

FIGURE 10-40   Using DECODE to determine sales tax rate by state

In Figure 10-40, the DECODE function begins on line 2. The State column is identified as the value to compare against the list. The state of California (CA) is the first item against which the value of the State column is compared. If the State column contains the value CA, a sales tax rate of .08 is returned, and the DECODE function is processed again for the next customer. If the value in the State column isn't CA, the value is compared against the next item in the list. (Note that the second listed item has been placed on line 3 to improve readability of the function.) If the value for the State column is equal to FL, a sales tax rate of .07 is returned. If the value in the State column isn't equal to the two items listed (CA or FL), the default value is assigned, which in this case is zero.

The CASE Expression

The **CASE** expression is similar to the DECODE function, in that it can perform IF … THEN … ELSE conditional processing. Both the simple CASE expression and the DECODE function evaluate equality conditions. However, the **searched CASE** expression gives you more flexibility because it allows other comparisons besides equality. For example, you might need to determine the retirement level assigned to each employee, based on the number of years employed at JustLee Books. The retirement levels are determined by using ranges of years. The Level 1 retirement category includes employees employed less than 4 years, Level 2 is less than 8 years, and so forth. The searched CASE expression in Figure 10-41 determines the retirement category for each employee based on years employed.

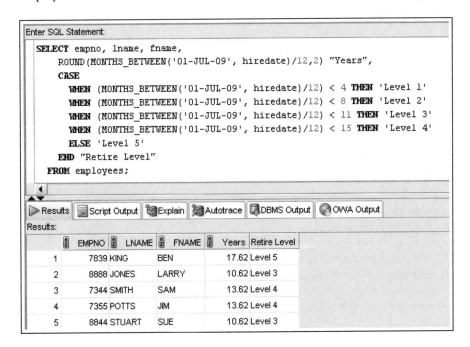

FIGURE 10-41 Using a searched CASE expression

The SOUNDEX Function

Some government agencies and organizations perform searches for information based on the phonetic pronunciation of words instead of their spelling. For example, in many states, the first four characters and numbers of a driver's license number represent the phonetic sound of the person's last name at the time the license was issued (but doesn't change if the person's last name changes). Oracle 11g can reference the phonetic sound or representation of words with the **SOUNDEX** function. The syntax of the function is SOUNDEX(c), where c is the character string being referenced.

Say that a customer has called to make an inquiry about an order. She doesn't have the order number, so you enter a search based on her last name, which is Smyth. No records are returned because no customers with this last name are recorded in the database. What do you do next? A secondary search with the SOUNDEX function can return any names recorded with a pronunciation similar to Smyth, as shown in Figure 10-42. Notice that two possible matches are found, both with the last name Smith. In this case, an employee mistakenly entered the last name with an "i" rather than a "y."

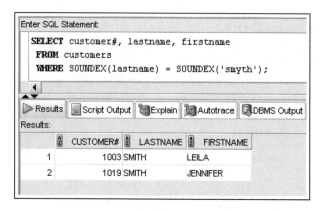

FIGURE 10-42   Using the SOUNDEX function

The TO_NUMBER Function

The **TO_NUMBER** function converts a value to a numeric datatype, if possible. For example, the string value 2009 stored in a date or character string could be converted to a numeric datatype to use in calculations. If the string being converted contains nonnumeric characters, the function returns an error.

For example, you need to calculate how old each book in the Computer category is in terms of years. You can subtract the current date and book publication date, divide by 365, and then use rounding. Another method is identifying the year of each date and using subtraction. Figure 10-43 shows using the TO_NUMBER function to convert the year value of each date to a numeric datatype to use in the subtraction operation.

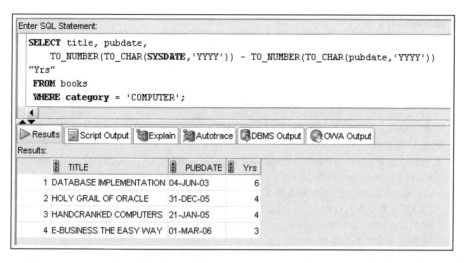

FIGURE 10-43   Using the TO_NUMBER function to convert a string to a numeric datatype

> **NOTE**
>
> Your output might vary because SYSDATE is used in the calculation.

THE DUAL TABLE

Any of the single-row functions covered in this chapter can be used with the DUAL table. Although the DUAL table is rarely used in the industry, it can be valuable for someone learning how to work with functions or testing new functions. For example, if you want to practice rounding numbers or determining the length of character strings, you can enter a specific value in the function you're practicing and reference the DUAL table in the FROM clause, as shown in Figure 10-44.

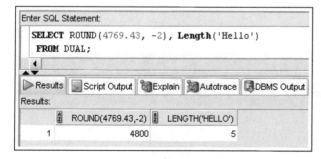

FIGURE 10-44   Practicing functions by using the DUAL table

Chapter Summary

- Single-row functions return a result for each row or record processed.
- Case conversion functions, such as UPPER, LOWER, and INITCAP, can be used to alter the letter case of character strings.
- Character manipulation functions can be used to extract substrings (portions of a string), identify the position of a substring in a string, replace occurrences of a string with another string, determine the length of a character string, and trim spaces or characters from strings.
- Nesting one function inside another allows performing multiple operations on data.
- Simple number functions, such as ROUND and TRUNC, can round or truncate a number on both the left and right side of a decimal.
- The MOD function is used to return the remainder of a division operation.
- Date functions can be used to perform calculations with dates or change the format of dates entered by a user.
- Regular expressions enable complex pattern-matching operations.
- The NVL, NVL2, and NULLIF functions are used to identify and manipulate NULL values.
- The TO_CHAR function enables you to display numeric data and dates in a specific format.
- The DECODE function allows determining the resulting value by testing for equality to a specific value.
- The searched CASE expression enables you to evaluate conditions to determine the resulting value.
- The SOUNDEX function looks for records based on the phonetic pronunciation of characters.
- The DUAL table can be helpful when testing functions.

Chapter 10 Syntax Summary

The following table summarizes the syntax you have learned in this chapter. You can use the table as a study guide and reference.

Syntax Guide

Function	Description	Syntax
Case Conversion Functions		
LOWER	Converts characters to lowercase letters.	LOWER (c) c = Character string or field to convert to lowercase
UPPER	Converts characters to uppercase letters.	UPPER (c) c = Character string or field to convert to uppercase

Function	Description	Syntax
Case Conversion Functions		
INITCAP	Converts words to mixed case (with initial capital letters).	INITCAP(c) c = Character string or field to convert to mixed case
Character Manipulation Functions		
SUBSTR	Returns a substring (portion of a string) in the output.	SUBSTR(c, p, l) c = Character string p = Position (beginning) for the extraction l = Length of output string
INSTR	Identifies the position of the search string.	INSTR(c, s, p, o) c = Character string to search s = Search string p = Search starting position o = Occurrence of search string to identify
LENGTH	Returns the numbers of characters in a string.	LENGTH(c) c = Character string to analyze
LPAD and RPAD	Pads, or fills in, the area to the left (or right) of a character string, using a specific character—or even a blank space.	LPAD(c, l, s) c = Character string to be pad l = Length of character string after padding s = Symbol or character used as padding
RTRIM and LTRIM	Trims, or removes, a specific string of characters from the right (or left) of data.	LTRIM(c, s) c = Characters to modify s = String to be removed from the left of data
REPLACE	Performs a search and replace of displayed results.	REPLACE(c, s, r) c = Data or column to search s = String of characters to find r = String of characters to substitute for s
TRANSLATE	Converts single characters to a substitution value.	TRANSLATE(c, s, r) c = Character string to search s = Search character r = Substitution character

Function	Description	Syntax
Character Manipulation Functions		
CONCAT	Used to concatenate two data items.	CONCAT($c1$, $c2$) $c1$ = First data item to concatenate $c2$ = Second data item to concatenate
Number Functions		
ROUND	Rounds numeric fields.	ROUND(n, p) n = Numeric data or field to round p = Position to which the data should be rounded
TRUNC	Truncates, or cuts, numbers to a specific position.	TRUNC(n, p) n = Numeric data or field to truncate p = Position to which the data should be truncated
MOD	Returns the remainder of a division operation.	MOD(n, d) n = Numerator d = Denominator
ABS	Returns the absolute value of a numeric value.	ABS(n) n = Numeric value
POWER	Raises a number to a specified power.	POWER(x, y) x = Number to raise y = Power to which the number should be raised
Date Functions		
MONTHS_BETWEEN	Determines the number of months between two dates.	MONTHS_BETWEEN($d1$, $d2$) $d1$ and $d2$ = Dates in question $d2$ is subtracted from $d1$
ADD_MONTHS	Adds months to a date to signal a target date in the future.	ADD_MONTHS(d, m) d = Date (beginning) for the calculation m = Months—number of months to add to the date
ROUND	Rounds date fields by month or year.	ROUND(d, u) d = Date value u = Date unit (YEAR or MONTH)

374

Function	Description	Syntax
Date Functions		
NEXT_DAY	Determines the next day—a specific day of the week after a given date.	`NEXT_DAY(d, DAY)` d = Date (starting) DAY = Day of the week to be identified
LAST_DAY	Determines the last day of the month for the month of a given date.	`LAST_DAY(d)` d = Date
TO_DATE	Converts a date in a specified format to the default date format.	`TO_DATE(d, 'f')` d = Date entered by the user f = Format argument to use
Regular Expressions		
REGEXP_LIKE	Searches for pattern values in a character string.	`REGEXP_LIKE(c, p)` c = Character string p = Pattern operators
REGEXP_SUBSTR	Extracts the part of a string that matches a pattern of values.	`REGEXP_SUBSTR(c, p)` c = Character string p = Pattern operators
Other Functions		
NVL	Solves problems caused by performing arithmetic operations with fields that might contain NULL values. When a NULL value is used in a calculation, the result is a NULL value. The NVL function is used to substitute a value for the existing NULL.	`NVL(x, y)` y = Value to be substituted if x is NULL
NVL2	Provides options based on whether a NULL value exists.	`NVL2(x, y, z)` y = What should be substituted if x is not NULL z = What should be substituted if x is NULL
NULLIF	Returns a NULL value if the given values equate; otherwise, returns the first given value.	`NULLIF(x, y)` x and y = Values to compare

375

Function	Description	Syntax
Other Functions		
TO_CHAR	Converts dates and numbers to a formatted character string.	TO_CHAR(n, 'f') n = Number or date to format f = Format argument to use
DECODE	Compares a given value to values in a list. If a match is found, the specified result is returned. If no match is found, a default result is returned. If no default result is defined, a NULL value is returned.	DECODE(V, L1, R1, L2, R2, ..., D) V = Value to search for L1 = First value in the list R1 = Result to return if L1 and V match D = Default result to return if no match is found
Searched CASE expression	Evaluates a given value with conditions to determine a resulting value.	CASE WHEN V1 cond THEN R1 WHEN V2 cond THEN R2 ELSE D END V1 = First value evaluated cond = Condition to evaluate R1 = Result to return if cond for V1 is TRUE D = Default result to return if no cond is TRUE
SOUNDEX	Converts alphabetic characters to their phonetic pronunciation, using an alphanumeric algorithm.	SOUNDEX(c) c = Characters to represent phonetically
TO_NUMBER	Converts numeric digits stored in a date or character value to a number.	TO_NUMBER(v) v = Value to convert

Review Questions

1. Why are functions in this chapter referred to as "single-row" functions?
2. What's the difference between the NVL and NVL2 functions?
3. What's the difference between the TO_CHAR and TO_DATE functions when working with date values?
4. How is the TRUNC function different from the ROUND function?
5. What functions can be used to search character strings for specific patterns of data?

6. What's the difference between using the CONCAT function and the concatenation operator (| |) in a SELECT clause?
7. Which functions can be used to convert the letter case of character values?
8. Describe a situation that calls for using the DECODE function.
9. What format model should you use to display the date 25-DEC-09 as Dec. 25?
10. Why does the function `NVL(shipdate, 'Not Shipped')` return an error message?

Multiple Choice

To answer the following questions, refer to the tables in the JustLee Books database.

1. Which of the following is a valid SQL statement?

    a. `SELECT SYSDATE;`

    b. `SELECT UPPER(Hello) FROM dual;`

    c. `SELECT TO_CHAR(SYSDATE, 'Month DD, YYYY')`
       `FROM dual;`

    d. all of the above

    e. none of the above

2. Which of the following functions can be used to extract a portion of a character string?

    a. EXTRACT

    b. TRUNC

    c. SUBSTR

    d. INITCAP

3. Which of the following determines how long ago orders that haven't shipped were received?

    a. `SELECT order#, shipdate-orderdate delay`
       `FROM orders;`

    b. `SELECT order#, SYSDATE - orderdate`
       `FROM orders`
       `WHERE shipdate IS NULL;`

    c. `SELECT order#, NVL(shipdate, 0)`
       `FROM orders`
       `WHERE orderdate is NULL;`

    d. `SELECT order#, NULL(shipdate)`
       `FROM orders;`

4. Which of the following SQL statements produces "Hello World" as the output?

    a. `SELECT "Hello World" FROM dual;`

    b. `SELECT INITCAP('HELLO WORLD') FROM dual;`

    c. `SELECT LOWER('HELLO WORLD') FROM dual;`

    d. both a and b

    e. none of the above

5. Which of the following functions can be used to substitute a value for a NULL value?

    a. NVL

    b. TRUNC

    c. NVL2

    d. SUBSTR

    e. both a and d

    f. both a and c

6. Which of the following is *not* a valid format argument for displaying the current time?

    a. `'HH:MM:SS'`

    b. `'HH24:SS'`

    c. `'HH12:MI:SS'`

    d. All of the above are valid.

7. Which of the following lists only the last four digits of the contact person's phone number at American Publishing?

    a.
```
SELECT EXTRACT (phone, -4, 1)
   FROM publisher
   WHERE name = 'AMERICAN PUBLISHING';
```

    b.
```
SELECT SUBSTR (phone, -4, 1)
   FROM publisher
   WHERE name = 'AMERICAN PUBLISHING';
```

    c.
```
SELECT EXTRACT (phone, -1, 4)
   FROM publisher
   WHERE name = 'AMERICAN PUBLISHING';
```

    d.
```
SELECT SUBSTR (phone, -4, 4)
   FROM publisher
   WHERE name = 'AMERICAN PUBLISHING';
```

8. Which of the following functions can be used to determine how many months a book has been available?

    a. MONTH

    b. MON

    c. MONTH_BETWEEN

    d. none of the above

9. Which of the following displays the order date for order 1000 as 03/31?

    a.
```
SELECT TO_CHAR (orderdate, 'MM/DD')
   FROM orders
   WHERE order# = 1000;
```

    b.
```
SELECT TO_CHAR (orderdate, 'Mth/DD')
   FROM orders
   WHERE order# = 1000;
```

c. SELECT TO_CHAR(orderdate, 'MONTH/YY')
    FROM orders
    WHERE order# = 1000;

d. both a and b

e. none of the above

10. Which of the following functions can produce different results, depending on the value of a specified column?

a. NVL

b. DECODE

c. UPPER

d. SUBSTR

11. Which of the following SQL statements is *not* valid?

a. SELECT TO_CHAR(orderdate, '99/9999')
    FROM orders;

b. SELECT INITCAP(firstname), UPPER(lastname)
    FROM customers;

c. SELECT cost, retail,
    TO_CHAR(retail-cost, '$999.99') profit
  FROM books;

d. all of the above

12. Which function can be used to add spaces to a column until it's a specific width?

a. TRIML

b. PADL

c. LWIDTH

d. none of the above

13. Which of the following SELECT statements returns 30 as the result?

a. SELECT ROUND(24.37, 2) FROM dual;

b. SELECT TRUNC(29.99, 2) FROM dual;

c. SELECT ROUND(29.01, -1) FROM dual;

d. SELECT TRUNC(29.99, -1) FROM dual;

14. Which of the following is a valid SQL statement?

a. SELECT TRUNC(ROUND(125.38, 1), 0) FROM dual;

b. SELECT ROUND(TRUNC(125.38, 0)
    FROM dual;

c. SELECT LTRIM(LPAD(state, 5, ' '), 4, -3, "*")
    FROM dual;

d. SELECT SUBSTR(ROUND(14.87, 2, 1), -4, 1)
    FROM dual;

15. Which of the following functions can't be used to convert the letter case of a character string?

    a. UPPER

    b. LOWER

    c. INITIALCAP

    d. All of the above can be used for case conversion.

16. Which of the following format elements causes months to be displayed as a three-letter abbreviation?

    a. MMM

    b. MONTH

    c. MON

    d. none of the above

17. Which of the following SQL statements displays a customer's name in all uppercase characters?

    a. `SELECT UPPER ('firstname', 'lastname')`
       `FROM customers;`

    b. `SELECT UPPER (firstname, lastname)`
       `FROM customers;`

    c. `SELECT UPPER (lastname, ',' firstname)`
       `FROM customers;`

    d. none of the above

18. Which of the following functions can be used to display the character string FLORIDA in the query results whenever FL is entered in the State field?

    a. SUBSTR

    b. NVL2

    c. REPLACE

    d. TRUNC

    e. none of the above

19. What's the name of the table provided by Oracle 11*g* for completing queries that don't involve a table?

    a. DUMDUM

    b. DUAL

    c. ORAC

    d. SYS

20. If an integer is multiplied by a NULL value, the result is:

    a. an integer

    b. a whole number

    c. a NULL value

    d. None of the above—a syntax error message is returned.

Hands-On Assignments

To perform the following assigments, refer to the tables in the JustLee Books database.

1. Produce a list of all customer names in which the first letter of the first and last names is in uppercase and the rest are in lowercase.

2. Create a list of all customer numbers along with text indicating whether the customer has been referred by another customer. Display the text "NOT REFERRED" if the customer wasn't referred to JustLee Books by another customer or "REFERRED" if the customer was referred.

3. Determine the amount of total profit generated by the book purchased on order 1002. Display the book title and profit. The profit should be formatted to display a dollar sign and two decimal places. Take into account that the customer might not pay the full retail price, and each item ordered can involve multiple copies.

4. Display a list of all book titles and the percentage of markup for each book. The percentage of markup should be displayed as a whole number (that is, multiplied by 100) with no decimal position, followed by a percent sign (for example, .2793 = 28%). (The percentage of markup should reflect the difference between the retail and cost amounts as a percent of the cost.)

5. Display the current day of the week, hour, minutes, and seconds of the current date setting on the computer you're using.

6. Create a list of all book titles and costs. Precede each book's cost with asterisks so that the width of the displayed Cost field is 12.

7. Determine the length of data stored in the ISBN field of the BOOKS table. Make sure each different length value is displayed only once (not once for each book).

8. Using today's date, determine the age (in months) of each book that JustLee sells. Make sure only whole months are displayed; ignore any portions of months. Display the book title, publication date, current date, and age.

9. Determine the calendar date of the next occurrence of Wednesday, based on today's date.

10. Produce a list of each customer number and the third and fourth digits of his or her zip code. The query should also display the position of the first occurrence of a 3 in the customer number, if it exists.

Advanced Challenge

To perform this activity, refer to the tables in the JustLee Books database.

Management is proposing to increase the price of each book. The amount of the increase will be based on each book's category, according to the following scale: Computer books, 10%; Fitness books, 15%; Self-Help books, 25%; all other categories, 3%. Create a list that displays each book's title, category, current retail price, and revised retail price. The prices should be displayed with two decimal places. The column headings for the output should be as follows: Title, Category, Current Price, and Revised Price. Sort the results by category. If there's more than one book in a category, a secondary sort should be performed on the book's title.

Create a document to show management the SELECT statement used to generate the results and the results of the statement.

Case Study: *City Jail*

Note: Make sure you have run the CityJail_8.sql script from Chapter 8. This script makes all database objects available for completing this case study.

The following list reflects current data requests from city managers. Provide the SQL statement to satisfy each request. Test the statements and show execution results.

1. List the following information for all crimes that have a period greater than 14 days between the date charged and the hearing date: crime ID, classification, date charged, hearing date, and number of days between the date charged and the hearing date.

2. Produce a list showing each active police officer and his or her community assignment, indicated by the second letter of the precinct code. Display the community description listed in the following chart, based on the second letter of the precinct code.

Second Letter of Precinct Code	Description
A	Shady Grove
B	Center City
C	Bay Landing

3. Produce a list of sentencing information to include criminal ID, name (displayed in all uppercase letters), sentence ID, sentence start date, and length in months of the sentence. The number of months should be shown as a whole number. The start date should be displayed in the format "December 17, 2009."

4. A list of all amounts owed is needed. Create a list showing each criminal name, charge ID, total amount owed (fine amount plus court fee), amount paid, amount owed, and payment due date. If nothing has been paid to date, the amount paid is NULL. Include only criminals who owe some amount of money. Display the dollar amounts with a dollar sign and two decimals.

5. Display the criminal name and probation start date for all criminals who have a probation period greater than two months. Also, display the date that's two months from the beginning of the probation period, which will serve as a review date.

6. An INSERT statement is needed to support users adding a new appeal. Create an INSERT statement using substitution variables. Note that users will be entering dates in the format of a two-digit month, a two-digit day, and a four-digit year, such as "12 17 2009." In addition, a sequence named APPEALS_ID_SEQ exists to supply values for the Appeal_ID column, and the default setting for the Status column should take effect (that is, the DEFAULT option on the column should be used). Test the statement by adding the following appeal: crime_ID = 25344031, filing date = 02 13 2009, and hearing date = 02 27 2009.

GROUP FUNCTIONS

INTRODUCTION

Group functions, also called **multiple-row functions,** return one result per group of rows processed. Multiple-row functions covered in this chapter include SUM, AVG, COUNT, MIN, MAX, STDDEV, and VARIANCE. This chapter also explains using the GROUP BY clause to identify groups of records to process and the HAVING clause to restrict groups returned in the query results. The last section of the chapter introduces enhanced aggregation capabilities with the GROUPING SETS, CUBE, and ROLLUP operations. Table 11-1 gives you an overview of this chapter's contents.

TABLE 11-1   Group Functions and GROUP BY Extensions Covered in This Chapter

Function and Syntax	Description	Example
Group Functions		
SUM([DISTINCT|ALL] *n*)	Returns the sum or total value of the selected numeric field. Ignores NULL values.	SELECT SUM(retail-cost) FROM books;
AVG([DISTINCT|ALL] *n*)	Returns the average value of the selected numeric field. Ignores NULL values.	SELECT AVG(cost) FROM books;
COUNT(* [DISTINCT|ALL] *c*)	Returns the number of rows containing a value in the identified field. Rows containing NULL values in the field aren't included in the results. To count rows containing NULL values, use an asterisk (*) rather than a field name.	SELECT COUNT(*) FROM books; *or* SELECT COUNT(shipdate) FROM orders;
MAX([DISTINCT|ALL] *c*)	Returns the highest (maximum) value from the selected field. Ignores NULL values.	SELECT MAX(customer#) FROM customers;
MIN([DISTINCT|ALL] *c*)	Returns the lowest (minimum) value from the selected field. Ignores NULL values.	SELECT MIN(retail-cost) FROM books;
STDDEV([DISTINCT|ALL] *n*)	Returns the standard deviation of the selected numeric field. Ignores NULL values.	SELECT STDDEV(retail) FROM books;

Function and Syntax	Description	Example
Group Functions		
`VARIANCE ([DISTINCT| ALL] n)`	Returns the variance of the selected numeric field. Ignores NULL values.	`SELECT VARIANCE(retail)` `FROM books;`
GROUP BY Extensions		
`GROUPING SETS`	Enables performing multiple GROUP BY clauses with a single query.	`SELECT name, category,` `    AVG(retail)` `FROM publisher` `JOIN books USING` `    (pubid)` `GROUP BY GROUPING SETS` `    (name, category,` `    (name,category),());`
`CUBE`	Performs aggregations for all possible combinations of columns included.	`SELECT name, category,` `    AVG(retail)` `FROM publisher` `JOIN books USING` `    (pubid)` `GROUP BY CUBE(name,` `    category)` `ORDER BY name,` `    category;`
`ROLLUP`	Performs increasing levels of cumulative subtotals, based on the provided column list.	`SELECT name, category,` `    AVG(retail)` `FROM publisher` `JOIN books USING` `    (pubid)` `GROUP BY ROLLUP(name,` `    category)` `ORDER BY name,` `    category;`

385

DATABASE PREPARATION

Before attempting to work through the examples in this chapter, make sure you have completed these two tasks: First, if you haven't already run the JLDB_Build_8.sql script from Chapter 8, execute this script to rebuild the JustLee Books database. Second, run the JLDB_Build_11.sql file in the Chapter11 folder of your student data files to make the necessary modifications to the JustLee Books database.

GROUP FUNCTIONS

Multiple-row functions are commonly referred to as group functions because they process groups of rows. Because these functions return only one result per group of data, they're also known as **aggregate functions**. For example, JustLee Books wants to determine the average retail price for all books currently in stock. For this task, the retail prices of all books are totaled and then divided by the number of books. This is what a group function—AVG, in this case—does. In other words, the group function processes the 14 rows in the BOOKS table and produces a single value (the average) as the result; this process is aggregation.

You can also define subsets or groups to use in an aggregate operation. JustLee Books might need to identify the average retail price for books by category instead of the overall average. In this case, you can use a GROUP BY clause in the query to define the groups. If you review the data in the BOOKS table, you'll notice that the 14 books are assigned to eight different book categories. Therefore, an aggregate query calculating the average by category produces eight averages: one for each category. You can also filter output based on aggregated results by using the HAVING clause. Figure 11-1 shows the position of the GROUP BY and HAVING clauses in the SELECT statement.

386

```
SELECT *|columnname, columnname...
FROM tablename
[WHERE condition]
[GROUP BY columnname, columnname...]
[HAVING group condition];
```

FIGURE 11-1   SELECT statement syntax

The SUM Function

The **SUM** function is used to calculate the total amount stored in a numeric field for a group of records. The syntax of the SUM function is SUM([DISTINCT|ALL] *n*), where *n* is a column containing numeric data. The optional **DISTINCT** keyword instructs Oracle 11*g* to include only *unique* numeric values in its calculation. The **ALL** keyword instructs Oracle 11*g* to include *multiple* occurrences of numeric values when totaling a field. If the DISTINCT or ALL keywords aren't included when using the SUM function, Oracle 11*g* assumes the ALL keyword by default and uses all the numeric values in the field when the query is executed, as shown in Figure 11-2.

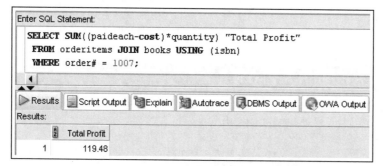

FIGURE 11-2   Using the SUM function to calculate order profit

In Figure 11-2, the query calculates the total profit from books sold in order 1007. The SUM function in the SELECT clause uses the argument `(paideach-cost)*quantity` to instruct Oracle 11g to calculate the profit each book item in the order generates before totaling the profit. The quantity must be included because multiple copies of a book might be purchased. For example, one customer's order might be for two copies of *Revenge of Mickey* and one copy of *Handcranked Computers*. Notice that the SELECT clause also includes a column alias, "Total Profit," to describe the profit output. As with single-row functions, if a group function is used in a SELECT clause, the actual function is displayed as the column header unless you assign a column alias.

The WHERE clause in Figure 11-2 restricts rows used in the calculation to only the books in order 1007. Because the books ordered and prices paid (the Paideach column) are stored in the ORDERITEMS table, and book costs are stored in the BOOKS table, the two tables are joined in the FROM clause. The difference between the Cost and Paideach values for each book is calculated, then these profits per book are totaled, and finally, the total profit for the order is returned as a single value of output.

Suppose management wants to determine total sales for one day—April 2, 2009. You might assume you could simply query the ORDERS table; however, this table doesn't include the total "amount due" for an order. The only way to calculate how much a customer owes for his or her order is to multiply the quantity of books purchased by the price paid for each book `(quantity*paideach)`. This calculation is commonly referred to as an "extended price." The extended prices are then totaled to yield a customer's total amount due. (You learn how to determine a customer's total amount due in "Grouping Data" later in this chapter.) However, in this case, management wants to know just the total sales for April 2, 2009. To calculate this total, simply add the extended prices for all orders placed on that day. The query in Figure 11-3 determines the day's total sales.

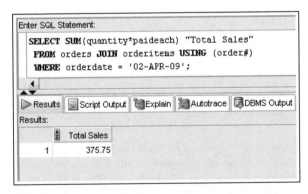

FIGURE 11-3 Using the SUM function to calculate total sales for a specified date

Review the clauses in Figure 11-3. The SELECT clause uses the SUM function to calculate the extended price for each book ordered. Notice that in the FROM clause, two tables must be joined. Why? The order date is stored in the ORDERS table and the quantity ordered and price paid are stored in the ORDERITEMS table. The WHERE clause is included to restrict the calculation to orders placed on April 2, 2009. So any rows not having the date April 2, 2009 are eliminated before the SUM function is calculated.

Keep the following two rules in mind as you work with group functions:

- Use the DISTINCT keyword to include only unique values. The *ALL keyword is the default*, and it instructs Oracle 11g to include all values (except nulls).
- All group functions, except the COUNT(*) function, ignore NULL values. To include NULL values, nest the NVL function inside the group function.

The AVG Function

The **AVG** function calculates the average of numeric values in a specified column. The syntax of the AVG function is AVG ([DISTINCT|ALL] *n*), where *n* is a column containing numeric data.

For example, if the management of JustLee Books wants to know the average profit generated by all books in the Computer category, you can use the WHERE clause to restrict the rows processed to those containing the value COMPUTER in the Category column. As with the SUM function, the profit for each book is calculated, and then all profits are totaled. This total is then divided by the number of records containing non-NULL values in the specified column, as shown in Figure 11-4.

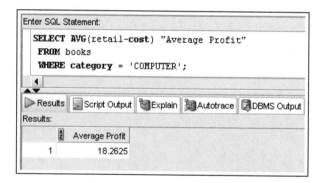

FIGURE 11-4   Using the AVG function to calculate average profit

As the query results show, the average profit of books in the Computer category is $18.2625. When a query includes division, the resulting display might include more than two decimal positions. Because management prefers displaying results with only two decimal positions, the TO_CHAR function is included in the modified query in Figure 11-5 to specify the correct format. When the TO_CHAR function is used on numeric data, any excess decimals are rounded (not truncated) to the specified number of digits. As shown in this example, group functions can be nested inside single-row functions.

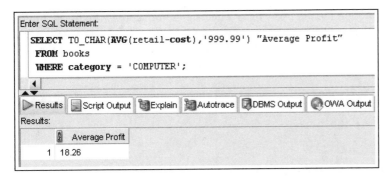

FIGURE 11-5 Embedding a group function in a single-row function

> **NOTE**
>
> If you're working with the SQL*Plus client tool, the entire column alias might not be displayed in Figure 11-5's output. Why? When the TO_CHAR function is included to apply a format argument to the average profit, numeric values are converted to characters for display purposes. Because the average profit is then considered a character string, Oracle 11g truncates the column alias to match the displayed column's width.

Managing NULL values might be an issue when calculating averages. For example, the EMPLOYEES table contains the monthly salary and current bonus amount for each employee. The Bonus column is NULL if no bonus has been earned so far. Review the employee data in Figure 11-6. Notice that the employee with the last name Stuart has a NULL value for the Bonus column.

Enter SQL Statement:

```
SELECT empno, lname, mthsal, bonus
FROM employees;
```

Results:

	EMPNO	LNAME	MTHSAL	BONUS
1	7839	KING	6000	3000
2	8888	JONES	4200	1200
3	7344	SMITH	4900	1500
4	7355	POTTS	4900	1900
5	8844	STUART	3700	(null)

FIGURE 11-6 Data in the EMPLOYEES table

Group Functions

How does the AVG function handle this value? The calculation ignores the row containing a NULL value for the Bonus column and uses only the first four rows of data to perform the calculation. Figure 11-7 shows the average bonus per employee, confirming that the NULL value is ignored.

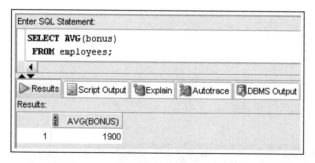

FIGURE 11-7　The AVG function ignores NULL values

However, if the NULL value actually represents a bonus of zero, it must be included in the calculation. In this case, the NVL function covered in Chapter 10 can be used to indicate that a NULL value should be treated as a zero. The query in Figure 11-8 includes this modification, and the average amount is lower now. The NVL function is embedded in the AVG function to substitute a zero value for a NULL before the average is calculated.

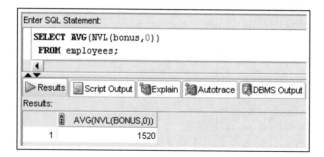

FIGURE 11-8　Embedding NVL in a group function

The COUNT Function

Depending on the argument used, the **COUNT** function can count the records having non-NULL values in a specified field or count the total records meeting a specific condition, including those containing NULL values. The syntax of the COUNT function is COUNT(* [DISTINCT|ALL] c), where c represents a numeric or nonnumeric column. The query in Figure 11-9 tells Oracle 11g to use the COUNT function to return the number of distinct categories represented by titles stored in the BOOKS table.

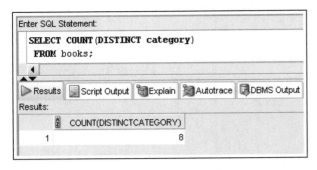

FIGURE 11-9 Using the COUNT function with the DISTINCT option

Notice that the DISTINCT keyword precedes the column name in the COUNT function's argument instead of being placed directly after the SELECT keyword. This placement instructs Oracle 11g to count each different value found in the Category column. If the DISTINCT keyword is entered directly after SELECT, it applies to the entire COUNT function and is interpreted to mean that only duplicate rows, not duplicate category values, should be suppressed. However, in Figure 11-9, the COUNT function returns only one value (or row), so the DISTINCT keyword has no effect on the results.

As shown in Figure 11-10, if the DISTINCT keyword is placed after the SELECT keyword, Oracle 11g returns a count of how many rows are in the BOOKS table. Why? Because the DISTINCT keyword applies to the results of the COUNT function—after all rows containing a value in the Category column have been counted.

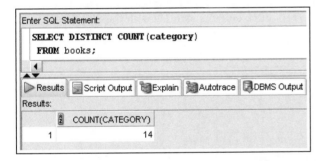

FIGURE 11-10 Flawed query: The COUNT function counts all rows with a Category value

However, in this case, management wants to know how many different categories are represented by books in the BOOKS table. Because the DISTINCT keyword should apply to the Category column's contents, the keyword must be placed immediately before the column name, inside the COUNT function's argument. As Figure 11-9 showed, the 14 titles in the BOOKS table represent eight different categories.

Now suppose management asks how many orders are currently outstanding—that is, they haven't been shipped to customers. One solution is to print a list of all orders having a NULL value for the ship date. However, you still need to count the records returned in the results. Figure 11-11 shows a simpler solution.

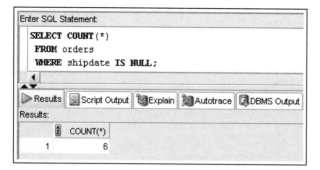

FIGURE 11-11   Using the COUNT(*) function to include NULL values

> **TIP**
>
> Remember that the equal sign (=) can't be used when searching for a NULL value. Use the correct comparison operator, IS NULL, for finding rows containing a NULL value.

When the argument supplied in the COUNT function is an asterisk (*), the entire record is counted, so a NULL value in one column doesn't cause the COUNT function to ignore a row. As Figure 11-11 showed, the WHERE clause restricts the rows that should be counted to only those without a value in the Shipdate column.

By contrast, look at the flawed query in Figure 11-12, which modifies the query in Figure 11-11. It illustrates a common error—replacing the asterisk with the Shipdate column in the COUNT argument.

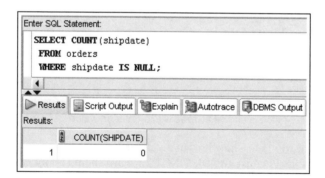

FIGURE 11-12   Flawed query: The COUNT function specifying a column ignores NULL values

Because the WHERE clause restricts the records counted to only those having a NULL value in the Shipdate column, the function returns a count of zero. Basically, because the specified column contains no value, there's nothing to count. Therefore, whenever NULL values could affect the COUNT function, you should use *an asterisk rather than a column name* as the argument.

The MAX Function

The **MAX** function returns the largest value stored in the specified column. The syntax of the MAX function is `MAX([DISTINCT|ALL] c)`, where `c` can represent any numeric, character, or date column. The query in Figure 11-13 retrieves the maximum profit generated by a book.

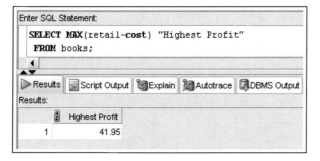

FIGURE 11-13   Using the MAX function on numeric data

As shown in the results, the largest profit earned by a single book is $41.95. The problem with this query is that you can't tell which book is generating the profit. Therefore, the result isn't helpful to management, who needs at least the title to identify which book is the most profitable. A common mistake is attempting to add a descriptive column in the query with aggregation, as shown in Figure 11-14.

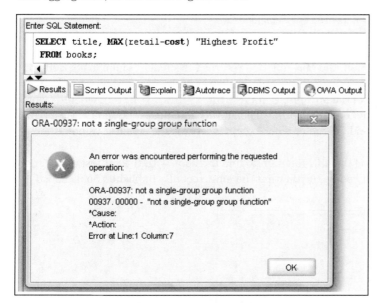

FIGURE 11-14   Common aggregation error

If the SELECT clause includes both aggregate (`MAX(retail-cost)`) and nonaggregate (Title) columns, a GROUP BY clause must be used and all the nonaggregate columns must be included. As you learn in "Grouping Data" later in this chapter, adding a GROUP BY clause to this query still produces incorrect results because the query calculates the profit generated by *each book* instead of the highest profit of *all books*. A subquery must be used to identify both the title and profit of the most profitable book in inventory. You learn how to work with subqueries in Chapter 12.

The MAX function can also be used with nonnumeric data. With nonnumeric data, the output shows the first value that occurs when a column is sorted in descending order. For example, if a column contains dates, the most recent date is considered to have the highest value (based on its Julian date, discussed in Chapter 10). If the MAX function is applied to a character column, the letter Z has a higher, or larger, "value" than the letter A.

Suppose you need to find the book title at the end of a list of book titles sorted alphabetically in ascending order (A to Z). In other words, the book title has the highest value of all books in the BOOKS table. As shown in Figure 11-15, when the MAX function is applied to the Title column, the book title *The Wok Way to Cook* is displayed.

FIGURE 11-15 Using the MAX function on character data

The MIN Function

In contrast to the MAX function, the **MIN** function returns the smallest value in a specified column. As with the MAX function, the MIN function works with any numeric, character, or date column. The syntax of the MIN function is MIN ([DISTINCT|ALL] c), where c represents any character, numeric, or date column. The MIN function uses the same logic as the MAX function for numeric and character data, except it returns the smallest value rather than the largest value.

Figure 11-16 shows using the MIN function to find the book with the earliest publication date in the BOOKS table. Of course, if you want the most recently published book, substitute the MAX function.

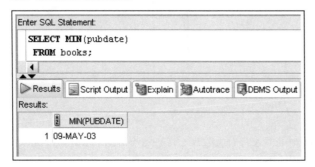

FIGURE 11-16 Using the MIN function on date data

In Figure 11-4, the SELECT query returns the average profit of all books in the Computer category. However, suppose the management of JustLee Books wants to know the average profit for each category of books. One solution is reissuing the query in Figure 11-4 once for each category, using the WHERE clause to restrict the query to a specific category each time. Another solution is dividing the records in the BOOKS table into groups, and then calculating the average for each group—preferably all in one query. For this solution, you can use with the **GROUP BY** clause. The syntax is GROUP BY *columnname* [, *columnname*, . . .], where *columnname* is the column used to create the groups or sets of data.

What happens when you attempt to create this query by adding the Category column in the SELECT clause without including the GROUP BY clause? In Figure 11-17, the SELECT clause includes both the nonaggregate column Category and the aggregate column using the AVG group function. However, the SELECT statement doesn't include a GROUP BY clause to specify that groups should be created by using the Category column values Therefore, an error message is returned.

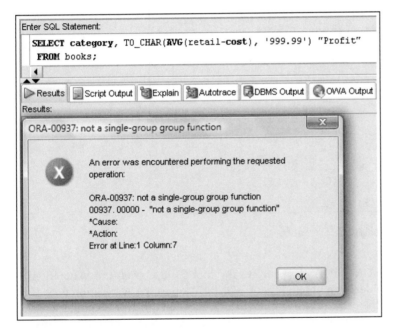

FIGURE 11-17 Flawed query: Including both aggregate and nonaggregate columns requires a GROUP BY clause

When using the GROUP BY clause, remember the following:

- If a group function is used in the SELECT clause, any single (nonaggregate) columns listed in the SELECT clause must also be listed in the GROUP BY clause.
- Columns used to group data in the GROUP BY clause don't have to be listed in the SELECT clause. They're included in the SELECT clause only to have these groups identified in the output.

- Column aliases can't be used in the GROUP BY clause.
- Results returned from a SELECT statement that includes a GROUP BY clause are displayed in ascending order of the columns listed in the GROUP BY clause. To have a different sort sequence, use the ORDER BY clause.

The required GROUP BY clause is added in Figure 11-18 to correct the error in Figure 11-17's query. As you can see, the nonaggregate column in the SELECT clause (Category) is included in the GROUP BY clause. When the query is executed, the records in the BOOKS table are grouped by category, and then the average profit for each category is calculated. Because the Category column is listed in the SELECT clause, each category is displayed in the results along with the average profit each category generates.

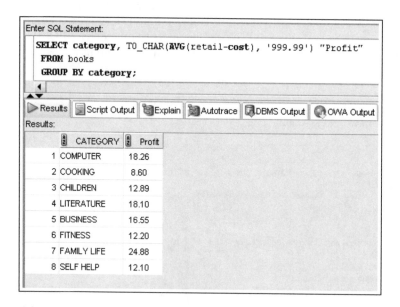

FIGURE 11-18 Adding the GROUP BY clause

Even though adding a GROUP BY clause solves the "not a single-group group function" error, it's not always suitable. Keep in mind that GROUP BY forces the aggregation to occur at the group level. Recall the error returned in Figure 11-14 when the Title column was added to the MAX function. Figure 11-19 corrects this error by adding a GROUP BY clause. The statement executes successfully; however, the results don't produce the needed data, which is the maximum profit for all books. Adding the GROUP BY clause means a maximum profit is determined for each group—or book, in this case.

N O T E

More options for adding descriptive data values in aggregation queries are explained in Chapter 12.

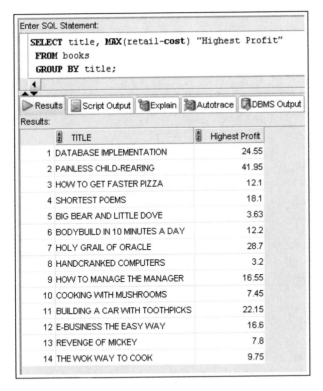

FIGURE 11-19 Inappropriate use of GROUP BY

A grouping action can include multiple column values. You used the SUM function earlier to calculate the total sales for a order. The SUM function is a group function and, therefore, returns one total for all rows processed. What if you need a list of all orders and the total amount due *by customer and order*? This task is perfect for the GROUP BY clause.

For example, if the Billing Department requests a list of the amount due from each customer for each order, use the GROUP BY clause to group rows for each order, and then use the SUM function to calculate the extended price for the items ordered and return the total amount due for each order. Figure 11-20 shows the SQL statement to create this list.

When the statement is executed, Oracle 11g displays each order, the customer number of the person who placed the order, and the total amount due. Because the SELECT clause in Figure 11-20 includes the Customer# and Order# columns, these columns must also be listed in the GROUP BY clause. You might wonder whether including the order number in the query is necessary. Suppose a customer placed two orders recently. If the order number isn't included in the query, the SQL statement returns the total amount due from each customer, not the amount due from a customer for each order. The customer number is included to identify who placed each order.

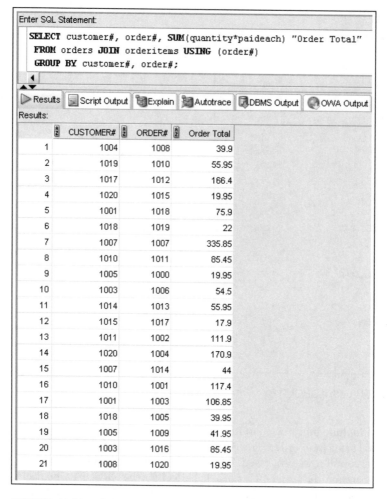

FIGURE 11-20 Calculate the total amount due by each customer and order

RESTRICTING AGGREGATED OUTPUT

The **HAVING** clause is used to restrict the groups returned by a query. If you need to use a group function to restrict groups, you must use the HAVING clause because the WHERE clause *can't contain group functions*. Although the WHERE clause restricts the records the query processes, the HAVING clause specifies which groups are displayed in the results. The syntax of the HAVING clause is `HAVING groupfunction comparisonoperator value`.

Suppose you want to display book categories with an average profit of more than $15.00. Figure 11-21 shows the query to perform this task. The HAVING clause serves as the WHERE clause for aggregated data. Referring to the syntax of the HAVING clause, notice that `groupfunction` is AVG, `comparisonoperator` is >, and `value` is 15.

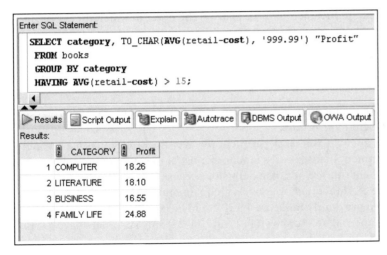

FIGURE 11-21   Using a HAVING clause to restrict which groups are displayed

The GROUP BY clause specifies calculating the average profit for each category. Then the HAVING clause checks each category average to see whether it's greater than $15.00. In Figure 11-21, only four categories return an average profit of more than $15.00. As with the WHERE clause, the logical operators NOT, AND, and OR can be used instead of comparison operators in the HAVING clause to join group conditions, if necessary.

Keep in mind that a WHERE clause can still be used to restrict specific rows in the query. In Figure 11-22, the WHERE clause restricts the records that are processed to only those with a publication date after January 1, 2005. The GROUP BY clause groups the records meeting this publication date restriction by the category to which each book is assigned. The HAVING clause restricts the group data displayed to categories with an average profit greater than $15.00.

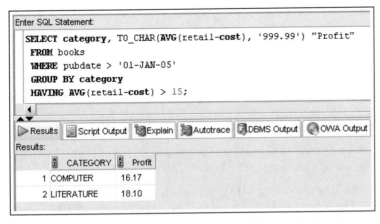

FIGURE 11-22   Using the WHERE, GROUP BY, and HAVING clauses

Group Functions

When a SELECT statement includes all three clauses, the order in which they're evaluated is as follows:

- The WHERE clause
- The GROUP BY clause
- The HAVING clause

In essence, the WHERE clause filters the data *before* grouping, and the HAVING clause filters the groups *after* the grouping occurs.

For another example, after the Billing Department receives the order totals list created in Figure 11-20, the department manager asks for another list of the amount due—but only for orders with a total amount due greater than $100.00. Because output is to be restricted based on the results of the SUM function, which is a group function, a HAVING clause is required. The query shown previously in Figure 11-20 could be modified by adding HAVING SUM(quantity*paideach) > 100. As shown in Figure 11-23, only six orders have a total amount due greater than $100.

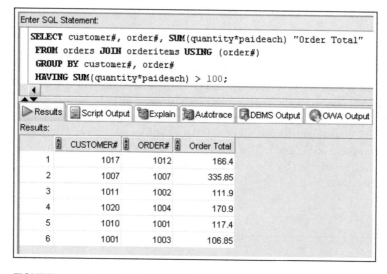

FIGURE 11-23   Using a HAVING clause to restrict grouped output

Including Category column filters in a HAVING clause is a common issue in data-filtering queries. For example, suppose you need to generate a list of each book category along with the average profit, as you did earlier. This query needs to include the following data filters:

- Show only book categories with an average profit greater than $15.00.
- Include only the categories Computer, Children, and Business.

The first filtering task should be done with a HAVING clause because it places a condition on an aggregated value. The second filtering task should be handled with a WHERE clause at the row level to include only the book rows in the specified categories before the aggregation is performed. Figure 11-24 shows the correct query.

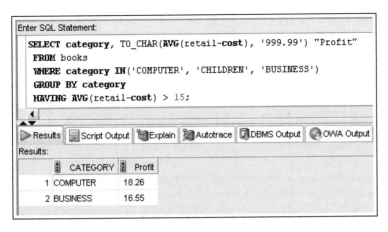

FIGURE 11-24 Filtering correctly with the WHERE and HAVING clauses

However, many new SQL users make the mistake of placing all filtering actions in the HAVING clause. Figure 11-25 shows the same query with all filtering action in the HAVING clause.

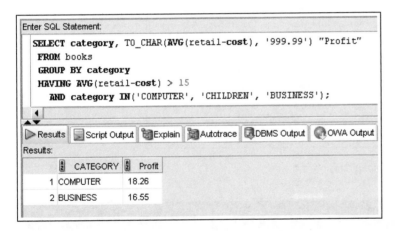

FIGURE 11-25 Filtering incorrectly with the HAVING clause

Notice that the same results are produced. The query can perform the category filtering in the HAVING clause, after the aggregation is performed, because the Category value is available as a grouping value. However, this method is quite inefficient and considered poor SQL programming practice. The statement must process all rows in the BOOKS table with the aggregated calculation and then eliminate categories. By using a WHERE clause instead, the rows not needed are eliminated at the beginning of the query process, before the aggregation.

NESTING FUNCTIONS

When group functions are **nested**, *the inner function is resolved first*, as with single-row functions. The result of the inner function is passed as input for the outer function. Unlike single-row functions that have no restriction on the number of nesting levels, group functions can be nested *only to a depth of two*. As you saw in Figure 11-25, group functions can be nested inside single-row functions (AVG embedded in a TO_CHAR function). In addition, single-row functions can be nested inside group functions, as shown previously in Figure 11-8 (NVL embedded in an AVG function). You can also nest a group function inside another group function.

Suppose you need to calculate the average sales amount per order. To do this, you need to calculate the total sales by order, and then compute the average of these amounts. In Figure 11-26, the SUM function is nested inside the AVG function to determine the average total amount for an order. First, the GROUP BY clause groups all records, based on the Order# column. Second, the SUM function calculates the total order amount for each group or order. Third, the AVG function calculates the average of the total order amounts calculated by the SUM function. The resulting output is the average total amount due for orders stored in the ORDERS table.

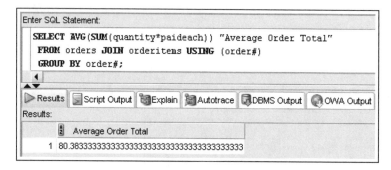

FIGURE 11-26   Nesting group functions

TIP

Don't forget to include two closing parentheses at the end of the function in the SELECT clause. The first parenthesis closes the SUM function, and the second closes the AVG function.

NOTE

A TO_CHAR function could be used to format the results with two decimal places, as in previous examples.

Keep in mind that group functions can be nested to only *two* levels, and the query *must* include a GROUP BY clause. The inner group function creates the aggregated result for each group. The outer group function performs an aggregation on the grouped results.

STATISTICAL GROUP FUNCTIONS

Oracle 11g provides **statistical group functions** to perform calculations for data analysis. In most organizations, marketing and accounting tasks require using data analysis to detect sales trends, price fluctuations, and so on. Oracle 11g provides functions to support basic statistical calculations, such as standard deviation and variance. Although these calculations are easy to do in Oracle 11g, most people need training in statistical analysis to interpret the calculations' results. This chapter is intended to give you an overview of the calculations' purposes, not to train you in statistical analysis. The statistical functions covered in this section are STDDEV and VARIANCE.

The STDDEV Function

The **STDDEV** function calculates the standard deviation for a specified field. A **standard deviation** calculation determines how close each value in a group of numbers is to the mean, or average, of the group. The syntax of the STDDEV function is STDDEV ([DISTINCT|ALL] *n*), where *n* represents a numeric column.

The SELECT statement in Figure 11-27 displays each book category in the JustLee Books database, the average profit for each category, the count (number of books in each category), and the standard deviation of the profit for each category.

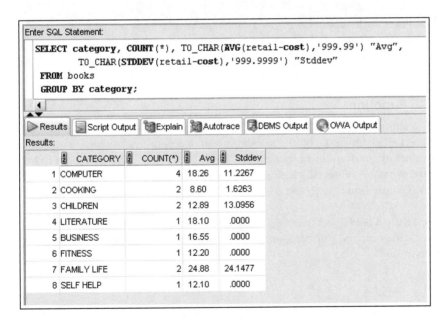

FIGURE 11-27   Using the STDDEV function

Now take a look at how to interpret these results. For the value calculated by the standard deviation to be useful, it must be compared to the calculated "average profit" for each category. For example, the average profit for books in the Computer category is $18.26. However, are most books in the category close to this average? Or do most books generate a small profit (perhaps only $1) and one book generates a large profit (perhaps $20)—which inflates the average? The standard deviation is a statistical approximation of how many books in a certain category fall within a certain range around the average.

The STDDEV function is based on the concept of **normal distribution**, which means that if you input many data values, they tend to cluster around an average value. The basic assumption is that as you move closer to the average value, more data values are clustered around the average. However, some values might be extreme and, therefore, be much larger or smaller than the average. Of course, each extreme data value affects the group's average. For example, calculating the average of 5, 6, 7, and 100 results in a larger value than if the 100 isn't included. When performing statistical analysis, the standard deviation is calculated to determine how closely data matches the average value for the group.

In a normal distribution, you can expect to find 68% of books within one standard deviation (plus or minus) of the average and 95% of books within two standard deviations (plus or minus) of the average. In simpler terms, again using the example in Figure 11-27, 68% of books in the Computer category have a profit between $7.03 ($18.26 - $11.23) and $29.49 ($18.26 + $11.23). The standard deviation can give management a quick picture of the range of profit values for books in a particular category, without having to examine each book—a time-consuming task if the category includes thousands of books.

> **TIP**
>
> Notice that some categories in Figure 11-27 have a standard deviation of zero. If the STDDEV function processes only one record per group, as shown in the COUNT column, the result is always zero.

The VARIANCE Function

The **VARIANCE** function determines how widely data is spread in a group. The variance of a group of records is calculated based on the minimum and maximum values for a specified field. If the data values are clustered together closely, the variance is small. However, if the data contains extreme values (unusually high or low values), the variance is larger. The syntax for the VARIANCE function is VARIANCE ([DISTINCT|ALL] n), where n represents a numeric field.

The query in Figure 11-28 lists the categories for all books in the BOOKS table, the profit variance of each category, and (for comparison purposes) the lowest and highest profit in each category.

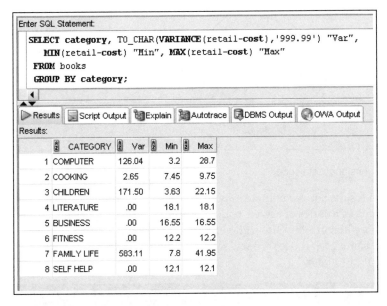

FIGURE 11-28 Using the VARIANCE function

As with the standard deviation, if a group of data consists of only one value (the Business, Fitness, Literature, and Self Help categories, for example), the calculated variance is zero. However, unlike standard deviation, variance isn't measured with the same units (for example, dollars) as the source data used for the calculation.

To interpret the results of a VARIANCE function, you must look at how large or small the value is. For example, the Cooking category has a smaller variance than in other categories, meaning that profits for books in the Cooking category are clustered tightly together (that is, the profit doesn't cover a wide range of values). In the Cooking category, notice that the profit range for all books is $2.30: $9.75 (maximum) - $7.45 (minimum). On the other hand, the Family Life category has the largest profit range of all the categories. This data should throw up a warning flag to management that some books might generate little profit, and others might return a large profit. Therefore, the average profit for books in the Family Life category shouldn't be the sole basis for decision making.

ENHANCED AGGREGATION FOR REPORTING

Oracle provides extensions to the GROUP BY clause that allow aggregating across multiple dimensions or generating increasing levels of subtotals with a single SELECT statement. A **dimension** is a term for describing any category used in analyzing data, such as time, geography, and product line. Each dimension could contain different levels of aggregation. For example, a time dimension might include aggregation by month, quarter, and year. Multi-dimensional analysis involves data aggregated across more than one dimension. A cube of data, as shown in Figure 11-29, is a common analogy used to help you visualize data with many dimensions.

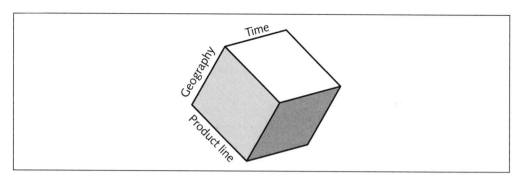

FIGURE 11-29   A multidimensional data cube

Producing aggregated results across multiple dimensions or increasing levels of aggregation in one dimension requires a series of aggregate queries joined with a UNION operation. However, this method isn't efficient. The GROUPING SETS expression is a much simpler method that achieves the same goal. It uses a single scan to compute all aggregates, which typically improves query performance.

NOTE

The advanced aggregation extensions of GROUP BY discussed in this section are widely used in online analytical processing (OLAP) and data warehousing.

A basic Microsoft Excel pivot table can help demonstrate the need for multidimensional data. A pivot table allows dragging dimensions of data to different areas on a spreadsheet to display different aggregations. For example, Figure 11-30 shows a pivot table with two dimensions: Publisher and Category. This table analyzes the total number of available books.

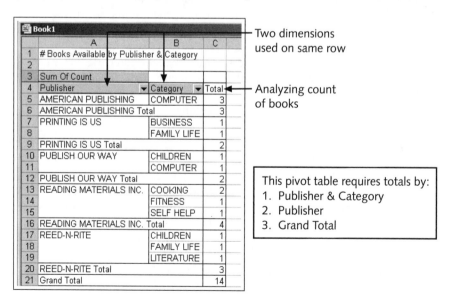

FIGURE 11-30   A pivot table with two dimensions on a row

The analysis requires three levels of aggregation, listed in Figure 11-30. What happens when you drag the Category dimension from the row area to the column area? Figure 11-31 shows the results of this action; four levels of aggregation are required now.

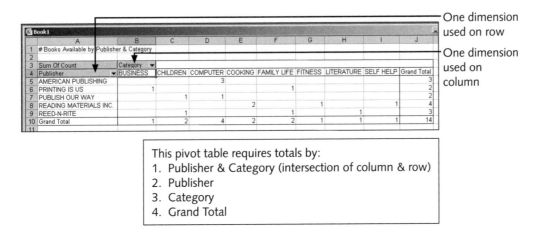

This pivot table requires totals by:
1. Publisher & Category (intersection of column & row)
2. Publisher
3. Category
4. Grand Total

FIGURE 11-31 A pivot table with one row and one column dimension

This type of analysis is possible by using multidimensional data or a data cube, which stores data with all possible aggregations for quicker and more versatile data analysis.

The GROUPING SETS Expression

The **GROUPING SETS** expression is the component on which the other GROUP BY extensions, ROLLUP and CUBE, are built. With this expression, you can use a single query statement to perform multiple GROUP BY clauses. Figure 11-32 shows the GROUPING SETS expression added to the GROUP BY clause. This single query produces the average retail price for books in four groupings: 1) publisher (the Name column) and category, 2) category, 3) publisher, and 4) overall average. Notice that a WHERE clause is used to filter data to make output shorter for the examples in this section. You could remove the WHERE clause filtering and perform the operations on all rows.

The column arguments in parentheses after the GROUPING SETS expression indicate the different GROUP BY operations to be performed—four different groupings, in this example. The ( ) column argument listed last indicates an overall total aggregation. The NULL values in the column output indicate a subtotal row. Without the GROUPING SETS expression, this task would require combining results of four separate queries with a UNION operation, as shown in Figure 11-33.

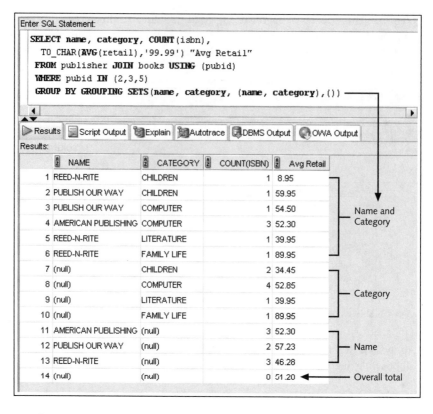

FIGURE 11-32 Using a GROUPING SETS expression in a GROUP BY clause

```
Enter SQL Statement:
 SELECT name, category, COUNT(isbn), TO_CHAR(AVG(retail),'99.99') "Avg Retail"
   FROM publisher JOIN books USING (pubid)
   WHERE pubid IN (2,3,5)
   GROUP BY (name, category)
  UNION
 SELECT NULL, category, COUNT(isbn), TO_CHAR(AVG(retail),'99.99') "Avg Retail"
   FROM publisher JOIN books USING (pubid)
   WHERE pubid IN (2,3,5)
   GROUP BY (NULL, category)
  UNION
 SELECT name, NULL, COUNT(isbn), TO_CHAR(AVG(retail),'99.99') "Avg Retail"
   FROM publisher JOIN books USING (pubid)
   WHERE pubid IN (2,3,5)
   GROUP BY (name, NULL)
  UNION
 SELECT NULL, NULL, COUNT(isbn), TO_CHAR(AVG(retail),'99.99') "Avg Retail"
   FROM publisher JOIN books USING (pubid)
   WHERE pubid IN (2,3,5);
```

FIGURE 11-33 Using a UNION operation to perform the same action as the GROUPING SETS expression

The GROUPING SETS expression gives you control over the specific aggregations to be performed. As mentioned, two extensions of the GROUP BY clause are used for additional GROUPING SETS actions: The CUBE extension performs cross-tabular aggregations, and the ROLLUP extension calculates subtotals.

The CUBE Extension

The **CUBE** extension of GROUP BY instructs Oracle to perform aggregations for all possible combinations of the specified columns. Figure 11-34 shows a statement using the CUBE option to perform all aggregate combinations on two columns: Name (publisher name) and Category.

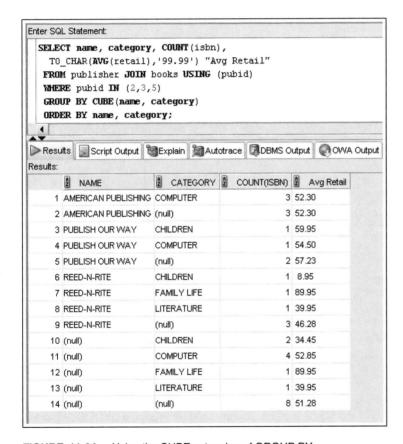

FIGURE 11-34   Using the CUBE extension of GROUP BY

This output matches all the aggregations performed in Figure 11-32 with the GROUP-ING SETS expression. If you need only a subset of the four aggregate levels calculated, you must use the GROUPING SETS expression because the CUBE extension *always* performs all aggregation levels.

Identifying subtotal rows is helpful in labeling, sorting, and restricting output. You can add a **GROUPING** function to the CUBE extension to identify subtotal rows in the results. This function returns a 1 to identify a row that displays a subtotal for a column, as shown in Figure 11-35.

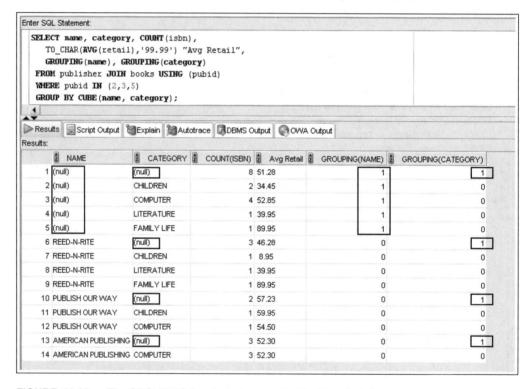

FIGURE 11-35 The GROUPING function returns a 1 to identify subtotal rows

In addition, the results of the GROUPING function could be used in a DECODE operation to label subtotal rows, as shown in Figure 11-36.

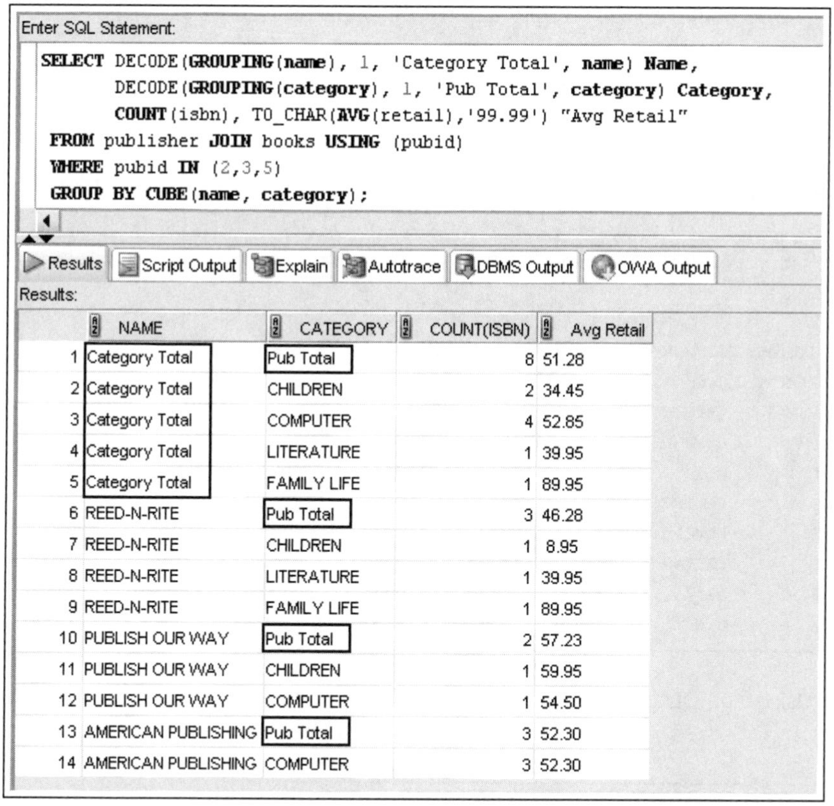

```
Enter SQL Statement:

SELECT DECODE(GROUPING(name), 1, 'Category Total', name) Name,
       DECODE(GROUPING(category), 1, 'Pub Total', category) Category,
       COUNT(isbn), TO_CHAR(AVG(retail),'99.99') "Avg Retail"
FROM publisher JOIN books USING (pubid)
WHERE pubid IN (2,3,5)
GROUP BY CUBE(name, category);
```

Results | Script Output | Explain | Autotrace | DBMS Output | OWA Output

Results:

	NAME	CATEGORY	COUNT(ISBN)	Avg Retail
1	Category Total	Pub Total	8	51.28
2	Category Total	CHILDREN	2	34.45
3	Category Total	COMPUTER	4	52.85
4	Category Total	LITERATURE	1	39.95
5	Category Total	FAMILY LIFE	1	89.95
6	REED-N-RITE	Pub Total	3	46.28
7	REED-N-RITE	CHILDREN	1	8.95
8	REED-N-RITE	LITERATURE	1	39.95
9	REED-N-RITE	FAMILY LIFE	1	89.95
10	PUBLISH OUR WAY	Pub Total	2	57.23
11	PUBLISH OUR WAY	CHILDREN	1	59.95
12	PUBLISH OUR WAY	COMPUTER	1	54.50
13	AMERICAN PUBLISHING	Pub Total	3	52.30
14	AMERICAN PUBLISHING	COMPUTER	3	52.30

FIGURE 11-36   Combining DECODE and the GROUPING function to label subtotal rows

The ROLLUP Extension

The **ROLLUP** extension of GROUP BY calculates cumulative subtotals for the specified columns. If multiple columns are indicated, subtotals are performed for each column in the argument list, except the one on the far right. A grand total is also calculated. Figure 11-37 shows a ROLLUP operation on the same two columns used in the CUBE operation: Name and Category. Three levels of increasing aggregation are performed: 1) combination of Name and Category columns, 2) Name subtotal, and 3) grand total.

You can also do a partial ROLLUP by including only a subset of the columns in the GROUP BY clause. For example, if you need only subtotals by Category and each Name in each Category, the statement in Figure 11-38 uses a partial ROLLUP for this task. In this example, the column outside the ROLLUP operation, Category, is considered the **aggregate value**. A subtotal is calculated for the aggregate value as well as for each unique value of the ROLLUP column in the aggregate value.

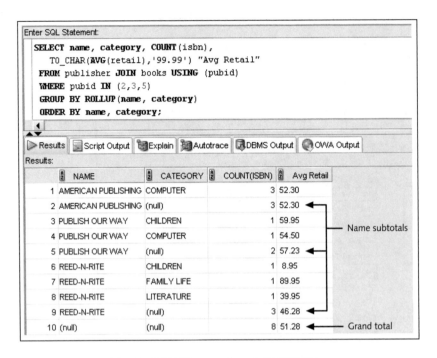

FIGURE 11-37 Using the ROLLUP extension of GROUP BY

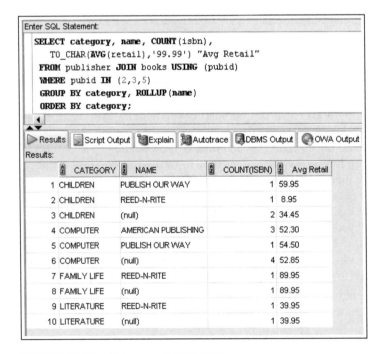

FIGURE 11-38 Using a partial ROLLUP

Using parentheses on a collection of columns creates a **composite column**, which is treated as a single unit in grouping operations. First, review a ROLLUP operation containing three columns, as shown in Figure 11-39. This ROLLUP operation performs four levels of increasing aggregation: 1) Name, Mth (the column alias for the combination of month and year), and Category, 2) Name and Mth, 3) Name, and 4) grand total.

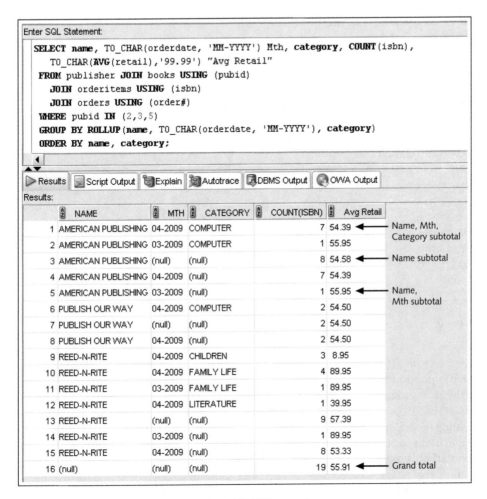

FIGURE 11-39   Using three columns in a ROLLUP operation

Next, modify this statement to combine Mth and Category as a composite column. Figure 11-40 shows the ROLLUP operation with the composite column, which is created by placing parentheses around the Mth and Category columns in the ROLLUP arguments.

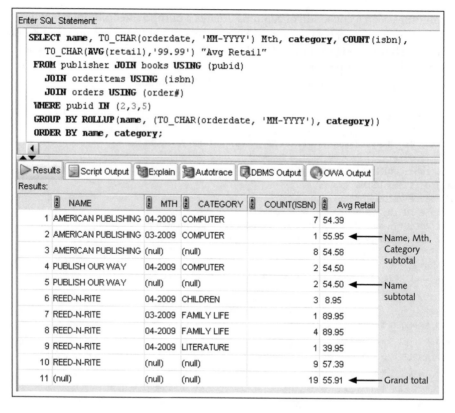

FIGURE 11-40   Using a composite column in a ROLLUP operation

Compared with Figure 11-39's operation without a composite column, one less level of aggregation is performed. The aggregation level of Name and Mth is no longer performed because the Mth column isn't considered a separate aggregation value in the ROLLUP operation. This ROLLUP operation is considered to have only two arguments: Name and the composite column of Mth and Category combined.

Combinations of groupings can be generated by using concatenated groupings. A **concatenated grouping** operation is created by listing multiple grouping sets in the grouping operation. Say you need accumulated aggregate totals of a book count and the average retail value of books by two different groupings: 1) Name and Category and 2) Mth and Yr. Figure 11-41 shows a concatenated grouping that includes two ROLLUP operations.

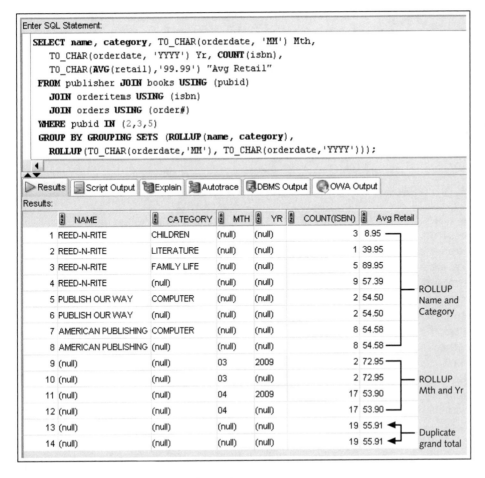

FIGURE 11-41 Using concatenated groupings

As grouping operations become more complex, sometimes duplicate grouping results are generated, as you can see in the duplicate grand totals in the last two rows of Figure 11-41's output. To eliminate duplicate grouping results, the **GROUP_ID** function is available. It returns a value of 1 for duplicate output rows. Figure 11-42 adds the GROUP_ID function to the previous statement.

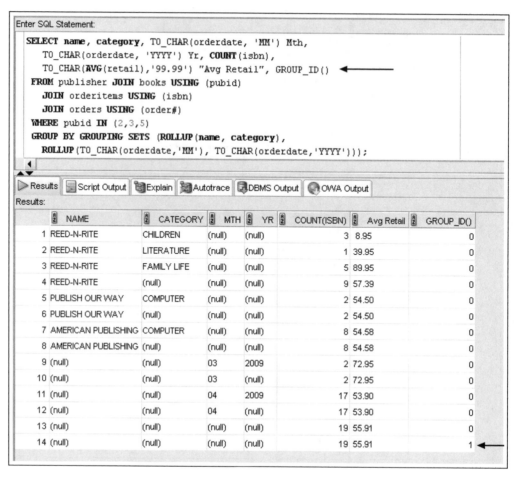

FIGURE 11-42 Using GROUP_ID to eliminate duplicate grouping results

Notice that the repeated grand total row has a GROUP_ID of 1. To eliminate the duplicate row in the output, add the HAVING clause HAVING GROUP_ID() = 0 to the query. The GROUP_ID value can be checked for any grouping level. The example in Figure 11-42 uses ( ) as the GROUP_ID argument, which represents the grand total grouping level.

TIP

Partial groupings, composite columns, and concatenated groupings can also be performed with GROUPING SETS and CUBE operations, using the same technique shown with the ROLLUP examples.

Chapter Summary

- The AVG, SUM, STDDEV, and VARIANCE functions are used only with numeric fields.
- The COUNT, MAX, and MIN functions can be applied to any datatype.
- The AVG, SUM, MAX, MIN, STDDEV, and VARIANCE functions ignore NULL values. The COUNT(*) function counts records containing NULL values. To include NULL values in other group functions, use the NVL function.
- By default, the AVG, SUM, MAX, MIN, COUNT, STDDEV, and VARIANCE functions include duplicate values. To include only unique values, use the DISTINCT keyword.
- The GROUP BY clause is used to divide query data into groups.
- If a SELECT clause contains both a nonaggregate column name and a group function, the column name must also be included in a GROUP BY clause.
- The HAVING clause is used to restrict groups, based on aggregated results.
- Group functions can be nested to a depth of only two. The inner function is always performed first, using the specified grouping. The results of the inner function are used as input for the outer function.
- The STDDEV and VARIANCE functions are used to perform statistical analysis on a set of data values.
- GROUPING SETS operations can be used to perform multiple GROUP BY aggregations with a single query.
- The CUBE extension of GROUP BY calculates aggregations for all possible combinations or groupings of specified columns.
- The ROLLUP extension of GROUP BY calculates increasing levels of accumulated subtotals for the specified column list.
- Composite columns and concatenated groupings can be used in GROUPING SETS, CUBE, and ROLLUP operations.
- The GROUP_ID function helps eliminate duplicate grouping results.

Chapter 11 Syntax Summary

The following table summarizes the syntax you have learned in this chapter. You can use the table as a study guide and reference.

Syntax Guide

Function and Syntax	Description	Example
Group Functions		
SUM([DISTINCT| ALL] n)	Returns the sum or total value of the selected numeric field. Ignores NULL values.	SELECT SUM(retail-cost) FROM books;

Function and Syntax	Description	Example
Group Functions		
AVG ([DISTINCT|ALL] n)	Returns the average value of the selected numeric field. Ignores NULL values.	SELECT AVG(cost) FROM books;

COUNT (*
[DISTINCT|ALL] c)
	Returns the number of rows containing a value in the identified field. Rows containing NULL values in the field aren't included in the results. To count all rows, including those with NULL values, use an asterisk (*) rather than a field name.	SELECT COUNT(*)
		  FROM books;
		or
		SELECT COUNT(shipdate)
		  FROM orders;

MAX ([DISTINCT| ALL] c)	Returns the highest (maximum) value from the selected field. Ignores NULL values.	SELECT MAX(customer#) FROM customers;
MIN ([DISTINCT| ALL] c)	Returns the lowest (minimum) value from the selected field. Ignores NULL values.	SELECT MIN(retail-cost) FROM books;
STDDEV ([DISTINCT| ALL] n)	Returns the standard deviation of the selected numeric field. Ignores NULL values.	SELECT STDDEV(retail) FROM books;
VARIANCE ([DISTINCT| ALL] n)	Returns the variance of the selected numeric field. Ignores NULL values.	SELECT VARIANCE(retail) FROM books;
Clauses		

GROUP BY
  columnname
  [, columnname, . . .]
	Divides data into sets or groups, based on the contents of specified columns.	SELECT AVG(cost)
		  FROM books
		  GROUP BY category;

HAVING
  groupfunction
  comparisonoperator
  value
	Restricts the groups displayed in query results.	SELECT AVG(cost)
		  FROM books
		  GROUP BY category
		  HAVING AVG(cost) > 21;

Function and Syntax	Description	Example
GROUP BY Extensions		

GROUPING SETS	Enables performing multiple GROUP BY operations with a single query.	```SELECT name, category,```
		 ``` AVG(retail)```
		 ```FROM publisher```
		 ```JOIN books USING (pubid)```
		 ```GROUP BY GROUPING SETS```
		 ``` (name, category,```
		 ``` (name,category),());```

CUBE	Performs aggregations for all possible combinations of columns included.	```SELECT name, category,```
		 ``` AVG(retail)```
		 ```FROM publisher```
		 ```JOIN books USING (pubid)```
		 ```GROUP BY CUBE(name,```
		 ``` category)```
		 ```ORDER BY name, category;```

ROLLUP	Performs increasing levels of cumulative subtotals, based on the specified column list.	```SELECT name, category,```
		 ``` AVG(retail)```
		 ```FROM publisher```
		 ```JOIN books USING (pubid)```
		 ```GROUP BY ROLLUP(name,```
		 ``` category)```
		 ```ORDER BY name, category;```

419

Review Questions

1. Explain the difference between single-row and group functions.
2. Which group function can be used to perform a count that includes NULL values?
3. Which clause can be used to restrict or filter the groups returned by a query based on a group function?
4. Under what circumstances *must* you include a GROUP BY clause in a query?
5. In which clause should you include the condition "pubid=4" to restrict the rows processed by a query?
6. In which clause should you include the condition MAX(cost) > 39 to restrict groups displayed in the query results?
7. What's the basic difference between the ROLLUP and CUBE extensions of the GROUP BY clause?
8. What's the maximum depth allowed when nesting group functions?
9. In what order are output results displayed if a SELECT statement contains a GROUP BY clause and no ORDER BY clause?
10. Which clause is used to restrict the records retrieved from a table? Which clause restricts groups displayed in the query results?

Multiple Choice

To answer these questions, refer to the tables in the JustLee Books database.

1. Which of the following statements is true?

    a. The MIN function can be used only with numeric data.

    b. The MAX function can be used only with date values.

    c. The AVG function can be used only with numeric data.

    d. The SUM function can't be part of a nested function.

2. Which of the following is a valid SELECT statement?

    a. ```
    SELECT AVG(retail-cost)
        FROM books
        GROUP BY category;
    ```

    b. ```
    SELECT category, AVG(retail-cost)
        FROM books;
    ```

    c. ```
    SELECT category, AVG(retail-cost)
        FROM books
        WHERE AVG(retail-cost) > 8.56
        GROUP BY category;
    ```

    d. ```
    SELECT category, AVG(retail-cost) Profit
        FROM books
        GROUP BY category
        HAVING profit > 8.56;
    ```

3. Which of the following statements is correct?

    a. The WHERE clause can contain a group function only if the function isn't also listed in the SELECT clause.

    b. Group functions can't be used in the SELECT, FROM, or WHERE clauses.

    c. The HAVING clause is always processed before the WHERE clause.

    d. The GROUP BY clause is always processed before the HAVING clause.

4. Which of the following is *not* a valid SQL statement?

    a. ```
    SELECT MIN(pubdate)
        FROM books
        GROUP BY category
        HAVING pubid = 4;
    ```

    b. ```
    SELECT MIN(pubdate)
        FROM books
        WHERE category = 'COOKING';
    ```

    c. ```
    SELECT COUNT(*)
        FROM orders
        WHERE customer# = 1005;
    ```
    ```

d.
```
SELECT MAX(COUNT(customer#))
   FROM orders
   GROUP BY customer#;
```

5. Which of the following statements is correct?

   a. The COUNT function can be used to determine how many rows contain a NULL value.

   b. Only distinct values are included in group functions, unless the ALL keyword is included in the SELECT clause.

   c. The HAVING clause restricts which rows are processed.

   d. The WHERE clause determines which groups are displayed in the query results.

   e. none of the above

6. Which of the following is a valid SQL statement?

   a.
   ```
   SELECT customer#, order#, MAX(shipdate-orderdate)
      FROM orders
      GROUP BY customer#
      WHERE customer# = 1001;
   ```

   b.
   ```
   SELECT customer#, COUNT(order#)
      FROM orders
      GROUP BY customer#;
   ```

   c.
   ```
   SELECT customer#, COUNT(order#)
      FROM orders
      GROUP BY COUNT(order#);
   ```

   d.
   ```
   SELECT customer#, COUNT(order#)
      FROM orders
      GROUP BY order#;
   ```

7. Which of the following SELECT statements lists only the book with the largest profit?

   a.
   ```
   SELECT title, MAX(retail-cost)
      FROM books
      GROUP BY title;
   ```

   b.
   ```
   SELECT title, MAX(retail-cost)
      FROM books
      GROUP BY title
      HAVING MAX(retail-cost);
   ```

   c.
   ```
   SELECT title, MAX(retail-cost)
      FROM books;
   ```

   d. none of the above

8. Which of the following is correct?

   a. A group function can be nested inside a group function.

   b. A group function can be nested inside a single-row function.

   c. A single-row function can be nested inside a group function.

d. a and b

e. a, b, and c

9. Which of the following functions is used to calculate the total value stored in a specified column?

   a. COUNT

   b. MIN

   c. TOTAL

   d. SUM

   e. ADD

10. Which of the following SELECT statements lists the highest retail price of all books in the Family category?

   a.
```
SELECT MAX (retail)
   FROM books
   WHERE category = 'FAMILY';
```

   b.
```
SELECT MAX (retail)
   FROM books
   HAVING category = 'FAMILY';
```

   c.
```
SELECT retail
   FROM books
   WHERE category = 'FAMILY'
   HAVING MAX (retail);
```

   d. none of the above

11. Which of the following functions can be used to include NULL values in calculations?

   a. SUM

   b. NVL

   c. MAX

   d. MIN

12. Which of the following is *not* a valid statement?

   a. You must enter the ALL keyword in a group function to include all duplicate values.

   b. The AVG function can be used to find the average calculated difference between two dates.

   c. The MIN and MAX functions can be used on any type of data.

   d. all of the above

   e. none of the above

13. Which of the following SQL statements determines how many total customers were referred by other customers?

   a.
```
SELECT customer#, SUM (referred)
   FROM customers
   GROUP BY customer#;
```

b.   SELECT COUNT (referred)
        FROM customers;

c.   SELECT COUNT (*)
        FROM customers;

d.   SELECT COUNT (*)
        FROM customers
        WHERE referred IS NULL;

Use the following SELECT statement to answer questions 14–18:

```
1    SELECT customer#, COUNT (*)
2    FROM customers JOIN orders USING (customer#)
3    WHERE orderdate > '02-APR-09'
4    GROUP BY customer#
5    HAVING COUNT (*) > 2;
```

14. Which line of the SELECT statement is used to restrict the number of records the query processes?

   a.   1
   b.   3
   c.   4
   d.   5

15. Which line of the SELECT statement is used to restrict groups displayed in the query results?

   a.   1
   b.   3
   c.   4
   d.   5

16. Which line of the SELECT statement is used to group data stored in the database?

   a.   1
   b.   3
   c.   4
   d.   5

17. Because the SELECT clause contains the Customer# column, which clause must be included for the query to execute successfully?

   a.   1
   b.   3
   c.   4
   d.   5

18. The COUNT(*) function in the SELECT clause is used to return:

    a. the number of records in the specified tables

    b. the number of orders placed by each customer

    c. the number of NULL values in the specified tables

    d. the number of customers who have placed an order

19. Which of the following functions can be used to determine the earliest ship date for all orders recently processed by JustLee Books?

    a. COUNT function

    b. MAX function

    c. MIN function

    d. STDDEV function

    e. VARIANCE function

20. Which of the following is *not* a valid SELECT statement?

    a.
```
SELECT STDDEV(retail)
    FROM books;
```

    b.
```
SELECT AVG(SUM(retail))
    FROM orders
    NATURAL JOIN orderitems NATURAL JOIN books
    GROUP BY customer#;
```

    c.
```
SELECT order#, TO_CHAR(SUM(retail),'999.99')
    FROM orderitems JOIN books USING (isbn)
    GROUP BY order#;
```

    d.
```
SELECT title, VARIANCE(retail-cost)
    FROM books
    GROUP BY pubid;
```

Hands-On Assignments

To perform these assignments, refer to the tables in the JustLee Books database.

1. Determine how many books are in the Cooking category.

2. Display the number of books with a retail price of more than $30.00.

3. Display the most recent publication date of all books sold by JustLee Books.

4. Determine the total profit generated by sales to customer 1017. *Note*: Quantity should be reflected in the total profit calculation.

5. List the retail price of the least expensive book in the Computer category.

6. Determine the average profit generated by orders in the ORDERS table. *Note*: The total profit by order must be calculated before finding the average profit.

7. Determine how many orders have been placed by each customer. Do not include in the results any customer who hasn't recently placed an order with JustLee Books.

8. Determine the average retail price of books by publisher name and category. Include only the categories Children and Computer and the groups with an average retail price greater than $50.

9. List the customers living in Georgia or Florida who have recently placed an order totaling more than $80.

10. What's the retail price of the most expensive book written by Lisa White?

Advanced Challenge

To perform this activity, refer to the tables in the JustLee Book database.

JustLee Books has a problem: Book storage space is filling up. As a solution, management is considering limiting the inventory to only those books returning at least a 55% profit. Any book returning less than a 55% profit would be dropped from inventory and not reordered.

This plan could, however, have a negative impact on overall sales. Management fears that if JustLee stops carrying the less profitable books, the company might lose repeat business from its customers. As part of management's decision-making process, it wants to know whether current customers purchase less profitable books frequently. Therefore, management wants to know how many times these less profitable books have been purchased recently.

Determine which books generate less than a 55% profit and how many copies of these books have been sold. Summarize your findings for management, and include a copy of the query used to retrieve data from the database tables.

Case Study: *City Jail*

Note: Make sure you have run the CityJail_8.sql script from Chapter 8. This script makes all database objects available to complete this case study.

The city's Crimes Analysis unit has submitted the following data requests. Provide the SQL statements to satisfy these requests. Test the statements and show the query results.

1. Show the average number of crimes reported by an officer.
2. Show the total number of crimes by status.
3. List the highest number of crimes committed by a person.
4. Display the lowest fine amount assigned to a crime charge.
5. List criminals (ID and name) who have multiple sentences assigned.
6. List the total number of crime charges successfully defended (guilty status assigned) by precinct. Include only precincts with at least seven guilty charges.
7. List the total amount of collections (fines and fees) and the total amount owed by crime classification.

   Use single queries to address the following requests:

8. List the total number of charges by crime classification and charge status. Include a grand total in the results.
9. Perform the same task as in Question #8 and add the following: a) a subtotal by each crime classification and b) a subtotal for each charge status. Provide two different queries to accomplish this task.
10. Perform the same task as in Question #8 and add a subtotal by each crime classification. Provide two different queries to accomplish this task.

SUBQUERIES AND MERGE STATEMENTS

LEARNING OBJECTIVES

After completing this chapter, you should be able to do the following:

- Determine when using a subquery is appropriate
- Identify which clauses can contain subqueries
- Distinguish between an outer query and a subquery
- Use a single-row subquery in a WHERE clause
- Use a single-row subquery in a HAVING clause
- Use a single-row subquery in a SELECT clause
- Distinguish between single-row and multiple-row comparison operators
- Use a multiple-row subquery in a WHERE clause
- Use a multiple-row subquery in a HAVING clause
- Use a multiple-column subquery in a WHERE clause
- Create an inline view by using a multiple-column subquery in a FROM clause
- Compensate for NULL values in subqueries
- Distinguish between correlated and uncorrelated subqueries
- Nest a subquery inside another subquery
- Use a subquery in a DML action
- Process multiple DML actions with a MERGE statement

INTRODUCTION

Suppose that the management of JustLee Books requests a list of every computer book that has a higher retail price than the book *Database Implementation*. In previous chapters, you would have followed a procedure including two separate queries: 1) Query the database to determine the retail price of *Database Implementation*, and then 2) create a second SELECT statement to find the titles of all computer books retailing for more than *Database Implementation*. In this chapter, you learn how to use an alternative approach, called a subquery, to get the same output by using a single SQL statement. A **subquery** is a nested query—one complete query inside another query.

The subquery's output can consist of a single value (a single-row subquery), several rows of values (a multiple-row subquery), or even multiple columns of data (a multiple-column subquery). This chapter addresses each type of subquery. In addition, the final section returns to the topic of DML and introduces advanced DML actions, including subqueries and the MERGE statement. With the MERGE statement, you can conditionally process multiple DML actions with a single SQL statement. Table 12-1 gives you an overview of this chapter's contents.

TABLE 12-1   Topics Covered in This Chapter

Subquery	Description
Single-row subquery	Returns to the outer query one row of results consisting of one column
Multiple-row subquery	Returns to the outer query more than one row of results
Multiple-column subquery	Returns to the outer query more than one column of results
Correlated subquery	References a column in the outer query and executes the subquery once for every row in the outer query
Uncorrelated subquery	Executes the subquery first and passes the value to the outer query
DML subquery	Uses a subquery to determine the rows affected by the DML action
MERGE statement	Conditionally processes a series of DML statements

DATABASE PREPARATION

Before attempting to work through the examples in this chapter, run the JLDB_Build_12.sql script to make the necessary additions to the JustLee Books database. You should have already executed the JLDB_Build_8.sql script as instructed in Chapter 8.

SUBQUERIES AND THEIR USES

As described earlier, getting an answer to a query sometimes requires a multistep operation. First, you must create a query to determine a value you don't know but that's stored in the database. This first query is the subquery. The subquery's results are passed as input to the **outer query** (also called the **parent query**). The outer query incorporates this value into its calculations to determine the final output.

Although subqueries are used most commonly in the WHERE or HAVING clause of a SELECT statement, at times using a subquery in the SELECT or FROM clause is appropriate. When the subquery is nested in a WHERE or HAVING clause, the results it returns are used as a condition in the outer query. Any type of subquery (single-row, multiple-row, or multiple-column) can be used in the WHERE, HAVING, or FROM clause of a SELECT statement. As you'll see, the only type of subquery that can be used in a SELECT clause is a single-row subquery.

NOTE

Using a subquery in a FROM clause has a specific purpose, as you learn later in "Multiple-Column Subquery in a FROM Clause."

Keep the following rules in mind when working with any type of subquery:

- A subquery must be a *complete query in itself*—in other words, it must have at least a SELECT and a FROM clause.
- A subquery, except one in the FROM clause, can't have an ORDER BY clause. If you need to display output in a specific order, include an ORDER BY clause as the outer query's last clause.
- A subquery must be *enclosed in parentheses* to separate it from the outer query.
- If you place a subquery in the outer query's WHERE or HAVING clause, you can do so only on the *right side* of the comparison operator.

SINGLE-ROW SUBQUERIES

A **single-row subquery** is used when the outer query's results are based on a single, unknown value. Although this query type is formally called "single-row," the name implies that the query returns multiple columns—but only one row—of results. However, a single-row subquery can return *only one row of results consisting of only one column* to the outer query. Therefore, this textbook refers to the output of a single-row subquery as a **single value**.

Single-Row Subquery in a WHERE Clause

To see how subqueries work, you can compare creating multiple queries, which you've studied in previous chapters, with creating one query containing a subquery. In this chapter's introduction, management requested a list of all computer books with a higher retail price than the book *Database Implementation*. As shown in Figure 12-1, the first step is to create a query to determine the book's retail price, which is $31.40.

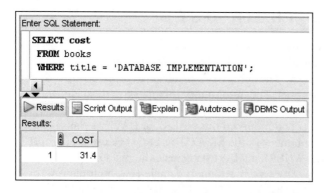

FIGURE 12-1 Query to determine the retail price of *Database Implementation*

To determine which computer books retail for more than $31.40, you must issue a second query stating the cost of *Database Implementation* in the WHERE clause condition, as shown in Figure 12-2. The WHERE clause includes the retail price of *Database Implementation*, which the query in Figure 12-1 found. The category condition also restricts records to those only in the Computer category.

```
Enter SQL Statement:

  SELECT category, title, cost
  FROM books
  WHERE cost > 31.4
    AND category = 'COMPUTER';
```

	CATEGORY	TITLE	COST
1	COMPUTER	HOLY GRAIL OF ORACLE	47.25
2	COMPUTER	E-BUSINESS THE EASY WAY	37.9

FIGURE 12-2 Query for computer books costing more than $31.40

You can get these same results with a single SQL statement by using a single-row subquery. A single-row subquery is appropriate in this example because 1) to get the results you need, an unknown value that's stored in the database must be found, and 2) only one value should be returned from the inner query (the retail price of *Database Implementation*).

In Figure 12-3, a single-row subquery is substituted for the `cost > 31.4` condition of the SELECT statement in Figure 12-2. This subquery is enclosed in parentheses to distinguish it from the outer query's clauses.

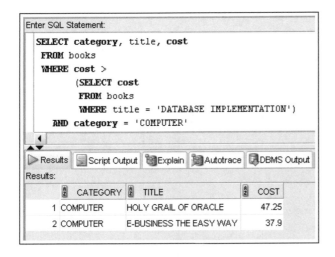

```
Enter SQL Statement:
    SELECT category, title, cost
    FROM books
    WHERE cost >
            (SELECT cost
             FROM books
             WHERE title = 'DATABASE IMPLEMENTATION')
        AND category = 'COMPUTER'
```

	CATEGORY	TITLE	COST
1	COMPUTER	HOLY GRAIL OF ORACLE	47.25
2	COMPUTER	E-BUSINESS THE EASY WAY	37.9

FIGURE 12-3   A single-row subquery

In Figure 12-3, the inner query is executed first, and this query's result, a single value of 31.4, is passed to the outer query. The outer query is then executed, and all books having a retail price greater than $31.40 and belonging to the Computer category are listed in the output. Using a single SQL statement prevents the need for user intervention to accomplish the task. In addition, the subquery enables this query to always reflect the current price of the *Database Implementation* book.

N O T E

The indentation in Figure 12-3's subquery and in other figures in this chapter is used only to improve readability; it isn't required by Oracle 11*g*.

Operators indicate to Oracle 11*g* whether you're creating a single-row subquery or a multiple-row subquery. The single-row operators are =, >, <, >=, <=, and <>. Although other operators, such as IN, are allowed, single-row operators instruct Oracle 11*g* that only one value is expected from the subquery. If more than one value is returned, the SELECT statement fails, and you get an error message.

Suppose management makes another request: the title of the most expensive book sold by JustLee Books. The MAX function covered in Chapter 11 handles this task. You might be tempted to create the query shown in Figure 12-4.

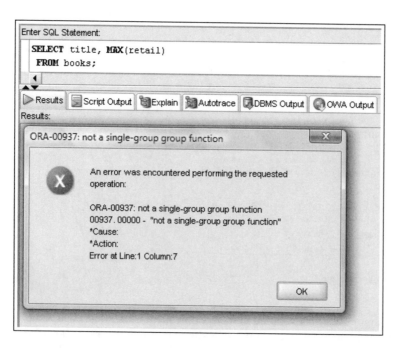

FIGURE 12-4 Flawed query: attempt to determine the book with the highest retail value

You encountered this problem when working with aggregate functions in Chapter 11. Recall this rule when working with group functions: If a nonaggregate field is listed with a group function in the SELECT clause, the field must also be listed in a GROUP BY clause. In this example, however, adding a GROUP BY clause doesn't make sense. If a GROUP BY clause containing the Title column is added, each book would be its own group because each title is different. In other words, the results would be the same as using SELECT title, retail in the query.

Therefore, to retrieve the title of the most expensive book, you can use a subquery to determine the highest retail price of any book. This retail price can then be returned to an outer query and displayed in the results.

As shown in Figure 12-5, the most expensive book sold by JustLee Books is *Painless Child-Rearing*. The book's retail price is included in the query output, but it isn't required. In this case, only one book matches the highest price of $89.95; however, multiple book titles could be displayed if more than one book matched the high price.

TIP

The query statement serving as the subquery should be created and executed first by itself. In this way, you can verify that the query produces the expected results before embedding it in another query as a subquery.

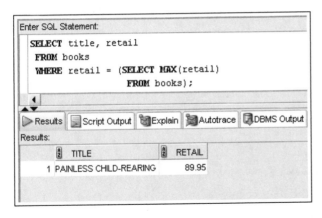

```
Enter SQL Statement:
  SELECT title, retail
  FROM books
  WHERE retail = (SELECT MAX(retail)
                      FROM books);
```

Results | Script Output | Explain | Autotrace | DBMS Output

Results:

	TITLE	RETAIL
1	PAINLESS CHILD-REARING	89.95

FIGURE 12-5   Query to determine the title of the most expensive book

CAUTION

If an error message is returned for the query in Figure 12-5, make sure the subquery contains four parentheses—one set around the `retail` argument for the MAX function and one set around the subquery.

You can include multiple subqueries in a SELECT statement. For example, suppose management needs to know the title of all books published by the publisher of *Big Bear and Little Dove* that generate more than the average profit returned by all books sold by JustLee Books. In this case, two values are unknown: the identity of the publisher of *Big Bear and Little Dove* and the average profit of all books. How might you create a query that extracts these values? The SELECT statement in Figure 12-6 uses two separate subqueries in the WHERE clause to find the information.

Notice that both subqueries in Figure 12-6 are complete because they contain a minimum of one SELECT clause and one FROM clause. Because they are subqueries, each one is enclosed in parentheses. The first subquery determines the publisher of *Big Bear and Little Dove* and returns the result to the first condition of the WHERE clause (WHERE pubid =). The second subquery finds the average profit of all books sold by JustLee Books by using the AVG function, and then passes this value to the second condition of the WHERE clause (AND retail-cost >) to be compared against the profit for each book. Because the two conditions of the outer query's WHERE clause are combined with the AND logical operator, both values returned by the subqueries must be met for a book to be listed in the outer query's output. In this example, the query finds two books published by the publisher of *Big Bear and Little Dove* that return more than the average profit.

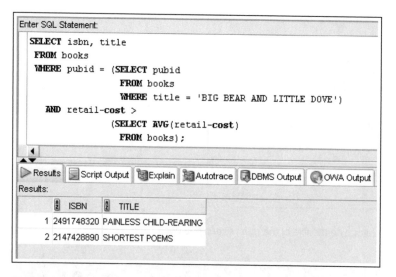

Enter SQL Statement:

```
SELECT isbn, title
 FROM books
 WHERE pubid = (SELECT pubid
                  FROM books
                  WHERE title = 'BIG BEAR AND LITTLE DOVE')
     AND retail-cost >
                  (SELECT AVG(retail-cost)
                     FROM books);
```

| ▶ Results | 🖹 Script Output | 🖼 Explain | 🖼 Autotrace | 🖥 DBMS Output | 🌐 OWA Output |

Results:

	ISBN	TITLE
1	2491748320	PAINLESS CHILD-REARING
2	2147428890	SHORTEST POEMS

FIGURE 12-6   SELECT statement with two single-row subqueries

Single-Row Subquery in a HAVING Clause

As mentioned, you can include a subquery in a HAVING clause. A HAVING clause is used when the group results of a query need to be restricted based on some condition. If a subquery's result must be compared with a group function, you must nest the inner query in the outer query's HAVING clause.

For example, if management needs a list of all book categories returning a higher average profit than the Literature category, you would follow these steps:

1. Calculate the average profit for all Literature books.
2. Calculate the average profit for each category.
3. Compare the average profit for each category with the average profit for the Literature category.

To perform these steps, you use the AVG function and include a subquery in the HAVING clause, as shown in Figure 12-7. The results are restricted to groups having a higher average profit than the Literature category. Because the subquery's results are applied to groups of data, nesting the subquery in the HAVING clause is necessary. (Recall from Chapter 11 that filtering by aggregate data occurs in the HAVING clause, not the WHERE clause).

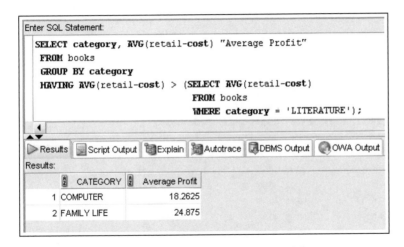

Enter SQL Statement:

```
SELECT category, AVG(retail-cost) "Average Profit"
FROM books
GROUP BY category
HAVING AVG(retail-cost) > (SELECT AVG(retail-cost)
                           FROM books
                           WHERE category = 'LITERATURE');
```

Results | Script Output | Explain | Autotrace | DBMS Output | OWA Output

Results:

	CATEGORY	Average Profit
1	COMPUTER	18.2625
2	FAMILY LIFE	24.875

FIGURE 12-7   Single-row subquery nested in a HAVING clause

As Figure 12-7 shows, the subquery in a HAVING clause must follow the same guidelines as for WHERE clauses: It must include at least a SELECT clause and a FROM clause, and it must be enclosed in parentheses.

Single-Row Subquery in a SELECT Clause

A single-row subquery can also be nested in the outer query's SELECT clause. In this case, the value the subquery returns is available for every row of output the outer query generates. Typically, this technique is used to perform calculations with a value produced from a subquery. For example, suppose management wants to compare the price of each book in inventory against the average price of all books in inventory. The query output must show each book's price and the amount above or below the average.

You can accomplish this task by using a subquery in a SELECT clause that calculates the average retail price of all books. When a single-row subquery is included in a SELECT clause, the subquery's results are displayed in the outer query's output. To include a subquery in a SELECT clause, you use a comma to separate the subquery from the table columns, as though you were listing another column. In fact, you can even give the subquery results a column alias. First, use the query in Figure 12-8 to verify that the same average amount the subquery returns is available for each row in the parent query. The TO_CHAR function is used to round the average amount to two decimal places.

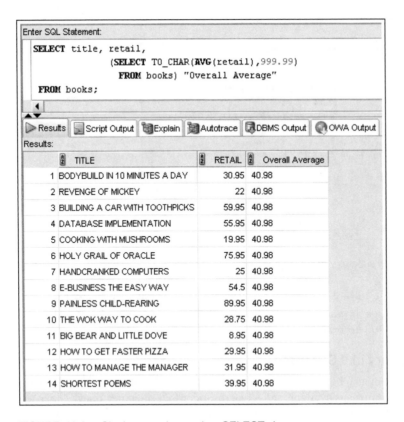

FIGURE 12-8   Single-row subquery in a SELECT clause

To calculate the average price of all books in inventory, the outer query's SELECT clause includes the Title and Retail columns as well as the subquery in the column list. The average calculated by the subquery is displayed for every book included in the output. The column alias, Overall Average, is assigned to the subquery's results to indicate the column's contents. If a column alias isn't used, the actual subquery shows as the column heading, which is somewhat unattractive and not very descriptive. Having the subquery in the SELECT clause enables management to compare each book's retail price to the average retail price for all books by looking at just one list.

Can this subquery be used in a calculation to determine the difference between each book price and the average? Absolutely. Simply move the subquery into a calculation, as shown in Figure 12-9, to calculate the difference between the retail price and the average price.

```
SELECT title, retail, retail-(SELECT TO_CHAR(AVG(retail),999.99)
                              FROM books) "Diff from AVG"
  FROM books;
```

◄

▲▼

▷ Results | ▤ Script Output | ▤ Explain | ▤ Autotrace | ▤ DBMS Output | ◎ OWA Output

Results:

	TITLE	RETAIL	Diff from AVG
1	BODYBUILD IN 10 MINUTES A DAY	30.95	-10.03
2	REVENGE OF MICKEY	22	-18.98
3	BUILDING A CAR WITH TOOTHPICKS	59.95	18.97
4	DATABASE IMPLEMENTATION	55.95	14.97
5	COOKING WITH MUSHROOMS	19.95	-21.03
6	HOLY GRAIL OF ORACLE	75.95	34.97
7	HANDCRANKED COMPUTERS	25	-15.98
8	E-BUSINESS THE EASY WAY	54.5	13.52
9	PAINLESS CHILD-REARING	89.95	48.97
10	THE WOK WAY TO COOK	28.75	-12.23
11	BIG BEAR AND LITTLE DOVE	8.95	-32.03
12	HOW TO GET FASTER PIZZA	29.95	-11.03
13	HOW TO MANAGE THE MANAGER	31.95	-9.03
14	SHORTEST POEMS	39.95	-1.03

FIGURE 12-9   Use a subquery in a calculation in the SELECT clause

TIP

Users learning to work with subqueries often don't have much confidence in the output when the subquery is in a WHERE clause because the value the subquery generates isn't displayed. However, if the subquery used in the WHERE clause is also included in the SELECT clause, the value the single-row subquery generates can be compared against the outer query's final output. This comparison is an easy way to validate the output. The subquery is removed from the SELECT clause after validation to generate the requested results. Validation gives you more confidence in the final results and reduces the risk of distributing errone-ous data. However, this method works only for single-row subqueries. For other types of subqueries, you must execute the subquery as a separate SELECT statement to determine the values it generates because a SELECT clause can process only single-row subqueries.

MULTIPLE-ROW SUBQUERIES

Multiple-row subqueries are nested queries that can return more than one row of results to the parent query. Multiple-row subqueries are used most commonly in WHERE and HAVING clauses. The main rule to keep in mind when working with multiple-row sub-queries is that you *must* use multiple-row operators. If a single-row operator is used with a subquery that returns more than one row of results, Oracle 11g returns an error message, and the SELECT statement fails. Valid multiple-row operators include IN, ALL, and ANY, discussed in the following sections.

The IN Operator

Of the three multiple-row operators, the IN operator is used most often. Figure 12-10 shows a multiple-row subquery with this operator. This query identifies books with a retail value matching the highest retail value for any book category.

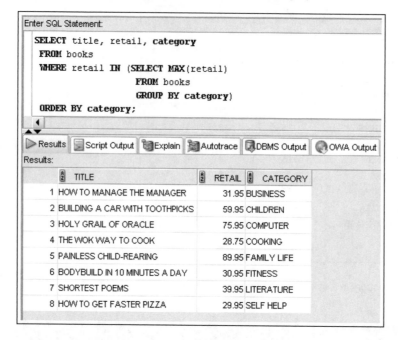

FIGURE 12-10 Multiple-row subquery with the IN operator

The IN operator in this subquery indicates that the records the outer query processes must match one of the values the subquery returns. (In other words, it creates an OR condition.) The order of execution in Figure 12-10 is as follows:

1. The subquery determines the price of the most expensive book in each category.
2. The maximum retail price in each category is passed to the WHERE clause of the outer query (a list of values).
3. The outer query compares each book's price to the prices generated by the subquery.
4. If a book's retail price matches one of the prices returned by the subquery, the book's title, retail price, and category are displayed in the output.

The ALL and ANY Operators

The ALL and ANY operators can be combined with other comparison operators to treat a subquery's results as a set of values instead of single values. Table 12-2 summarizes the use of the ALL and ANY operators with other comparison operators.

TABLE 12-2   ALL and ANY Operator Combinations

Operator	Description
>ALL	More than the highest value returned by the subquery
<ALL	Less than the lowest value returned by the subquery
<ANY	Less than the highest value returned by the subquery
>ANY	More than the lowest value returned by the subquery
=ANY	Equal to any value returned by the subquery (same as IN)

The ALL operator is fairly straightforward:

- If the ALL operator is combined with the "greater than" symbol (>), the outer query searches for all records with a value higher than the highest value returned by the subquery (in other words, more than ALL the values returned).
- If the ALL operator is combined with the "less than" symbol (<), the outer query searches for all records with a value lower than the lowest value returned by the subquery (in other words, less than ALL the values returned).

To examine the impact of using the ALL comparison operator, look at the query in Figure 12-11, which will be used later as a subquery. It returns the retail prices for two books in the Cooking category. The lowest value returned is $19.95, and the highest value is $28.75.

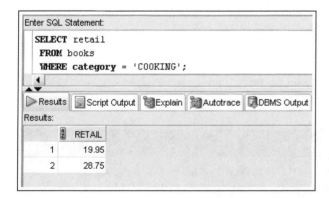

```
Enter SQL Statement:

SELECT retail
FROM books
WHERE category = 'COOKING';
```

	RETAIL
1	19.95
2	28.75

FIGURE 12-11   Retail price of books in the Cooking category

Suppose you want to know the titles of all books having a retail price greater than the most expensive book in the Cooking category. One approach is using the MAX function in a subquery to find the highest retail price. Another approach is using the >ALL operator, as shown in Figure 12-12.

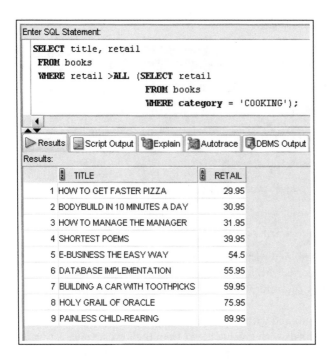

```
Enter SQL Statement:

  SELECT title, retail
  FROM books
  WHERE retail >ALL (SELECT retail
                     FROM books
                     WHERE category = 'COOKING');
```

	TITLE		RETAIL
1	HOW TO GET FASTER PIZZA		29.95
2	BODYBUILD IN 10 MINUTES A DAY		30.95
3	HOW TO MANAGE THE MANAGER		31.95
4	SHORTEST POEMS		39.95
5	E-BUSINESS THE EASY WAY		54.5
6	DATABASE IMPLEMENTATION		55.95
7	BUILDING A CAR WITH TOOTHPICKS		59.95
8	HOLY GRAIL OF ORACLE		75.95
9	PAINLESS CHILD-REARING		89.95

FIGURE 12-12   Using the >ALL operator

The Oracle 11g strategy for processing the SELECT statement in Figure 12-12 is as follows:

- The subquery passes the retail prices of the two books in the Cooking category ($19.95 and $28.75) to the outer query.
- Because the >ALL operator is used in the outer query, Oracle 11g is instructed to list all books with a retail price higher than the largest value returned by the subquery ($28.75).

In this example, nine books have a higher price than the most expensive book in the Cooking category.

> **TIP**
>
> You could get the same results as in Figure 12-12 by using the MAX function in the subquery. In this case, a single value for the highest priced book in the Cooking category is returned from the subquery, and a multiple-row operator isn't required.

Similarly, the <ALL operator is used to determine records with a value less than the lowest value returned by a subquery. Therefore, if you need to find books priced lower than the least expensive book in the Cooking category, first formulate a subquery that identifies books in the Cooking category. Then you can compare the retail price of books in the BOOKS table against the values returned by a subquery by using the <ALL operator.

As in the previous query, the subquery in Figure 12-13 first finds the two books in the Cooking category (shown in Figure 12-11). The retail prices of these books ($19.95 and $28.75) are then passed to the outer query. Because $19.95 is the lowest retail price of all books in the Cooking category, only books with a retail price lower than $19.95 are displayed in the output. In this case, Oracle 11g found only one book with a retail price lower than the least expensive book in the Cooking category: *Big Bear and Little Dove*.

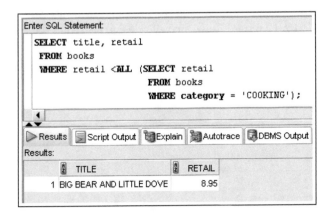

FIGURE 12-13 Using the <ALL operator

By contrast, the <ANY operator is used to find records with a value less than the highest value returned by a subquery. To determine which books cost less than the most expensive book in the Cooking category, evaluate the subquery's results by using the <ANY operator, as shown in Figure 12-14.

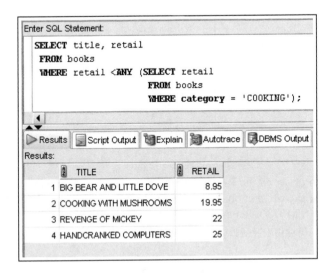

FIGURE 12-14 Using the <ANY operator

The outer query finds four books with a retail price lower than the most expensive book in the Cooking category. Notice, however, that the results also include *Cooking with Mushrooms*, a book in the Cooking category. Because the outer query compares the records to the highest value in the Cooking category, any other book in the Cooking category is also displayed in the query results. To eliminate any book in the Cooking category from appearing in the output, simply add the condition AND category < > 'COOKING' to the outer query's WHERE clause.

The >ANY operator is used to return records with a value greater than the lowest value returned by the subquery. In Figure 12-15, 12 records have a retail price greater than the lowest retail price returned by the subquery ($19.95).

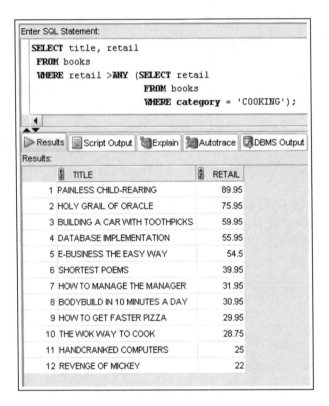

FIGURE 12-15   Using the >ANY operator

The =ANY operator works the same way as the IN comparison operator. For example, the query in Figure 12-16 searches for the titles of books purchased by customers who also purchased the book with the ISBN 0401140733. Because this book could have appeared on more than one order, and the query is supposed to identify all these orders, the =ANY operator is used.

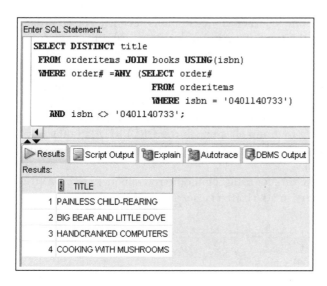

```
Enter SQL Statement:

  SELECT DISTINCT title
  FROM orderitems JOIN books USING(isbn)
  WHERE order# =ANY (SELECT order#
                        FROM orderitems
                        WHERE isbn = '0401140733')
    AND isbn <> '0401140733';
```

Results | Script Output | Explain | Autotrace | DBMS Output

Results:

	TITLE
1	PAINLESS CHILD-REARING
2	BIG BEAR AND LITTLE DOVE
3	HANDCRANKED COMPUTERS
4	COOKING WITH MUSHROOMS

FIGURE 12-16   Using the =ANY operator

This query would have yielded the same results if the IN operator had been used instead of the =ANY operator. The DISTINCT keyword in the outer query's SELECT clause is included because, as mentioned, a title could have been ordered by more than one customer and would have multiple listings in the output.

TIP

If you don't get the same results as in Figure 12-16, make sure the closing parenthesis for the subquery is placed before the last line beginning with the AND keyword, which is part of the outer query.

Also, notice in Figure 12-16 that the columns needed to complete the outer query are in two different tables: ORDERITEMS and BOOKS. A join is required in the outer query to combine the rows of these two tables. Because the columns needed to perform the inner query are contained only in the ORDERITEMS table, no join is required in the subquery.

NOTE

Another operator, EXISTS, is available to handle multiple-row subqueries and is discussed later in "Correlated Subqueries."

Multiple-Row Subquery in a HAVING Clause

So far, you have seen multiple-row subqueries in a WHERE clause, but they can also be included in a HAVING clause. When the subquery's results are compared to grouped data in the outer query, the subquery *must* be nested in a HAVING clause in the outer query.

For example, you need to determine whether any customer's recently placed order has a total amount due greater than the total amount due for every order placed recently by customers in Florida. Getting this output requires determining the total amount due for each order placed by a Florida customer, and then comparing these totals with every order total. The order totals for Florida customers can be calculated in a subquery, but because one value is returned for each order, you need a multiple-row subquery. These totals need to be compared with the total amount due for each order, which requires the outer query to group all items in the ORDERITEMS table by the Order# column. Therefore, the outer query must use a HAVING clause because the comparison is based on grouped data.

As shown in Figure 12-17, the structure for using a multiple-row subquery in a HAVING clause is the same as using the subquery in a WHERE clause.

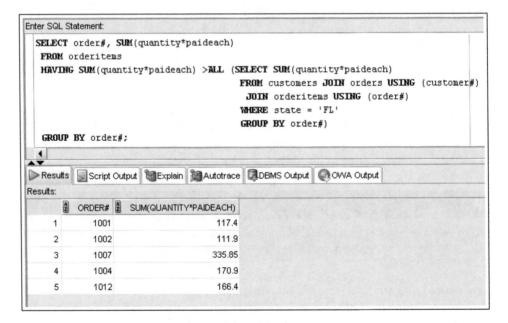

FIGURE 12-17   Multiple-row subquery in a HAVING clause

Single-row and multiple-row subqueries might look the same in terms of the subqueries themselves; however, a single-row subquery can return only *one* data value, whereas a multiple-row subquery can return *several* values. Therefore, if you execute a subquery that returns more than one data value and the comparison operator is intended to be used only with single-row subqueries, you get an error message and the query isn't executed, as shown in Figure 12-18.

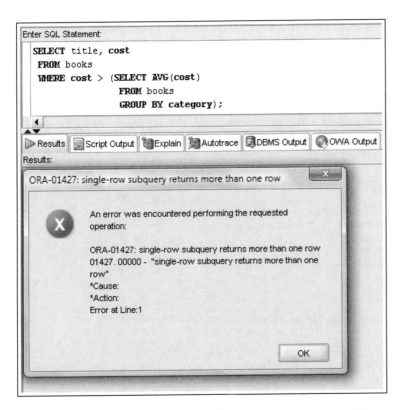

```
Enter SQL Statement:
  SELECT title, cost
  FROM books
  WHERE cost > (SELECT AVG(cost)
                FROM books
                GROUP BY category);
```

Results ▸ | Script Output | Explain | Autotrace | DBMS Output | OWA Output

Results:

ORA-01427: single-row subquery returns more than one row   ✕

An error was encountered performing the requested
operation:

ORA-01427: single-row subquery returns more than one row
01427. 00000 - "single-row subquery returns more than one
row"
*Cause:
*Action:
Error at Line:1

OK

FIGURE 12-18   Flawed query: using a single-row operator for a multiple-row subquery

445

MULTIPLE-COLUMN SUBQUERIES

Now that you've examined multiple-row subqueries, this section explores multiple-column
subqueries. A **multiple-column subquery** returns more than one column to the outer query
and can be listed in the outer query's FROM, WHERE, or HAVING clause.

Multiple-Column Subquery in a FROM Clause

When a multiple-column subquery is used in the outer query's FROM clause, it creates a
temporary table that can be referenced by other clauses of the outer query. This temporary
table is more formally called an **inline view**. The subquery's results are treated like any
other table in the FROM clause. If the temporary table contains grouped data, the grouped
subsets are treated as separate rows of data in a table.

NOTE

Views are covered in Chapter 13.

Suppose you need a list of all books in the BOOKS table that have a higher-than-average selling price compared with other books in the same category. For each book, you need to display the title, retail price, category, and average selling price of books in that category. Because the average selling price is based on grouped data, this query presents a problem. How might you solve it?

In Figure 12-19, a multiple-column subquery is nested in the outer query's FROM clause. The subquery creates a temporary table, including a column for the category and a column for the category average. The subquery determines the categories in the BOOKS table and the average selling price of every book in each category.

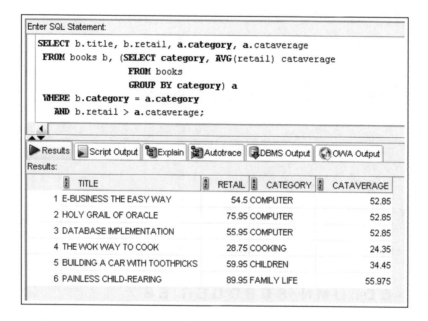

FIGURE 12-19   Multiple-column subquery in a FROM clause

However, how do you display the title of each book in the BOOKS table along with its retail price, its category, and the average price of all books in the same category? The BOOKS table contains the data for each book, and the subquery creates a temporary table that stores the grouped data. Notice in Figure 12-19 that the table alias "a" has been assigned to the subquery's results, so the columns in the subquery (Category and Cataverage) can be referenced by other clauses in the outer SELECT statement. Essentially, this alias assigns a table name to the subquery's results.

The query is referencing, or finding, data from two different tables, and one just happens to be created at runtime by the subquery. The tables have been joined by using the traditional approach—the outer query's WHERE clause. The problem with the traditional approach is that both tables contain a column called Category, which creates an ambiguity problem if the Category column is referenced anywhere in the outer query. To avoid this problem, the Category column needs a column qualifier to identify which table contains the category data to be displayed. Therefore, table aliases are used in the SELECT and WHERE clauses to identify the table containing the column being referenced.

As shown in Figure 12-20, the query could have also been created by using an ANSI JOIN operation supported by Oracle 11g. Because both tables (BOOKS and the temporary table created by the subquery) contain a column named Category, the tables are linked in the FROM clause with a join using the Category field. Because column qualifiers aren't allowed with the JOIN statement, the temporary table created by the subquery isn't assigned a table alias.

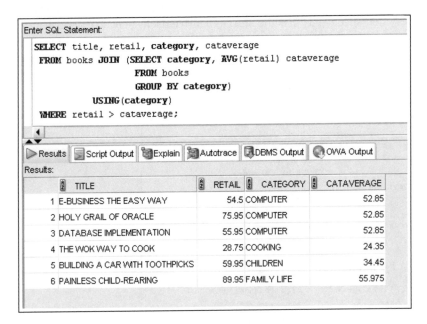

FIGURE 12-20 Using a join with a multiple-column subquery in the FROM clause

Multiple-Column Subquery in a WHERE Clause

When a multiple-column subquery is included in the outer query's WHERE or HAVING clause, the outer query uses the IN operator to evaluate the subquery's results. The subquery's results consist of more than one column of results.

The syntax of the outer WHERE clause is WHERE (`columnname, columnname, . . .`) IN `subquery`. Keep these rules in mind:

- Because the WHERE clause contains more than one column name, the column list must be enclosed in parentheses.
- Column names listed in the WHERE clause must be in the same order as they're listed in the subquery's SELECT clause.

TIP

Double-check that the column list in the outer query's WHERE clause is enclosed in parentheses and is in the same order as the column list in the subquery's SELECT clause.

In Figure 12-10 shown earlier, the subquery returned the price of the most expensive book in each category, and the outer query generated a list of the title, retail price, and category of books matching the retail price returned by the subquery. The overall result of the outer query was to display the title, retail price, and category for the most expensive book in each category. However, the query results could be misleading because a book retail value that matches a MAX value of any category is included in the output. So books in the output might be matches of the category MAX value of a different category than their own.

To create a query that specifically lists the most expensive books in each category, a multiple-column subquery is more suitable. Look at the example in Figure 12-21. The subquery finds the highest retail value in each category and passes both the category names and retail prices to the outer query.

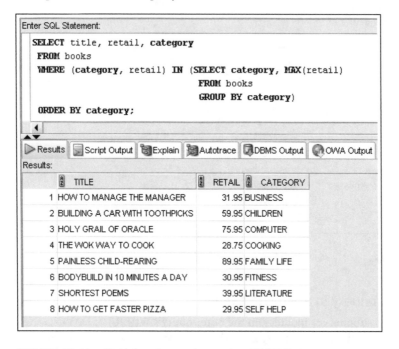

FIGURE 12-21   Multiple-column subquery in a WHERE clause

> **NOTE**
>
> Although a multiple-column subquery can be used in the outer query's HAVING clause, it's usually used only when analyzing extremely large sets of grouped numeric data. Generally, this method is discussed in more advanced courses focusing on quantitative methods.

NULL VALUES

As with everything else, NULL values present a challenge when using subqueries. Because a NULL value is the same as the absence of data, a NULL can't be returned to an outer query for comparison purposes; it's not equal to anything, not even another NULL. Therefore, if a NULL value is passed from a subquery, the results of the outer query are "no rows selected." Although the statement doesn't fail (it doesn't generate an Oracle 11*g* error message), you don't get the expected results, as you can see from the example in Figure 12-22.

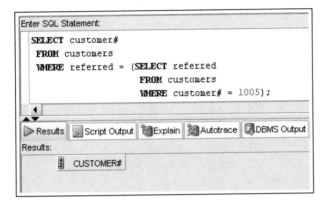

FIGURE 12-22   Flawed query: NULL results from a subquery

In Figure 12-22, the user is trying to determine whether the customer who referred customer 1005 has referred any other customers to JustLee Books. The problem is that no rows are listed as output from the outer query. Are no rows listed because the customer who referred customer 1005 hasn't referred any other customers or because customer 1005 wasn't referred to JustLee Books (in which case the Referred column is NULL)? If no one referred customer 1005, should the outer query's output be a list of all customers who weren't referred by other customers?

In this case, customer 1005 wasn't referred by any other customer; therefore, the Referred column is NULL. A NULL value is passed to the outer query, so no matches are found because the condition is WHERE referred = NULL. The IS NULL operator is required to identify NULL values in a conditional clause.

What if customer 1005 wasn't referred by another customer and you want a list of all customers who weren't referred by other customers? As always, it's the NVL function to the rescue.

NVL in Subqueries

If it's possible for a subquery to return a NULL value to the outer query for comparison, the NVL function should be used to substitute an actual value for the NULL. However, keep these two rules in mind:

- The substitution of the NULL value must occur for the NULL value in both the subquery and the outer query.
- The value substituted for the NULL value must be one that couldn't possibly exist anywhere else in that column.

Figure 12-23 uses the same premise as Figure 12-22 and shows an example of these two rules. The NVL function is included whenever the Referred column is referenced—in both the subquery and the outer query. In this example, a zero is substituted for a NULL value.

Because the value in the Referred column is actually a customer number in the CUSTOMERS table, and no customer has the customer number zero, substituting a zero for the NULL value doesn't accidentally make a NULL record equivalent to a non-NULL record.

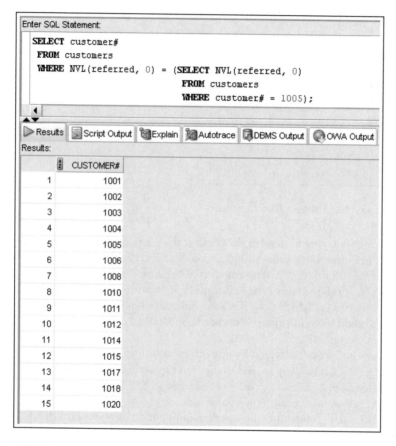

FIGURE 12-23   Using the NVL function to handle NULL values

When you substitute a value for a NULL, make sure no other record contains the substituted value. For example, use ZZZ for a customer name; in a date field, use a date that absolutely couldn't exist in the database.

IS NULL in Subqueries

Although passing a NULL value from a subquery to an outer query can be challenging, searches for NULL values are allowed in a subquery. As with regular queries, you can still search for NULL values with the IS NULL comparison operator.

For example, you need to find the title of all books that have been ordered but haven't shipped yet. The subquery in Figure 12-24 identifies the orders that haven't shipped—in other words, the ship date is NULL.

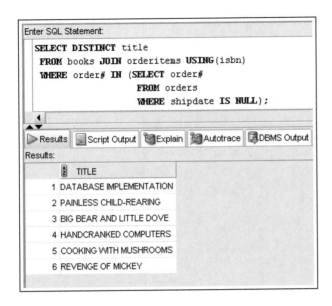

```
Enter SQL Statement:
  SELECT DISTINCT title
  FROM books JOIN orderitems USING(isbn)
  WHERE order# IN (SELECT order#
                   FROM orders
                   WHERE shipdate IS NULL);
```

	TITLE
1	DATABASE IMPLEMENTATION
2	PAINLESS CHILD-REARING
3	BIG BEAR AND LITTLE DOVE
4	HANDCRANKED COMPUTERS
5	COOKING WITH MUSHROOMS
6	REVENGE OF MICKEY

FIGURE 12-24   Using IS NULL in a subquery

As shown, the order number for each order is passed to the outer query, and the title for each book is displayed. The DISTINCT keyword is used to prevent duplicate titles from being listed. Although the subquery searches for records containing NULL values, it's the Order# column that's passed to the outer query. The Order# column is the primary key for the ORDERS table, and no NULL values can exist in this field. Therefore, there's no need to use the NVL function in this example.

CORRELATED SUBQUERIES

So far you have studied mostly **uncorrelated subqueries**: The subquery is executed first, its results are passed to the outer query, and then the outer query is executed. In a correlated subquery, Oracle 11g uses a different procedure to execute a query. A **correlated subquery** references one or more columns in the outer query, and the EXISTS operator is used to test whether the relationship or link is present.

Figure 12-25 shows an example of identifying books that have been ordered recently. Although this query is a multiple-row subquery, execution of the entire query requires processing each row in the BOOKS table to determine whether it also exists in the ORDERITEMS table.

Oracle 11g executes the outer query first, and when it encounters the outer query's WHERE clause, it's evaluated to determine whether that row is TRUE (whether it exists in the ORDERITEMS table). If it *is* TRUE, the book's title is displayed in the results. The outer query is executed again for the next book in the BOOKS table and compared to the ORDERITEMS table's contents, and so on, for each row of the BOOKS table. In other words, a correlated subquery is *processed, or executed, once for each row in the outer query*.

How does Oracle 11g distinguish between an uncorrelated and a correlated subquery? Simply speaking, if a subquery *references a column from the outer query*, it's a correlated

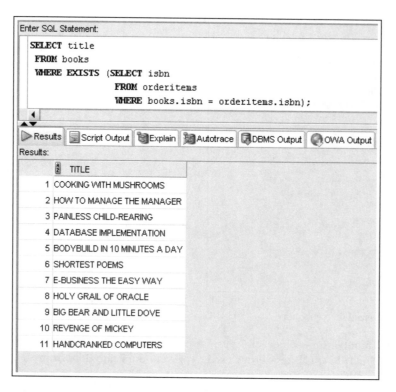

FIGURE 12-25 Correlated subquery

subquery. Notice that in the subquery in Figure 12-25, the WHERE clause specifies the ISBN column of the BOOKS table. Because the BOOKS table isn't included in the subquery's FROM clause, it's forced to use data processed by the outer query (the ISBN of books processed during that execution of the outer query). With an uncorrelated subquery, the subquery is executed first, and then the results are passed to the outer query. Because the subquery is used to identify every ISBN stored in the ORDERITEMS table, each ISBN listed in the table is returned to the outer query.

A join operation could also be used to tackle this query. Notice that in Figure 12-26, a join is used to identify the titles of books that have been ordered recently. Keep in mind that nonmatching rows are dropped from the results automatically in an equijoin.

Only partial results are included in Figure 12-26 because the output includes duplicates. Adding DISTINCT suppresses duplicates in the output so that it matches the results of the correlated subquery in Figure 12-25.

NOTE

You'll continue to discover that several techniques are available to solve query requests. Being familiar with different options is beneficial, as some techniques execute more efficiently in certain situations. Tuning is an advanced topic beyond this textbook's scope; however, the concept is introduced in Appendix E.

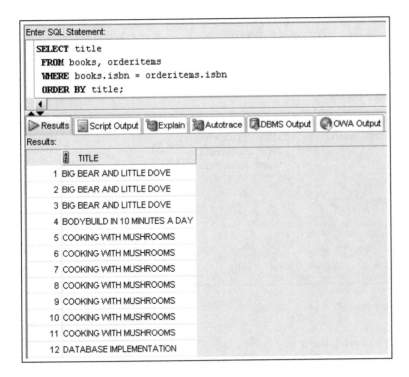

Enter SQL Statement:

```
SELECT title
FROM books, orderitems
WHERE books.isbn = orderitems.isbn
ORDER BY title;
```

Results | Script Output | Explain | Autotrace | DBMS Output | OWA Output

Results:

	TITLE
1	BIG BEAR AND LITTLE DOVE
2	BIG BEAR AND LITTLE DOVE
3	BIG BEAR AND LITTLE DOVE
4	BODYBUILD IN 10 MINUTES A DAY
5	COOKING WITH MUSHROOMS
6	COOKING WITH MUSHROOMS
7	COOKING WITH MUSHROOMS
8	COOKING WITH MUSHROOMS
9	COOKING WITH MUSHROOMS
10	COOKING WITH MUSHROOMS
11	COOKING WITH MUSHROOMS
12	DATABASE IMPLEMENTATION

FIGURE 12-26   Using a join rather than a correlated subquery

NESTED SUBQUERIES

You can nest subqueries inside the FROM, WHERE, or HAVING clauses of other subqueries. In Oracle 11g, subqueries in a WHERE clause can be nested to a depth of 255 subqueries, and there's no depth limit when subqueries are nested in a FROM clause. When nesting subqueries, you might want to use the following strategy:

- Determine exactly what you're trying to find—in other words, the goal of the query.
- Write the innermost subquery first.
- Next, look at the value you can pass to the outer query. If it isn't the value the outer query needs (for example, it references the wrong column), analyze how you need to convert the data to get the correct rows. If necessary, use another subquery between the outer query and the nested subquery. In some cases, you might need to create several layers of subqueries to link the value the innermost subquery returns to the value the outer query needs.

The most common reason for nesting subqueries is to create a chain of data. For example, you need to find the name of the customer who has ordered the most books from JustLee Books (not including multiple quantities of the same book) on *one* order. Figure 12-27 shows a query that returns these results.

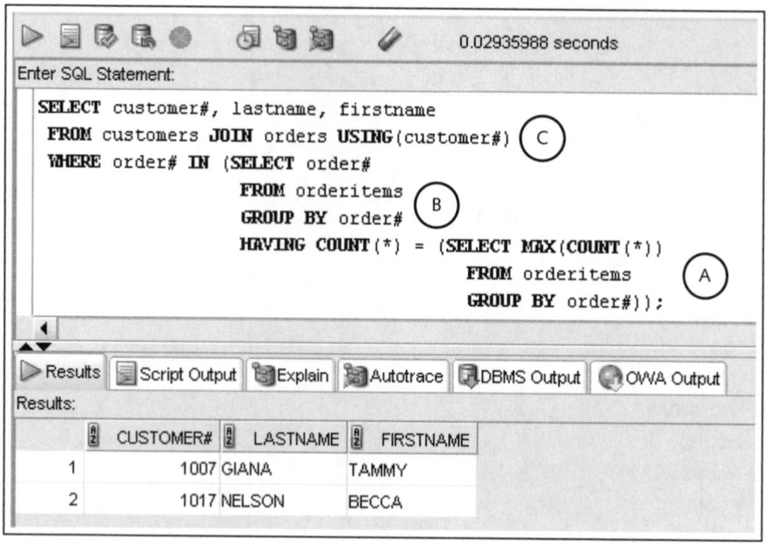

FIGURE 12-27 Nested subqueries

Here are the steps for creating the query in Figure 12-27:

1. The goal of the query is to count the number of items placed on each order and identify the order—or orders, in case of a tie—with the most items. The nested subquery identified by A in Figure 12-27 finds the highest count of books in any order.

2. The value of the highest count of items ordered is then passed to the outer subquery, B.

3. The outer subquery, B, is then used to identify which orders have the same number of items as the highest number of items that the innermost subquery, A, found.

4. After the order numbers have been identified, they are then passed to the outer query, C, which determines the customer number and name of the person who placed the orders. In this case, two customers tied for placing an order with the most items.

The statement uses the IN operator in the outer query's WHERE clause because the subquery, B, might return multiple rows.

TIP

Don't forget to include the extra set of parentheses for the nested group functions in the innermost subquery; if you do, you'll get an error message.

DML ACTIONS USING SUBQUERIES

In Chapter 5, you discovered you can insert data from existing tables into another table by using a subquery in the INSERT statement. You can also perform UPDATE and DELETE statements by using subqueries. For example, employee Sue Stuart needs her bonus set to equal the average bonus of all employees. The SET clause needs a subquery to determine the current average bonus of all employees, as shown in Figure 12-28.

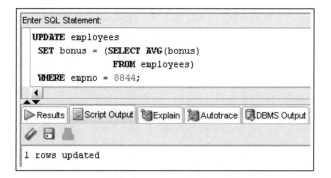

FIGURE 12-28   An UPDATE statement using a subquery

A subquery can also be used in the WHERE clause of a DELETE statement to determine which rows are deleted, based on the value the subquery returns. For example, the DEPARTMENT table contains a list of all departments initially established for JustLee Books; however, some departments might never have been used. JustLee management wants to eliminate any departments that currently have no employees. A subquery identifying all departments with employees can be used to accomplish this task with a DELETE statement, as shown in Figure 12-29. Notice that the WHERE clause uses the NOT IN operator to ensure that departments with employees aren't deleted.

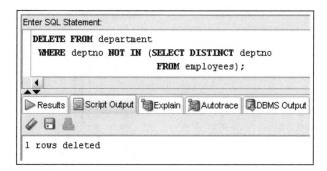

FIGURE 12-29   A DELETE statement using a subquery

Subqueries and MERGE Statements

MERGE STATEMENTS

With a **MERGE** statement, a series of DML actions can occur with a single SQL statement. The DML statements INSERT, UPDATE, and DELETE were covered in Chapter 5. However, conditionally updating one data source based on another wasn't covered. Now that you have an understanding of more complex SQL statements, this topic can be introduced.

In a data warehousing environment, often you need to conditionally update one table based on another table. For example, a BOOKS table might be used in the JustLee Books production system for recording orders. Any book price changes, category changes, and new book additions are entered in this table. Another copy of the BOOKS table could be kept for querying and reporting. Many organizations don't want to slow down the production system, so copies of tables are maintained on separate servers to handle querying and reporting requests. In this situation, the tables used for reporting need periodic updating. The MERGE statement assists in this task, as it can compare two data sources or tables and determine which rows need updating and which need inserting.

Take a look at an example involving two BOOKS tables. A table named BOOKS_1 serves as the reporting table, and BOOKS_2 serves as the production table. In this case, the BOOKS_2 table is the input source, and BOOKS_1 is the target. If a book exists in both tables, an UPDATE is needed to capture any changes in retail price or category assignments. If a book is in the BOOKS_2 table but not in the BOOKS_1 table, an INSERT is needed to add the book to BOOKS_1. First, query both tables to review the existing data, as shown in Figure 12-30.

The first three rows of the BOOKS_2 table, shown in Figure 12-30, are used to update the existing rows of the BOOKS_1 table because these three match based on ISBN. The last two rows of BOOKS_2 are added (using an INSERT) to the BOOKS_1 table, as these books don't currently exist in this table.

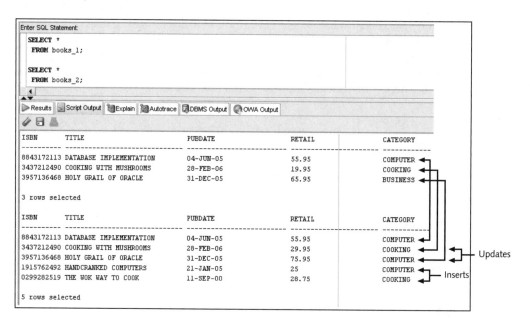

FIGURE 12-30 Current contents of the BOOKS_1 and BOOKS_2 tables

Figure 12-31 shows a MERGE statement that conditionally performs the UPDATEs and INSERTs.

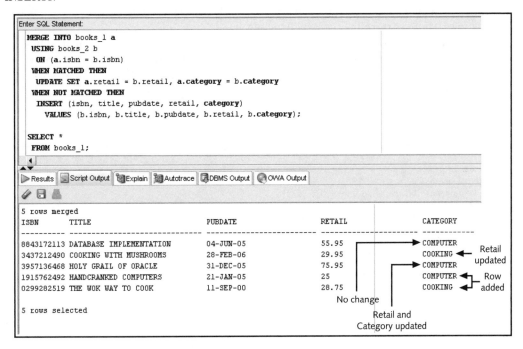

FIGURE 12-31 MERGE statement with UPDATE and INSERT

The following explains each part of this MERGE statement:

- `MERGE INTO books_1 a`: The BOOKS_1 table is to be changed, and a table alias of "a" is assigned to this table.
- `USING books_2 b`: The BOOKS_2 table provides the data to update or insert into BOOKS_1, and a table alias of "b" is assigned to this table.
- `ON (a.isbn = b.isbn)`: The rows of the two tables are joined or matched based on ISBN.
- `WHEN MATCHED THEN`: If a row match based on ISBN is discovered, execute the UPDATE action in this clause. The UPDATE action instructs Oracle to modify only two columns (Retail and Category).
- `WHEN NOT MATCHED THEN`: If no match is found based on the ISBN (a book exists in BOOKS_2 that isn't in BOOKS_1), perform the INSERT action in this clause.

NOTE

A MERGE statement containing an UPDATE and an INSERT clause is also called an UPSERT statement.

Including both WHEN MATCHED and WHEN NOT MATCHED isn't required. If only a particular DML operation is needed, you include only the corresponding clause.

Next, execute a ROLLBACK statement so that the BOOKS_1 data is set to the original three rows before performing the next example.

You can also include a WHERE condition in the matching clauses of a MERGE statement to conditionally perform the DML action based on a data value. Return to the previous example, but add a condition to update or insert only rows with the Computer category assigned in the BOOKS_2 table. Figure 12-32 shows WHERE clauses added to the previous MERGE statement.

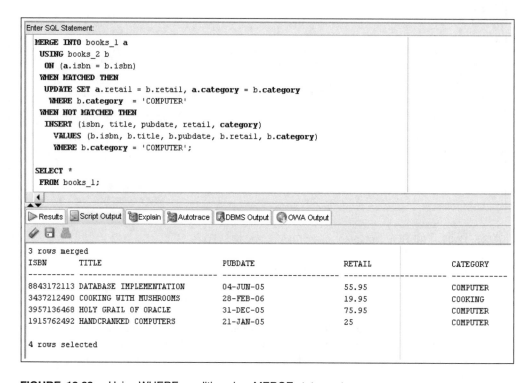

FIGURE 12-32   Using WHERE conditions in a MERGE statement

Recall that the BOOKS_2 table contains two rows with books in the Cooking category. These two rows are no longer processed because of the added WHERE conditions. Therefore, the retail price of the book *Cooking with Mushrooms* isn't updated, and the book *The Wok Way to Cook* isn't inserted. Also, the book *Holy Grail of Oracle* is originally assigned the category Business in the BOOKS_1 table and Computer in the BOOKS_2 table. The WHERE clause condition checks the category data in the BOOKS_2 table, which is Computer, so the MERGE statement updates this row in the BOOKS_1 table.

Execute another ROLLBACK statement so that the BOOKS_1 data is set to the original three rows before performing the next example.

When a match is found during a MERGE statement, a DELETE statement can also be conditionally processed. For example, assume the reporting table requires data only for books with a retail price of at least $50. Figure 12-33 shows a conditional DELETE action added to the WHEN MATCHED clause.

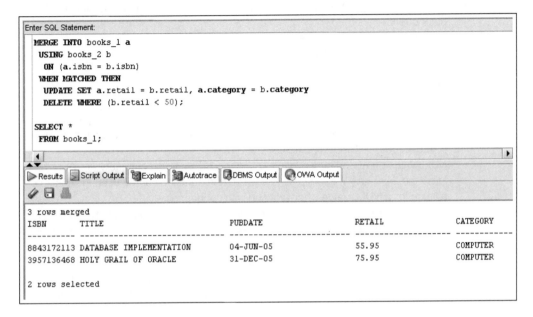

FIGURE 12-33   Conditional DELETE in a MERGE statement

The BOOKS_1 table contains only two rows instead of three because the DELETE action removed *Cooking with Mushrooms*. This row is deleted because the retail amount is $29.95 in the BOOKS_2 table, which meets the DELETE condition of costing less than $50. This MERGE statement processes only matched rows, so the other two rows in the BOOKS_2 table with retail amounts below $50 aren't processed.

Chapter Summary

- A subquery is a complete query nested in the SELECT, FROM, HAVING, or WHERE clause of another query. The subquery must be enclosed in parentheses and have a SELECT and a FROM clause, at a minimum.
- Subqueries are completed first. The result of the subquery is used as input for the outer query.
- A single-row subquery can return a maximum of one value.
- Single-row operators include =, >, <, >=, <=, and <>.
- Multiple-row subqueries return more than one row of results.
- Operators that can be used with multiple-row subqueries include IN, ALL, ANY, and EXISTS.
- Multiple-column subqueries return more than one column to the outer query. The columns of data are passed to the outer query in the same order in which they're listed in the subquery's SELECT clause.
- NULL values returned by a multiple-row or multiple-column subquery aren't a problem if the IN or =ANY operator is used. The NVL function can be used to substitute a value for a NULL value when working with subqueries.
- Correlated subqueries reference a column contained in the outer query. When using correlated subqueries, the subquery is executed once for each row the outer query processes.
- The EXISTS operator is used to formulate a correlated subquery.
- Subqueries can be nested to a maximum depth of 255 subqueries in the outer query's WHERE clause. The depth is unlimited for subqueries nested in the outer query's FROM clause.
- With nested subqueries, the innermost subquery is executed first, then the next highest level subquery is executed, and so on, until the outermost query is reached.
- DML actions can include subqueries to determine which rows are processed.
- A MERGE statement allows performing multiple DML actions conditionally while comparing data of two tables.

Chapter 12 Syntax Summary

The following table summarizes the syntax you have learned in this chapter. You can use the table as a study guide and reference.

Syntax Guide

Subquery Processing	Example
| Correlated subquery: References a column in the outer query. Executes the subquery once for every row in the outer query. | ```
SELECT title
FROM books b
WHERE b.isbn IN
    (SELECT isbn
     FROM orderitems o
     WHERE b.isbn = o.isbn);
``` |
```

Subquery Processing	Example

Uncorrelated subquery: Executes the subquery first and passes the value to the outer query.	`SELECT title`
	`FROM books b, orderitems o`
	`WHERE books isbn IN`
	`   (SELECT isbn`
	`   FROM orderitems)`
	`AND b.isbn = o.isbn;`

Multiple-Row Comparison Operators

Operator	Description
`>ALL`	More than the highest value returned by the subquery
`<ALL`	Less than the lowest value returned by the subquery
`<ANY`	Less than the highest value returned by the subquery
`>ANY`	More than the lowest value returned by the subquery
`=ANY`	Equal to any value returned by the subquery (same as IN)
`[NOT] EXISTS`	Row must match a value in the subquery

461

DML Action with a MERGE Statement	Example

Conditionally performs a series of DML actions	`MERGE INTO books_1 a`
	`   USING books_2 b`
	`   ON (a.isbn = b.isbn)`
	`WHEN MATCHED THEN`
	`   UPDATE SET a.retail = b.retail,`
	`   a.category = b.category`
	`WHEN NOT MATCHED THEN`
	`   INSERT (isbn, title, pubdate, retail,`
	`   category)`
	`   VALUES (b.isbn, b.title, b.pubdate,`
	`   b.retail, b.category);`

Review Questions

1. What's the difference between a single-row subquery and a multiple-row subquery?
2. What comparison operators are required for multiple-row subqueries?
3. What happens if a single-row subquery returns more than one row of results?
4. Which SQL clause(s) can't be used in a subquery in the WHERE or HAVING clauses?

5. If a subquery is used in the FROM clause of a query, how are the subquery's results referenced in other clauses of the query?
6. Why might a MERGE statement be used?
7. How can Oracle 11*g* determine whether clauses of a SELECT statement belong to an outer query or a subquery?
8. When should a subquery be nested in a HAVING clause?
9. What's the difference between correlated and uncorrelated subqueries?
10. What type of situation requires using a subquery?

Multiple Choice

To answer these questions, refer to the tables in the JustLee Books database.

1. Which query identifies customers living in the same state as the customer named Leila Smith?

   a. ```
      SELECT customer# FROM customers
         WHERE state = (SELECT state FROM customers
            WHERE lastname = 'SMITH');
      ```

   b. ```
      SELECT customer# FROM customers
         WHERE state = (SELECT state FROM customers
            WHERE lastname = 'SMITH'
               OR firstname = 'LEILA');
      ```

   c. ```
      SELECT customer# FROM customers
         WHERE state = (SELECT state FROM customers
            WHERE lastname = 'SMITH'
               AND firstname = 'LEILA'
            ORDER BY customer);
      ```

   d. ```
      SELECT customer# FROM customers
         WHERE state = (SELECT state FROM customers
            WHERE lastname = 'SMITH'
               AND firstname = 'LEILA');
      ```

2. Which of the following is a valid SELECT statement?

   a. ```
      SELECT order# FROM orders
         WHERE shipdate = SELECT shipdate FROM orders
            WHERE order# = 1010;
      ```

   b. ```
      SELECT order# FROM orders
         WHERE shipdate = (SELECT shipdate FROM orders)
            AND order# = 1010;
      ```

   c. ```
      SELECT order# FROM orders
         WHERE shipdate = (SELECT shipdate FROM orders
            WHERE order# = 1010);
      ```

   d. ```
      SELECT order# FROM orders
         HAVING shipdate = (SELECT shipdate FROM orders
            WHERE order# = 1010);
      ```

3. Which of the following operators is considered a single-row operator?

    a. IN

    b. ALL

    c. <>

    d. <>ALL

4. Which of the following queries determines which customers have ordered the same books as customer 1017?

    a.
```
SELECT order# FROM orders
  WHERE customer# = 1017;
```

    b.
```
SELECT customer# FROM orders
  JOIN orderitems USING(order#)
    WHERE isbn = (SELECT isbn FROM orderitems
      WHERE customer# = 1017);
```

    c.
```
SELECT customer# FROM orders
  WHERE order# = (SELECT order# FROM orderitems
    WHERE customer# = 1017);
```

    d.
```
SELECT customer# FROM orders
  JOIN orderitems USING(order#)
    WHERE isbn IN (SELECT isbn FROM orderitems
      JOIN orders USING(order#)
      WHERE customer# = 1017);
```

5. Which of the following statements is valid?

    a.
```
SELECT title FROM books
  WHERE retail < (SELECT cost FROM books
    WHERE isbn = '9959789321');
```

    b.
```
SELECT title FROM books
  WHERE retail = (SELECT cost FROM books
    WHERE isbn = '9959789321' ORDER BY cost);
```

    c.
```
SELECT title FROM books
  WHERE category IN (SELECT cost FROM orderitems
    WHERE isbn = '9959789321');
```

    d. none of the above statements

6. Which of the following statements is correct?

    a. If a subquery is used in the outer query's FROM clause, the data in the temporary table can't be referenced by clauses used in the outer query.

    b. The temporary table created by a subquery in the outer query's FROM clause must be assigned a table alias, or it can't be joined with another table by using the JOIN keyword.

    c. If a temporary table is created through a subquery in the outer query's FROM clause, the data in the temporary table can be referenced by another clause in the outer query.

    d. none of the above

7. Which of the following queries identifies other customers who were referred to JustLee Books by the same person who referred Jorge Perez?

   a. ```
   SELECT customer# FROM customers
      WHERE referred = (SELECT referred FROM customers
         WHERE firstname = 'JORGE'
            AND lastname = 'PEREZ');
   ```

   b. ```
   SELECT referred FROM customers
      WHERE (customer#, referred) = (SELECT customer#
         FROM customers WHERE firstname = 'JORGE'
            AND lastname = 'PEREZ');
   ```

   c. ```
   SELECT referred FROM customers
      WHERE (customer#, referred) IN (SELECT customer#
         FROM customers WHERE firstname = 'JORGE'
            AND lastname = 'PEREZ');
   ```

   d. ```
   SELECT customer# FROM customers
      WHERE customer# = (SELECT customer#
         FROM customers WHERE firstname = 'JORGE'
            AND lastname = 'PEREZ');
   ```

8. In which of the following situations is using a subquery suitable?

   a. when you need to find all customers living in a particular region of the country

   b. when you need to find all publishers who have toll-free telephone numbers

   c. when you need to find the titles of all books shipped on the same date as an order placed by a particular customer

   d. when you need to find all books published by Publisher 4

9. Which of the following queries identifies customers who have ordered the same books as customers 1001 and 1005?

   a. ```
   SELECT customer# FROM orders
      JOIN books USING(isbn)
         WHERE isbn = (SELECT isbn FROM orderitems
      JOIN books USING(isbn)
         WHERE customer# = 1001 OR customer# = 1005));
   ```

   b. ```
   SELECT customer# FROM orders
      JOIN books USING(isbn)
         WHERE isbn <ANY (SELECT isbn FROM orderitems
      JOIN books USING(isbn)
         WHERE customer# = 1001 OR customer# = 1005));
   ```

   c. ```
   SELECT customer# FROM orders
      JOIN books USING(isbn)
         WHERE isbn = (SELECT isbn FROM orderitems
      JOIN orders USING(order#)
         WHERE customer# = 1001 OR 1005));
   ```
   ```

d. `SELECT customer# FROM orders`
   `JOIN orderitems USING(order#)`
   `   WHERE isbn IN (SELECT isbn FROM orders`
   `JOIN orderitems USING(order#)`
   `   WHERE customer# IN (1001, 1005));`

10. Which of the following operators is used to find all values greater than the highest value returned by a subquery?

   a. >ALL

   b. <ALL

   c. >ANY

   d. <ANY

   e. IN

11. Which query determines the customers who have ordered the most books from JustLee Books?

   a. `SELECT customer# FROM orders`
       `JOIN orderitems USING(order#)`
       `   HAVING SUM(quantity) = (SELECT`
       `     MAX(SUM(quantity)) FROM orders`
       `JOIN orderitems USING(order#)`
       `GROUP BY customer#) GROUP BY customer#;`

   b. `SELECT customer# FROM orders`
       `JOIN orderitems USING(order#)`
       `   WHERE SUM(quantity) = (SELECT`
       `     MAX(SUM(quantity)) FROM orderitems`
       `GROUP BY customer#);`

   c. `SELECT customer# FROM orders`
       `   WHERE MAX(SUM(quantity)) = (SELECT`
       `     MAX(SUM(quantity) FROM orderitems`
       `GROUP BY order#);`

   d. `SELECT customer# FROM orders`
       `   HAVING quantity = (SELECT MAX(SUM(quantity))`
       `     FROM orderitems`
       `GROUP BY customer#);`

12. Which of the following statements is correct?

   a. The IN comparison operator can't be used with a subquery that returns only one row of results.

   b. The equals (=) comparison operator can't be used with a subquery that returns more than one row of results.

   c. In an uncorrelated subquery, statements in the outer query are executed first, and then statements in the subquery are executed.

   d. A subquery can be nested only in the outer query's SELECT clause.

13. What is the purpose of the following query?

```
SELECT isbn, title FROM books
  WHERE (pubid, category) IN (SELECT pubid, category
    FROM books WHERE title LIKE '%ORACLE%');
```

    a. It determines which publisher published a book belonging to the Oracle category and then lists all other books published by that same publisher.

    b. It lists all publishers and categories containing the value ORACLE.

    c. It lists the ISBN and title of all books belonging to the same category and having the same publisher as any book with the phrase ORACLE in its title.

    d. None of the above. The query contains a multiple-row operator, and because the inner query returns only one value, the SELECT statement will fail and return an error message.

14. A subquery must be placed in the outer query's HAVING clause if:

    a. The inner query needs to reference the value returned to the outer query.

    b. The value returned by the inner query is to be compared to grouped data in the outer query.

    c. The subquery returns more than one value to the outer query.

    d. None of the above. Subqueries can't be used in the outer query's HAVING clause.

15. Which of the following SQL statements lists all books written by the author of *The Wok Way to Cook*?

    a.
```
SELECT title FROM books
    WHERE isbn IN (SELECT isbn FROM bookauthor
      HAVING authorid IN 'THE WOK WAY TO COOK);
```

    b.
```
SELECT isbn FROM bookauthor
    WHERE authorid IN (SELECT authorid FROM books
      JOIN bookauthor USING(isbn)
        WHERE title = 'THE WOK WAY TO COOK');
```

    c.
```
SELECT title FROM bookauthor
    WHERE authorid IN (SELECT authorid FROM books
      JOIN bookauthor USING(isbn)
        WHERE title = 'THE WOK WAY TO COOK);
```

    d.
```
SELECT isbn FROM bookauthor
    HAVING authorid = SELECT authorid FROM books
      JOIN bookauthor USING(isbn)
        WHERE title = 'THE WOK WAY TO COOK';
```

16. Which of the following statements is correct?

    a. If the subquery returns only a NULL value, the only records returned by an outer query are those containing an equivalent NULL value.

    b. A multiple-column subquery can be used only in the outer query's FROM clause.

    c. A subquery can contain only one condition in its WHERE clause.

d. The order of columns listed in the SELECT clause of a multiple-column subquery must be in the same order as the corresponding columns listed in the outer query's WHERE clause.

17. In a MERGE statement, an INSERT is placed in which conditional clause?

   a. USING

   b. WHEN MATCHED

   c. WHEN NOT MATCHED

   d. INSERTs aren't allowed in a MERGE statement.

18. Given the following query, which statement is correct?

```
SELECT order# FROM orders
  WHERE order# IN (SELECT order# FROM orderitems
    WHERE isbn = '9959789321');
```

   a. The statement doesn't execute because the subquery and outer query don't reference the same table.

   b. The outer query removes duplicates in the subquery's Order# list.

   c. The query fails if only one result is returned to the outer query because the outer query's WHERE clause uses the IN comparison operator.

   d. No rows are displayed because the ISBN in the WHERE clause is enclosed in single quotation marks.

19. Given the following SQL statement, which statement is most accurate?

```
SELECT customer# FROM customers
  JOIN orders USING(customer#)
    WHERE shipdate-orderdate IN
      (SELECT MAX(shipdate-orderdate) FROM orders
      WHERE shipdate IS NULL);
```

   a. The SELECT statement fails and returns an Oracle error message.

   b. The outer query displays no rows in its results because the subquery passes a NULL value to the outer query.

   c. The customer number is displayed for customers whose orders haven't yet shipped.

   d. The customer number of all customers who haven't placed an order are displayed.

20. Which operator is used to process a correlated subquery?

   a. EXISTS

   b. IN

   c. LINK

   d. MERGE

Hands-On Assignments

To perform these activities, refer to the tables in the JustLee Books database. Use a subquery to accomplish each task. Make sure you execute the query you plan to use as the subquery to verify the results before writing the entire query.

1. List the book title and retail price for all books with a retail price lower than the average retail price of all books sold by JustLee Books.

2. Determine which books cost less than the average cost of other books in the same category.

3. Determine which orders were shipped to the same state as order 1014.

4. Determine which orders had a higher total amount due than order 1008.

5. Determine which author or authors wrote the books most frequently purchased by customers of JustLee Books.

6. List the title of all books in the same category as books previously purchased by customer 1007. Don't include books this customer has already purchased.

7. List the shipping city and state for the order that had the longest shipping delay.

8. Determine which customers placed orders for the least expensive book (in terms of regular retail price) carried by JustLee Books.

9. Determine the number of different customers who have placed an order for books written or cowritten by James Austin.

10. Determine which books were published by the publisher of *The Wok Way to Cook*.

Advanced Challenge

To perform this activity, refer to the tables in the JustLee Books database.

Currently, JustLee Books bills customers for orders by enclosing an invoice with each order when it's shipped. A customer then has 10 days to send in the payment. Of course, this practice has resulted in the company having to list some debts as "uncollectible." By contrast, most other online booksellers receive payment through a customer's credit card at the time of purchase. With this method, although payment would be deposited within 24 hours into JustLee's bank account, there's a downside. When a merchant accepts credit cards for payment, the company processing the credit card sales (usually called a "credit card clearinghouse") deducts a 1.5% processing fee from the total amount of the credit card sale.

The management of JustLee Books is trying to determine whether the surcharge for credit card processing is more than the amount usually deemed uncollectible when customers are sent an invoice. Historically, the average amount that JustLee Books has lost is about 4% of the total amount due from orders with a higher-than-average amount due. In other words, usually customers who have an order with a larger-than-average invoice total default on payments.

To determine how much money would be lost or gained by accepting credit card payments, management has requested that you do the following:

1. Determine how much the surcharge would be for all recently placed orders if payment had been made by a credit card.

2. Determine the total amount that can be expected to be written off as uncollectible based on recently placed orders with an invoice total more than the average of all recently placed orders.

Based on the results of these two calculations, you should determine whether the company will lose money by accepting payments via credit card. State your findings in a memo to management. Include the SQL statements for calculating the expected surcharge and the expected amount of uncollectible payments.

Case Study: *City Jail*

Make sure you have run the CityJail_8.sql script from Chapter 8. This script makes all database objects available for completing this case study.

The city's Crime Analysis unit has submitted the following data requests. Provide the SQL statements using subqueries to satisfy the requests. Test the statements and show execution results.

1. List the name of each officer who has reported more than the average number of crimes officers have reported.

2. List the criminal names for all criminals who have a less than average number of crimes and aren't listed as violent offenders.

3. List appeal information for each appeal that has a less than average number of days between the filing and hearing dates.

4. List the names of probation officers who have had a less than average number of criminals assigned.

5. List each crime that has had the highest number of appeals recorded.

6. List the information on crime charges for each charge that has had a fine above average and a sum paid below average.

7. List all the names of all criminals who have had any of the crime code charges involved in crime ID 10089.

8. Use a correlated subquery to determine which criminals have had at least one probation period assigned.

9. List the names of officers who have booked the highest number of crimes. Note that more than one officer might be listed.

Note: Use a MERGE statement to satisfy the following request:

10. The criminal data warehouse contains a copy of the CRIMINALS table that needs to be updated periodically from the production CRIMINALS table. The data warehouse table is named CRIMINALS_DW. Use a single SQL statement to update the data warehouse table to reflect any data changes for existing criminals and to add new criminals.

VIEWS

LEARNING OBJECTIVES

After completing this chapter, you should be able to do the following:

- Create a view by using the CREATE VIEW or CREATE OR REPLACE VIEW command
- Use the FORCE and NOFORCE options
- State the purpose of the WITH CHECK OPTION constraint
- Explain the effect of the WITH READ ONLY option
- Update a record in a simple view
- Re-create a view
- Explain the implication of using an expression in a view for DML operations
- Update a record in a complex view
- Identify problems associated with adding records through a complex view
- Identify the key-preserved table underlying a complex view
- Drop a view
- Explain inline views and the use of ROWNUM to perform a TOP-N analysis
- Create a materialized view to replicate data

INTRODUCTION

Views are database objects that store a SELECT statement and allow using a query's results as a table. Views have two purposes:

- Simplify issuing complex SQL queries
- Restrict users' access to sensitive data

Although views are database objects, they don't actually store data. A view stores a query and is used to access data in the underlying tables. You can think of a view as the result of a stored query: The results are given a name that allows using them as the source for queries, just as you would use a table. In fact, you can reference a view in the FROM clause of a SELECT statement, just as you reference any table. Views can use all the features of a query, including specifying columns, restricting rows, and aggregating data.

Figure 13-1 shows the basic processing of a view. When a query references a view, the query in the view is processed, and the results are treated as a virtual or temporary table. In this figure, any query referencing the BOOK_VU view can examine only books in the Cooking category because of the condition in the WHERE clause.

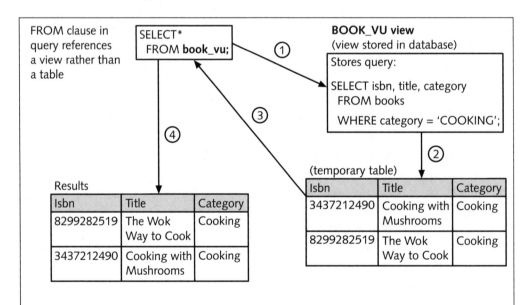

Processing steps:
1. A query references a view rather than a table in the FROM clause.
2. The query stored in the view is executed.
3. The results from the view query are stored in temporary storage.
4. The original query uses the temporary query results of the view as a table and completes execution.

FIGURE 13-1   View processing

Next, take a look at some examples of how views can help simplify complex queries. First, nontechnical users might not be familiar with SQL coding, and data retrieval from JustLee Books tables can require fairly complex queries. For example, an employee needs to find details for a specific order, including customer information, order date, book titles, prices paid, and calculated total item amount. To accomplish this task, the employee must query the CUSTOMERS, ORDERS, ORDERITEMS, and BOOKS tables simultaneously. The average employee probably lacks the training needed to create a query that joins multiple tables and performs the necessary calculations. To simplify the task, one option is creating a view that contains all the necessary information. So instead of teaching users how to create queries with joins and calculations, you can show them how to perform simple queries on a virtual table that includes customer names, order numbers, order dates, book titles, quantity of books ordered, prices paid, and calculated extended prices.

Application development can also be simplified by using views. For example, a developer might have several screens requiring customer order details that involve using a complex query. Instead of programming the query in multiple places in the application, the developer could create a view that stores the complex query. Then the developer just needs to perform a simple query on the view in the application code.

A view can also be used to restrict access to what management considers sensitive data. For example, the BOOKS table contains both the cost and retail price of each book in inventory. What happens if management decides book costs shouldn't be accessed by every employee in the company? Do you delete the column from the BOOKS table? If so, how would you calculate the profit for each book sold? Instead of giving users access to the actual table storing all data for books, you can give them access to data via a view that includes only the data they need, based on their job duties.

Views are used most commonly to query data. Some developers, however, might want to perform DML activities on data accessed via a view. This chapter explains the commands and guidelines regulating DML operations on data accessed by views. The last section of this chapter introduces materialized views, which are views that store data permanently. Table 13-1 gives you an overview of this chapter's contents.

TABLE 13-1 Overview of View Concepts

View Type	Description
Simple view	A view based on a subquery that references only one table and doesn't include group functions, expressions, or GROUP BY clauses
Complex view	A view based on a subquery that retrieves or derives data from one or more tables and can contain functions or grouped data
Inline view	A subquery used in the FROM clause of a SELECT statement to create a "temporary" table that can be referenced by the outer query's SELECT and WHERE clauses
Materialized view	A view that replicates data by physically storing the view query's results

TABLE 13-1   Overview of View Concepts (continued)

Task	Command Syntax or Example

Create a view	```CREATE [OR REPLACE] [FORCE|NOFORCE]```
	``` VIEW viewname (columnname, . . .)```
	```AS subquery```
	```[WITH CHECK OPTION [CONSTRAINT constraintname]]```
	```[WITH READ ONLY];```

Drop a view	```DROP VIEW viewname;```

Create an inline view	```SELECT columnname, . . .```
	```FROM (subquery)```
	``` WHERE ROWNUM <= n;```

Create a materialized view	```CREATE MATERIALIZED VIEW custbal_mv```
	``` REFRESH COMPLETE```
	``` START WITH SYSDATE NEXT SYSDATE + 7```
	``` AS SELECT customer#, city, state, order#,```
	``` SUM(quantity*retail) Amtdue```
	``` FROM customers JOIN orders```
	``` USING (customer#)```
	``` JOIN orderitems USING (order#)```
	``` JOIN books USING (isbn)```
	``` GROUP BY customer#, city, state, order#;```

DATABASE PREPARATION

Before attempting to work through the examples in this chapter, make sure you have executed the JLDB_Build_8.sql script, as instructed in Chapter 8.

CREATING A VIEW

A view is created with the **CREATE VIEW** command, using the syntax shown in Figure 13-2.

```
CREATE [OR REPLACE] [FORCE|NOFORCE] VIEW
    viewname (columnname, ...)
AS SELECT statement
[WITH CHECK OPTION [CONSTRAINT constraintname]]
[WITH READ ONLY];
```

FIGURE 13-2   Syntax of the CREATE VIEW command

The following is an overview of the syntax elements shown in Figure 13-2:

- *CREATE VIEW/CREATE OR REPLACE VIEW*: You use the CREATE VIEW keywords to create a view, choosing a name that no other database object in the current schema is using. *There's no way to modify or change an existing view*, so if you need to change a view, you must use the **CREATE OR REPLACE VIEW** keywords. The OR REPLACE option notifies Oracle 11g that a view with the same name might already exist; if it does, the view's previous version should be replaced with the one defined in the new command.

- *FORCE/NOFORCE*: **NOFORCE** is the default mode for the CREATE VIEW command, which means all tables and columns must be valid, or the view isn't created. So if you attempt to create a view based on a table that doesn't exist or is currently unavailable (for example, offline), Oracle 11g returns an error message, and the view isn't created. However, if you include the **FORCE** keyword in the CREATE clause, Oracle 11g creates the view in spite of the absence of any referenced tables. This approach is commonly used when a new database is being developed and the data hasn't yet been loaded, or entered, into database objects.

- *View name*: As mentioned, you should give each view a name that isn't already assigned to another database object in the same schema.

- *Column names*: If you want to assign new names for columns the view displays, list them after the VIEW keyword inside parentheses. The number of names listed *must match the number of columns returned by the SELECT statement*. An alternative is using column aliases in the query. In this case, Oracle 11g uses the aliases as column names in the view that's created.

- *AS clause*: The query listed after the **AS** keyword must be a complete SELECT statement (including both SELECT and FROM clauses) and can reference more than one table. The query can also include single-row and group functions, WHERE and GROUP BY clauses, nested subqueries, and so on. However, as with subqueries, the query can't include the ORDER BY clause. The query results are the content of the view that's created.

- *WITH CHECK OPTION constraint*: The **WITH CHECK OPTION** constraint ensures that any DML operations performed on the view (such as adding rows or changing data) don't prevent the view from accessing the row because it no longer meets the condition in the WHERE clause. For example, if a view consists of books only in the Cooking category, and the user attempts to change the category of a book in the view to Family Life, the change isn't allowed if WITH CHECK OPTION was included when the view was created. Why? The change would mean that the book is no longer listed in the view, which consists of books only in the Cooking category. If WITH CHECK OPTION is omitted when the view is created, any valid DML operation is allowed, even if the result is that rows being changed are no longer included in the view. However, if you're creating a view with the sole purpose of displaying data, the WITH READ ONLY option can be used instead to ensure that data can't be changed.

- *WITH READ ONLY option*: The **WITH READ ONLY** option prevents performing any DML operations on the view. This option is used often when it's important that users can only query data, not make any changes to it.

The following sections cover these operations in more depth. First, you see how to create a simple view, then how to change a simple view, and finally how to create a complex view.

Creating a Simple View

You create a **simple view** from a subquery that references only one table and doesn't include a group function, an expression (such as `retail-cost`), or a GROUP BY clause. For example, when JustLee Books customer service representatives assist customers with orders, they need to access the ISBN, title, and retail price of every book in JustLee's inventory. However, management doesn't want representatives to view the books' actual cost. The solution: Create a simple view that allows them to access only the data needed to assist customers—and not access irrelevant columns, such as Pubid, Cost, and so on. Figure 13-3 shows the command for creating the view customer service representatives need.

```
CREATE VIEW inventory
  AS SELECT isbn, title, retail price
     FROM books
  WITH READ ONLY;
```

FIGURE 13-3   Command to create the INVENTORY view

As indicated in the CREATE VIEW clause, the name of the new view is INVENTORY. Note these other elements in Figure 13-3:

- Because no other view with this name exists, the OR REPLACE clause isn't necessary.
- The only columns included in the view are the ISBN, Title, and Retail columns from the BOOKS table. Notice that the Retail column has been assigned the column alias "Price." Therefore, whenever the Retail column is referenced in a query on the INVENTORY view, it must be called Price in the query.
- The WITH READ ONLY option is used so that no customer service representative can change a book's ISBN, title, or price accidentally. Any changes made to data in a simple view created without the WITH READ ONLY option update the underlying BOOKS table automatically.

TIP

It's a common concern that users might accidentally, or even intentionally, change data when accessing it through views.

As shown in Figure 13-4, after the INVENTORY view is created, users can reference it in the FROM clause of a SELECT statement in the same manner as they would any table.

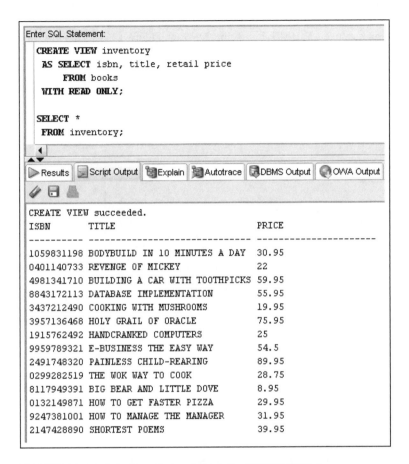

FIGURE 13-4   Selecting all records from the INVENTORY view

NOTE

If the command in Figure 13-4 returns an error message, make sure no view with the same name exists already. If it does, add the keywords OR REPLACE to the CREATE VIEW clause, and then execute the command again. In addition, if you get an error stating that you have insufficient privileges, you need to grant the CREATE VIEW privilege to your user account. Refer to your installation instructions or contact your instructor.

Keep in mind that the query defining the INVENTORY view didn't filter rows with a WHERE clause, so all book rows are accessible through the view. However, row filtering is a popular reason for creating views. For example, JustLee Books might want to limit regional managers' access to customer data based on their region. To limit their access to rows with a specific region value, you can use a WHERE clause in the view. Figure 13-5 shows creating this view for the Northeast region.

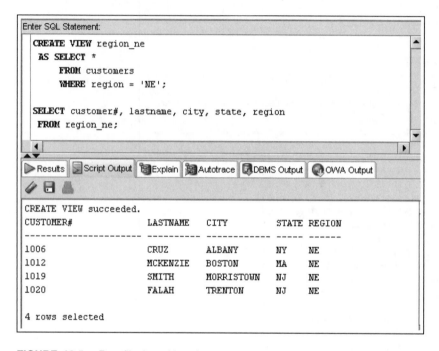

FIGURE 13-5 Row filtering with a view

Return to the view created in Figure 13-4. Although the INVENTORY view can be retrieved with a SELECT statement as though it were a regular table, the view in Figure 13-4 is created with a WITH READ ONLY option, which prevents performing any DML operations on the data. In Figure 13-6, the user is unsuccessfully attempting to update data in the Price and Title columns of the INVENTORY view.

First, the user attempts to change the retail price of the book *Shortest Poems*. Notice that the SET clause references the Price column (the column alias for the Retail column, specified in the view's SELECT statement). Also, because the user couldn't remember the book's exact title, a search pattern is used in the WHERE clause to identify the book being updated. The percent signs (%) indicate that characters might appear before and after the word *Poems*, but the book's title must contain the word *Poems*. However, Oracle 11*g* returns an error message indicating that DML operations can't be performed on read-only views.

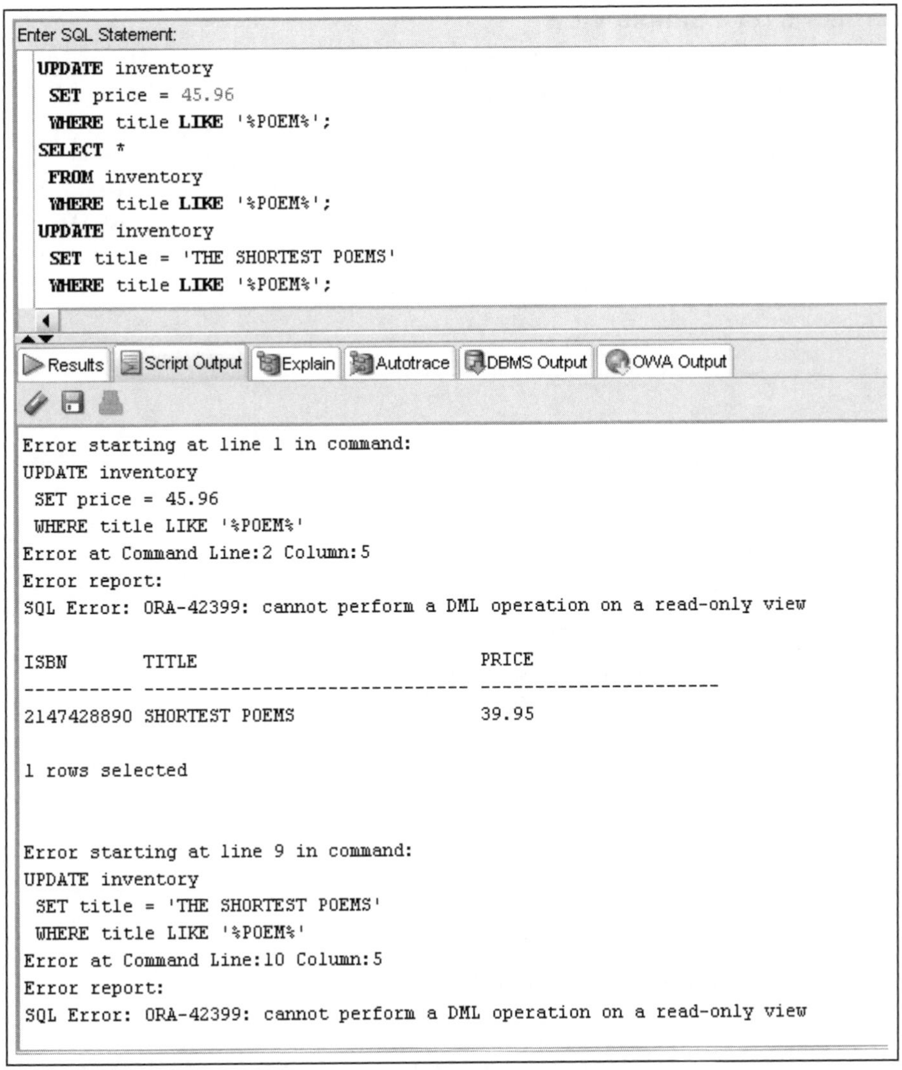

```
Enter SQL Statement:
  UPDATE inventory
   SET price = 45.96
   WHERE title LIKE '%POEM%';
  SELECT *
   FROM inventory
   WHERE title LIKE '%POEM%';
  UPDATE inventory
   SET title = 'THE SHORTEST POEMS'
   WHERE title LIKE '%POEM%';
```

Results | Script Output | Explain | Autotrace | DBMS Output | OWA Output

```
Error starting at line 1 in command:
UPDATE inventory
 SET price = 45.96
 WHERE title LIKE '%POEM%'
Error at Command Line:2 Column:5
Error report:
SQL Error: ORA-42399: cannot perform a DML operation on a read-only view

ISBN       TITLE                           PRICE
---------- ------------------------------- ----------------------
2147428890 SHORTEST POEMS                  39.95

1 rows selected


Error starting at line 9 in command:
UPDATE inventory
 SET title = 'THE SHORTEST POEMS'
 WHERE title LIKE '%POEM%'
Error at Command Line:10 Column:5
Error report:
SQL Error: ORA-42399: cannot perform a DML operation on a read-only view
```

FIGURE 13-6   Failed updates on the INVENTORY view

Second, the user attempts to change the book title to see whether the problem is caused by the column alias used for the Retail column, but the same error message is returned. Both UPDATE attempts failed because the view was created with the WITH READ ONLY option, so no DML operations are allowed.

DML Operations on a Simple View

If the technical staff decides that the INVENTORY view should be used to alter the retail prices of books currently in inventory, the view can be re-created with the CREATE OR REPLACE VIEW command—without the WITH READ ONLY option. Because the INVENTORY view already exists in the database, the OR REPLACE keywords must be included, or Oracle 11g returns an error message stating that the view already exists.

In Figure 13-7, the INVENTORY view has been re-created—without a WITH READ ONLY option, so updates are allowed. After this view is re-created, anyone with access to it can change the ISBN, title, or retail price of any book in the BOOKS table.

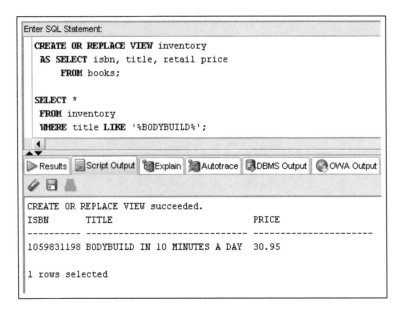

FIGURE 13-7   Re-create the view to allow DML activity

For example, the original retail price of *Bodybuild in 10 Minutes a Day* was $30.95. Figure 13-8 shows an UPDATE command issued on the view to change the book's retail price to $49.95. The previous version of the view wouldn't have allowed changing this data because WITH READ ONLY was used. As shown, now the view accepts the change, so the price in the underlying BOOKS table is altered.

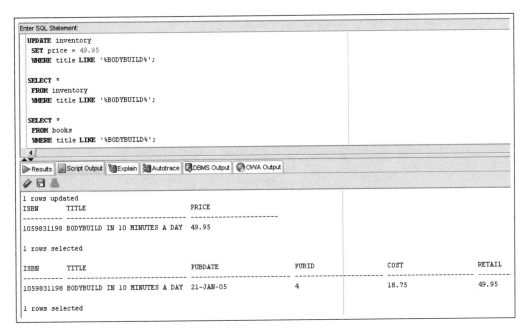

FIGURE 13-8 Issue a DML command on a simple view

The basic rule for DML operations on a simple view is this: As long as the view isn't created with the WITH READ ONLY option, any DML operation is allowed if it doesn't violate an existing constraint on the underlying table. In essence, you can add, modify, and even delete data in an underlying table as long as one of the following constraints doesn't prevent the operation:

- PRIMARY KEY
- NOT NULL
- UNIQUE
- FOREIGN KEY
- WITH CHECK OPTION

NOTE

If the SELECT statements in Figures 13-7 or 13-8 return an error message or don't display any rows, make sure the book title in the WHERE clause is enclosed in single quotation marks and includes the % signs in the search pattern for BODYBUILD.

Now take a look at another view example. As shown in Figure 13-9, the OUTSTANDING view has been created to display all orders in the ORDERS table that haven't been shipped yet.

```
Enter SQL Statement:
  CREATE VIEW outstanding
    AS SELECT customer#, order#, orderdate, shipdate
      FROM orders
      WHERE shipdate IS NULL
  WITH CHECK OPTION;

  SELECT *
  FROM outstanding;
```

Results | Script Output | Explain | Autotrace | DBMS Output | OWA Output

```
CREATE VIEW succeeded.
CUSTOMER#              ORDER#                ORDERDATE              SHIPDATE
--------------------  --------------------  ---------------------  -------------------------

1017                  1012                  03-APR-09
1020                  1015                  04-APR-09
1003                  1016                  04-APR-09
1001                  1018                  05-APR-09
1018                  1019                  05-APR-09
1008                  1020                  05-APR-09

6 rows selected
```

FIGURE 13-9   Create a view with the WITH CHECK OPTION constraint

Because the WITH CHECK OPTION constraint is included, the OUTSTANDING view can't be used to update the ship date of any of the six records that haven't been shipped because changing this field would remove the record from the view. Any attempt to change an order's ship date to a non-NULL value returns an error message, and the update fails, as shown in Figure 13-10.

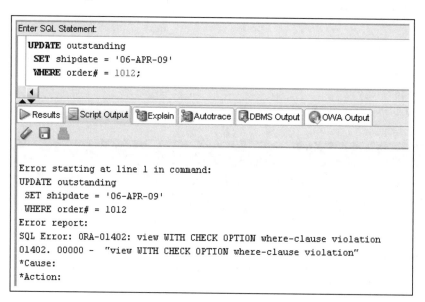

```
Enter SQL Statement:
  UPDATE outstanding
    SET shipdate = '06-APR-09'
    WHERE order# = 1012;
```

Results | Script Output | Explain | Autotrace | DBMS Output | OWA Output

```
Error starting at line 1 in command:
UPDATE outstanding
  SET shipdate = '06-APR-09'
  WHERE order# = 1012
Error report:
SQL Error: ORA-01402: view WITH CHECK OPTION where-clause violation
01402. 00000 -  "view WITH CHECK OPTION where-clause violation"
*Cause:
*Action:
```

FIGURE 13-10   Error returned on an update that violates WITH CHECK OPTION

If the purpose of the OUTSTANDING view is to allow users to enter an order's ship date when it's shipped, it should be re-created *without* the WITH CHECK OPTION constraint, as shown in Figure 13-11. Next, a command is issued to update order 1012's ship date to April 6, 2009. Because the WITH CHECK OPTION constraint is no longer used on the OUTSTANDING view, the update is allowed. However, the record isn't included in the view after the change occurs because the Shipdate field is no longer NULL.

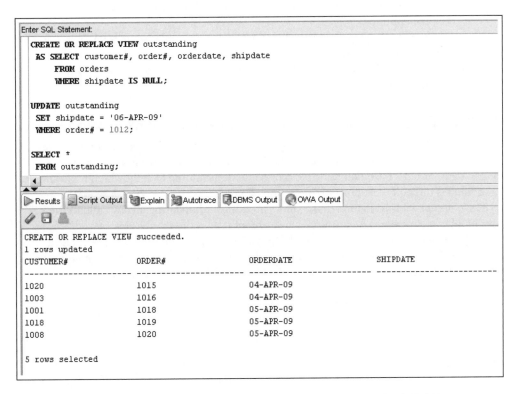

FIGURE 13-11   Update succeeds on the view created without WITH CHECK OPTION

> **TIP**
>
> You should attempt the DML operations in this chapter and experiment with variations of the examples. You can "undo" any DML operation by issuing the ROLLBACK command so that you don't have to rebuild the views. In a real-world situation, however, you would store copies of the tables in a test environment and experiment on the copies rather than the real data.

CREATING A COMPLEX VIEW

You create a **complex view** with the same CREATE VIEW command you use for a simple view. However, the SELECT statement in a complex view retrieves or derives data from one or more tables and can contain functions or grouped data. The main difference in functionality between simple and complex views is that *certain DML operations aren't permitted with complex views*. To explain how complex views react to DML operations, this section uses three different views:

- The first complex view is based on one table, but it uses an expression for one of the columns.
- The second complex view is based on two tables and also uses an expression for one of the columns.
- The third complex view is derived from four tables, and it includes a group function and a GROUP BY clause.

DML Operations on a Complex View with an Arithmetic Expression

As you'll see, different factors affect the type of DML operations allowed on complex views. For example, if a view contains a column that's the result of an arithmetic expression or grouped data, or if it's based on multiple tables and determining exactly which table should be modified is difficult, certain DML operations don't work.

The complex view in Figure 13-12 shows creating and updating a view called PRICES. This complex view seems like a simple view, except it uses the expression `retail-cost` to calculate the Profit column. It also includes an UPDATE command, using `SET retail = 29.95` to change the retail price of *Revenge of Mickey* from $22.00 to $29.95. Again, the view acts like a simple view because the UPDATE command to make the change works.

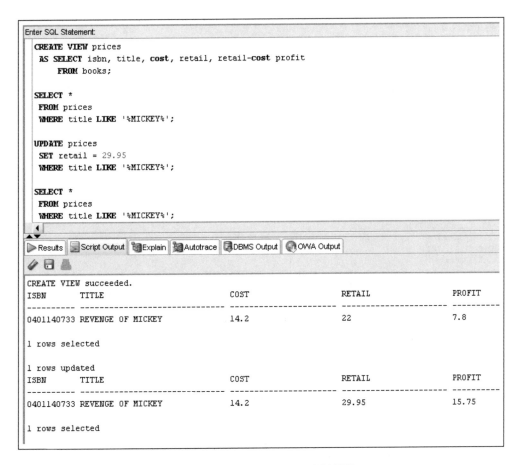

```
Enter SQL Statement:
  CREATE VIEW prices
   AS SELECT isbn, title, cost, retail, retail-cost profit
      FROM books;

  SELECT *
   FROM prices
   WHERE title LIKE '%MICKEY%';

  UPDATE prices
   SET retail = 29.95
   WHERE title LIKE '%MICKEY%';

  SELECT *
   FROM prices
   WHERE title LIKE '%MICKEY%';
```

Results | Script Output | Explain | Autotrace | DBMS Output | OWA Output

```
CREATE VIEW succeeded.
ISBN       TITLE                    COST                    RETAIL                  PROFIT
---------- ----------------------   ----------------------  ----------------------  ----------
0401140733 REVENGE OF MICKEY        14.2                    22                      7.8

1 rows selected

1 rows updated
ISBN       TITLE                    COST                    RETAIL                  PROFIT
---------- ----------------------   ----------------------  ----------------------  ----------
0401140733 REVENGE OF MICKEY        14.2                    29.95                   15.75

1 rows selected
```

FIGURE 13-12   Create and update a complex view named PRICES

What about removing rows in the view, however? In Figure 13-13, the first DELETE command raises a constraint error because it's attempting to remove a row from the BOOKS table that has FOREIGN KEY dependencies.

The next two commands disable the FOREIGN KEY constraints, and then the DELETE command is tried again. The second DELETE successfully removes the book *Revenge of Mickey* from the PRICES view, which actually removes the book from the BOOKS table. So the DELETE command works for this view, as long as it doesn't violate any constraints on the underlying table.

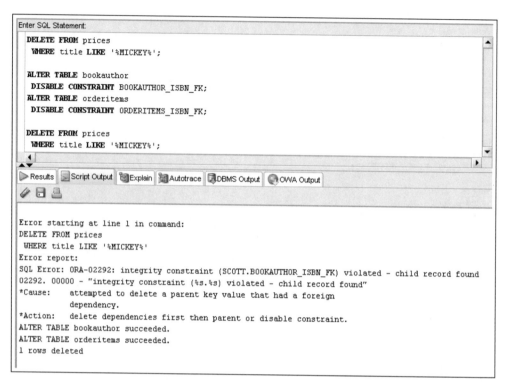

```
Enter SQL Statement:
DELETE FROM prices
 WHERE title LIKE '%MICKEY%';

ALTER TABLE bookauthor
 DISABLE CONSTRAINT BOOKAUTHOR_ISBN_FK;
ALTER TABLE orderitems
 DISABLE CONSTRAINT ORDERITEMS_ISBN_FK;

DELETE FROM prices
 WHERE title LIKE '%MICKEY%';
```

Results | Script Output | Explain | Autotrace | DBMS Output | OWA Output

```
Error starting at line 1 in command:
DELETE FROM prices
 WHERE title LIKE '%MICKEY%'
Error report:
SQL Error: ORA-02292: integrity constraint (SCOTT.BOOKAUTHOR_ISBN_FK) violated - child record found
02292. 00000 -  "integrity constraint (%s.%s) violated - child record found"
*Cause:    attempted to delete a parent key value that had a foreign
           dependency.
*Action:   delete dependencies first then parent or disable constraint.
ALTER TABLE bookauthor succeeded.
ALTER TABLE orderitems succeeded.
1 rows deleted
```

FIGURE 13-13   Deleting a book via the PRICES view

Now what about adding a record to the view? Before attempting it, remember that the Profit column is based on the expression `retail-cost` to calculate the profit generated by the book. So when you add a new book to the PRICES view, do you enter the profit generated, or should you let Oracle 11g calculate the profit? First, attempt an INSERT that includes a profit amount, and then try excluding it. The INSERT statements in Figure 13-14 shows the outcome of these two attempts.

The first attempt includes the profit that would be generated by the new book added to the view. However, Oracle 11g doesn't accept the calculated profit that's entered. The error message "virtual column not allowed here" is one you might see often until you learn the rules for DML operations on complex views. In this case, it means that because a column in the view is based on an arithmetic expression, a value can't be inserted into this column. In other words, a Profit column doesn't exist in the underlying BOOKS table, so this value has no place to be stored in the database.

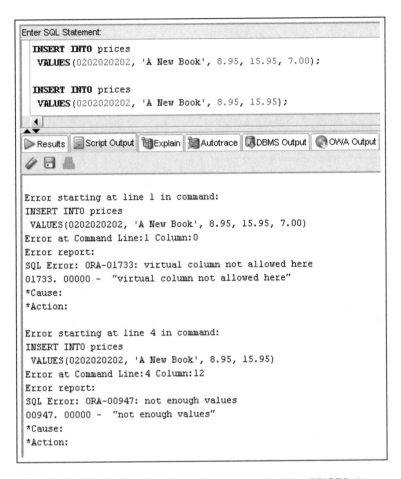

FIGURE 13-14   Failed attempts to add a new book via the PRICES view

The second INSERT attempt in Figure 13-14 tries to add a new record to the PRICES view and exclude the profit value. This attempt returns the error message "not enough values." This message is simple enough; it indicates that the view contains five values, but only four values are included in the INSERT statement. Therefore, the only way to add a new book to the PRICES view is to use a column list in the INSERT statement, as shown in Figure 13-15.

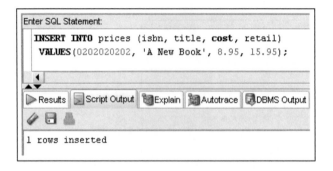

FIGURE 13-15   Successful attempt to add a new book via the PRICES view

You have discovered one of the rules governing DML operations in complex views: *Values can't be inserted into columns based on arithmetic expressions.*

Another consideration is NOT NULL constraints on the underlying table. What if the BOOKS table contains a NOT NULL constraint on the Pubid column? Would the INSERT command via the PRICES view still work? No! The PRICES view doesn't contain the Pubid column and, therefore, doesn't allow indicating a value for this column in an INSERT statement. Figure 13-16 shows adding a NOT NULL constraint on the Pubid column and then another attempt of the previously successful INSERT command. First, the row that was added previously must be deleted because it doesn't contain a Pubid value and would violate the NOT NULL constraint.

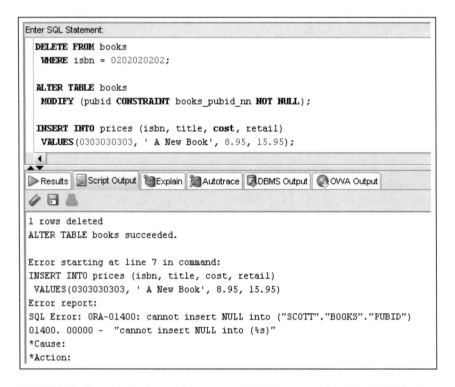

FIGURE 13-16   Constraint violation with an INSERT command via the PRICES view

DML Operations on a Complex View Containing Data from Multiple Tables

To add a little more complexity to the complex view called PRICES, the view has been re-created in Figure 13-17 to include the names of publishers from the PUBLISHER table.

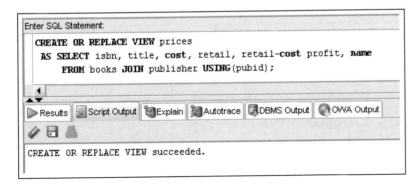

```
Enter SQL Statement:
  CREATE OR REPLACE VIEW prices
  AS SELECT isbn, title, cost, retail, retail-cost profit, name
      FROM books JOIN publisher USING(pubid);
```

CREATE OR REPLACE VIEW succeeded.

FIGURE 13-17   PRICES view with a table join

Now try to perform the same type of DML operations as you did on the previous version of the PRICES view. First, attempt to update the price of a book, as shown in Figure 13-18. The retail price of the book *Big Bear and Little Dove* has been changed from $8.95 to $13.95. As with the previous PRICES view, the DML operation to modify a record works.

```
Enter SQL Statement:
  SELECT *
  FROM prices
  WHERE title LIKE '%BEAR%';

  UPDATE prices
  SET retail = 13.95
  WHERE title LIKE '%BEAR%';

  SELECT *
  FROM prices
  WHERE title LIKE '%BEAR%';
```

ISBN	TITLE	COST	RETAIL	PROFIT	NAME
8117949391	BIG BEAR AND LITTLE DOVE	5.32	8.95	3.63	REED-N-RITE

1 rows selected

1 rows updated

ISBN	TITLE	COST	RETAIL	PROFIT	NAME
8117949391	BIG BEAR AND LITTLE DOVE	5.32	13.95	8.63	REED-N-RITE

1 rows selected

FIGURE 13-18   Updating the Retail column via the PRICES view

However, the change made to the view includes the publisher's name from the PUBLISHER table, so what happens if the Name column is updated? Look at Figure 13-19. When Oracle 11g attempts to update the name of the publisher of *Big Bear and Little Dove*, the UPDATE command fails, and you get the error message "cannot modify a column which maps to a non key-preserved table."

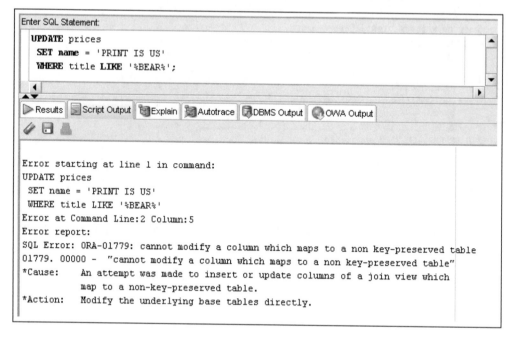

FIGURE 13-19   Failed attempt to update the publisher name via the PRICES view

Taking a step back and analyzing the underlying tables can help you understand the error message's meaning and what caused the error to occur. The PRICES view was built with columns from the BOOKS and PUBLISHER tables. When a view includes columns from more than one table, updates can be applied to only *one* table. The table that can be updated is the one including the primary key of an underlying table and is basically used as the primary key for the view. The PRICES view includes the primary key for the BOOKS table, so any UPDATE command can be performed on columns from the BOOKS table—if the change doesn't violate any constraints on that table. (For example, you can't change the primary key if it's used as a reference for a FOREIGN KEY constraint or if it would no longer be unique.)

In the PRICES view, the BOOKS table is known as the **key-preserved table**. In essence, a key-preserved table is the table containing the primary key that the view is using to uniquely identify each record it displays. By contrast, the Name column is from the PUBLISHER table. The primary key for this table is the Pubid column, and it's not included in the view. However, even if this column is included, Oracle 11g doesn't consider it the primary key for the PRICES view because it could have appeared more than once in the view's contents. Therefore, Oracle 11g treats data from the PUBLISHER table as coming from a **non-key-preserved table** because it doesn't uniquely identify records in the PRICES view.

One way to make sense of this problem is to remember that the BOOKS table actually stores publishers' ID numbers. Therefore, if Oracle 11*g* changes the publisher's name, as in Figure 13-19's UPDATE command, does it mean you want the Pubid column updated in the BOOKS table as well? Or do you change the publisher's name in the PUBLISHER table? If the name is changed in the PUBLISHER table, every book with the same Pubid as *Big Bear and Little Dove* would be published by the publisher Print Is Us, and this isn't the command's intention.

Now you have discovered a second rule that applies to complex views: *DML operations can't be performed on a non-key-preserved table*. To test this rule, execute the DELETE command on the PRICES view shown in Figure 13-20.

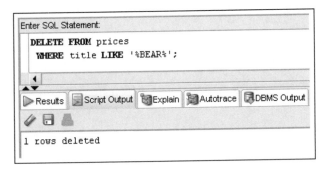

FIGURE 13-20   Deleting a book via the PRICES view

Given the results of this DELETE command, it might seem as though there's a problem with the second rule. You weren't allowed to update the publisher's name for the book, but you *were* allowed to delete the row for the book. However, what actually occurred is that *Big Bear and Little Dove* was deleted from the BOOKS table, which is the key-preserved table. No change occurred in the PUBLISHER table, so technically, the command didn't perform a DML operation on the non-key-preserved table; therefore, the rule was not violated.

N O T E

Keep in mind that the FOREIGN KEY constraints on the BOOKS table were disabled earlier.

DML Operations on a Complex View Containing Functions or Grouped Data

Views are also considered complex if they contain a function or GROUP BY clause. To determine what effect they have on DML commands, create a BALANCEDUE view that displays the total balance due for each order placed by a customer, as shown in Figure 13-21.

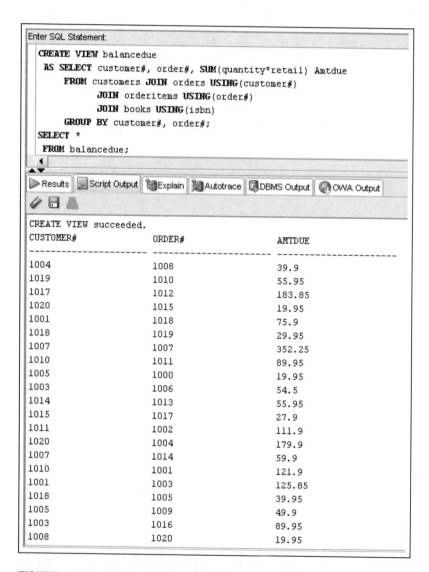

Enter SQL Statement:

```
CREATE VIEW balancedue
  AS SELECT customer#, order#, SUM(quantity*retail) Amtdue
     FROM customers JOIN orders USING(customer#)
          JOIN orderitems USING(order#)
          JOIN books USING(isbn)
     GROUP BY customer#, order#;
SELECT *
  FROM balancedue;
```

Results Script Output Explain Autotrace DBMS Output OWA Output

```
CREATE VIEW succeeded.
CUSTOMER#              ORDER#                AMTDUE
--------------------  --------------------  --------------------
1004                  1008                  39.9
1019                  1010                  55.95
1017                  1012                  183.85
1020                  1015                  19.95
1001                  1018                  75.9
1018                  1019                  29.95
1007                  1007                  352.25
1010                  1011                  89.95
1005                  1000                  19.95
1003                  1006                  54.5
1014                  1013                  55.95
1015                  1017                  27.9
1011                  1002                  111.9
1020                  1004                  179.9
1007                  1014                  59.9
1010                  1001                  121.9
1001                  1003                  125.85
1018                  1005                  39.95
1005                  1009                  49.9
1003                  1016                  89.95
1008                  1020                  19.95
```

FIGURE 13-21   View including grouped data

The BALANCEDUE view groups the items on each customer's order, and then calculates the total amount due, based on the number of books ordered and each book's retail price. Adding a record to this view isn't allowed because the Amtdue column is derived from a function (treated the same way as an expression). Even without the SUM function, Oracle 11g still wouldn't allow adding a record because of the GROUP BY clause. The data being displayed is grouped, so adding a single record to the view isn't possible.

In addition, the function and GROUP BY clause *prevent changing the displayed data* because each record might represent more than one row in the underlying key-preserved table. (Try a little experiment on your own, and see whether you can add or modify a record, but don't be too disappointed if you get an error message.) Therefore, the question is "Will Oracle 11g allow deleting a row from the view?" Figure 13-22 shows an attempt.

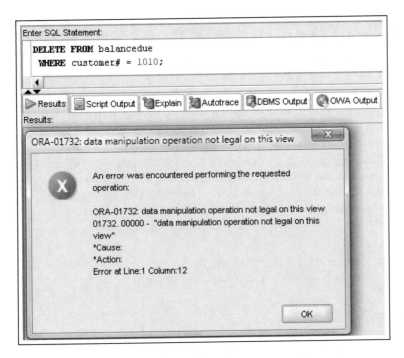

FIGURE 13-22 Failed DELETE command on a view with grouped data

Apparently, the answer is no. The rationale behind not allowing a record to be deleted from the view is that the data is grouped, so clarifying exactly what to delete is hard. Instead of allowing a user to mistakenly delete what could be several rows in the underlying key-preserved table, the operation is simply not allowed. Therefore, if you really want to delete a particular customer's orders, you have to delete them directly from the ORDERS table. (Of course, if this deletion violates any existing constraints between the ORDERS and ORDERITEMS tables, it might not be possible without including an ON CASCADE DELETE option.) Now a third rule has been identified: *DML operations aren't permitted if the view includes a group function or a GROUP BY clause.*

DML Operations on a Complex View Containing DISTINCT or ROWNUM

You must also consider the impact of using the DISTINCT keyword or the ROWNUM pseudocolumn in a view. Recall that the DISTINCT keyword is used to prevent duplicates in the results. Similarly, the **ROWNUM** pseudocolumn is used to limit the rows a query returns.

When using the DISTINCT keyword in a subquery to create a view, remember that in a SELECT clause, the keyword instructs Oracle 11g to suppress duplicates. In other words, if more than one row of a table contains the same data, Oracle displays the row only once in the view. If you consider each unique row as one group, the DISTINCT keyword acts almost like a GROUP BY clause. Therefore, a fourth rule has emerged: *DML operations on a view created with the DISTINCT keyword aren't permitted.*

The second point applies to using ROWNUM, which is a pseudocolumn that applies to every query, even though it isn't displayed with a SELECT * command. When a query is processed, each row returned is assigned a number called the ROWNUM. This value is assigned before any sorting or aggregation. The query in Figure 13-23 instructs Oracle 11g to list each customer's last name and the record's position in the query. However, the query also requires displaying last names in alphabetical order. As shown in the query results, the customer with the last name Cruz is listed first because the list is sorted alphabetically. However, this customer record is actually the sixth row returned by the query.

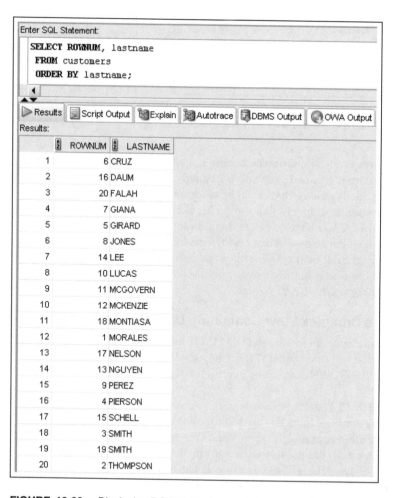

Enter SQL Statement:

```
SELECT ROWNUM, lastname
  FROM customers
  ORDER BY lastname;
```

Results | Script Output | Explain | Autotrace | DBMS Output | OWA Output

Results:

	ROWNUM	LASTNAME
1	6	CRUZ
2	16	DAUM
3	20	FALAH
4	7	GIANA
5	5	GIRARD
6	8	JONES
7	14	LEE
8	10	LUCAS
9	11	MCGOVERN
10	12	MCKENZIE
11	18	MONTIASA
12	1	MORALES
13	17	NELSON
14	13	NGUYEN
15	9	PEREZ
16	4	PIERSON
17	15	SCHELL
18	3	SMITH
19	19	SMITH
20	2	THOMPSON

FIGURE 13-23 Displaying ROWNUMs in a sorted customer list

So what does ROWNUM have to do with DML operations on a complex view? If the query of a complex view includes ROWNUM as one of the columns, no DML operation is allowed on the view. Because ROWNUM is a pseudocolumn that Oracle 11g uses to assign a value to each row, Oracle 11g doesn't allow any additions, deletions, or modifications on data displayed in the view. This results in the final rule: *DML operations aren't allowed on views that include the ROWNUM pseudocolumn.*

Summary Guidelines for DML Operations on a Complex View

The following list summarizes the guidelines regulating DML operations on complex views:

- DML operations that violate a constraint aren't permitted.
- A value can't be added to a column containing an arithmetic expression.
- DML operations aren't permitted on non-key-preserved tables.
- DML operations aren't permitted on views that include group functions, a GROUP BY clause, the DISTINCT keyword, or the ROWNUM pseudocolumn.

DROPPING A VIEW

A view can be dropped or deleted with the **DROP VIEW** command. Figure 13-24 shows the syntax of this command.

```
DROP VIEW viewname;
```

FIGURE 13-24   Syntax of the DROP VIEW command

The command to drop the PRICES view used earlier is shown in Figure 13-25.

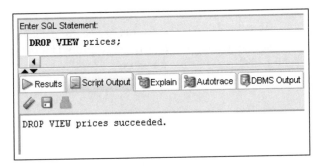

FIGURE 13-25   Command to drop the PRICES view

After the command is executed successfully, Oracle 11g returns a message stating that the view has been dropped. However, the data the view displayed is still available in the underlying tables used to create the view. All that has been deleted is the database object named PRICES that pointed to data stored in the underlying tables.

CREATING AN INLINE VIEW

In Chapter 12, you used a subquery in the FROM clause of a SELECT statement to create a "temporary" table that could be referenced by the SELECT and WHERE clauses. It was considered temporary because a copy of the data the subquery returned wasn't stored in the database. This temporary table is similar to what's called an **inline view** in Oracle 11g. The main difference between an inline view and the other views discussed is that an inline view exists *only while the command is being executed*. It's not a permanent database object and can't be referenced again by a subsequent query. This view is used most often to provide a temporary data source while a command is being executed. One of the most common uses for an inline view is performing a TOP-N analysis.

TOP-N Analysis

Suppose you want to find the five books that generate the most profit. In Chapter 11, you used the MAX group function to find the most profitable book. However, using this function yields only the highest value in a column. How do you find the five highest values? You use **TOP-N analysis**, in which the concepts of an inline view and the ROWNUM pseudocolumn are merged to create a temporary list of records in a sorted order, and then the top *N*, or number, of records are retrieved. An inline view must be used for this analysis because the subquery must use an ORDER BY clause to put records in the correct order before passing the results to the outer query—and ORDER BY clauses aren't allowed in the CREATE VIEW command. Figure 13-26 shows the syntax for performing TOP-N analysis.

```
SELECT columnname, ...
  FROM (subquery)
  WHERE ROWNUM <= N;
```

FIGURE 13-26　Syntax of TOP-N analysis

To determine the five books that generate the most profit, the subquery needs to calculate the profit for each book and then sort the results in descending order by profit before passing the values to the outer query. Figure 13-27 shows this subquery.

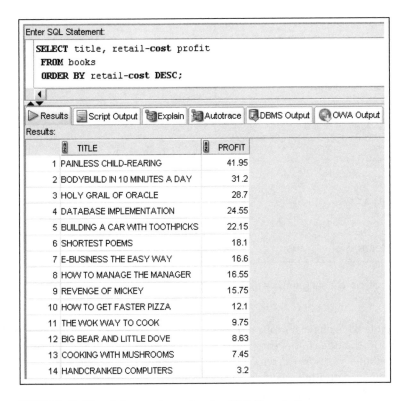

FIGURE 13-27 Subquery for performing TOP-N analysis

To perform the analysis, the subquery must be nested in the FROM clause of a SELECT statement to create the inline view. When the sorted results are passed from the subquery, each row is assigned a ROWNUM to identify its position in the results. Keep in mind that the ROWNUM is assigned before any sorting; however, in this case, the sorting has already occurred in the subquery. Then a WHERE clause is added to the outer query to select only those books with a ROWNUM less than or equal to *N*. In this case, *N* is 5 because you're looking for the five most profitable books.

Figure 13-28 shows the command to determine the five most profitable books. When the command is executed, Oracle 11*g* displays only books with a ROWNUM less than or equal to 5. Because the outer query receives data in descending order based on profit, the top five most profitable books are assigned ROWNUMs 1 through 5.

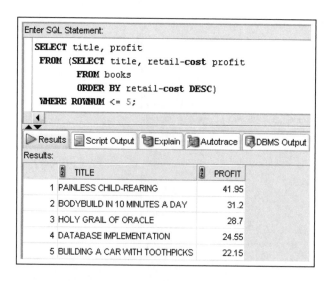

FIGURE 13-28 TOP-N analysis to identify the five most profitable books

The five most profitable books have a profit range between $22.15 and $41.95. What if management wants to know the titles of the three least profitable books, however? The simplest solution is using a subquery to sort the data, so the least profitable book receives the first ROWNUM, the second least profitable book receives the second ROWNUM, and so on. This query and output are shown in Figure 13-29.

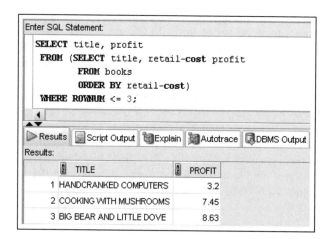

FIGURE 13-29 TOP-N analysis to determine the three least profitable books

The subquery in Figure 13-29 has been modified so that data is sorted in ascending order before it's passed to the outer query. The WHERE clause has also been changed to select only the first three books shown in the results. In this case, the lowest profit generated is $3.20, the next lowest profit is $7.45, and the third lowest profit is $8.63. Although the query actually returns the lowest *N* values, it's still considered TOP-N analysis because the results consist of the top *N* ROWNUMs.

CREATING A MATERIALIZED VIEW

A **materialized view** enables you to store data retrieved by the view query and reuse this data without executing the view query again. In other words, a materialized view allows replicating data. These views are often referred to as "snapshots," as they take a picture or capture a set of data at a specific point in time. The data already exists in the underlying tables, so why would you want to replicate it? Several business needs make materialized views useful:

- Complex queries or queries on large databases often require a lot of processing, which can affect system users. Most businesses want to maintain optimal performance for transactional processing. Replicating data for reporting and analysis allows dedicating system resources to transactional processes.
- Remote users could improve query performance by replicating data to a local database. Instead of transferring data across long distances, a local copy could be used for satellite offices.
- Data analysis needs might require freezing data for a specific time for comparison purposes.

Materialized views also have some disadvantages. First, additional storage space is needed for the copied data. Second, if modifications are made to data via the materialized view, these changes must be synchronized with underlying tables. Finally, if data in the materialized view is updated often, the reduction in processing overhead might be minimal because a materialized view must physically store the results.

Creating a materialized view, shown in Figure 13-30, is similar to creating the views you've worked with in this chapter. A number of additional options are available, such as defining the data refresh schedule and determining whether data in the view should be read only.

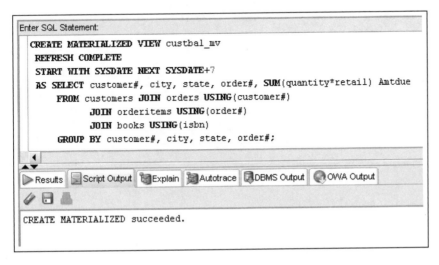

```
Enter SQL Statement:
    CREATE MATERIALIZED VIEW custbal_mv
    REFRESH COMPLETE
    START WITH SYSDATE NEXT SYSDATE+7
    AS SELECT customer#, city, state, order#, SUM(quantity*retail) Amtdue
        FROM customers JOIN orders USING(customer#)
            JOIN orderitems USING(order#)
            JOIN books USING(isbn)
        GROUP BY customer#, city, state, order#;
```

Results | Script Output | Explain | Autotrace | DBMS Output | OWA Output

```
CREATE MATERIALIZED succeeded.
```

FIGURE 13-30   Creating a materialized view named CUSTBAL_MV

The REFRESH clause in this example uses the COMPLETE option, which indicates that the data should be reconstructed from scratch. Other options are available, such as FAST, which applies only data changes. The START WITH clause indicates that the initial materialized view should be created immediately (the SYSDATE option). It also establishes a schedule to rebuild the materialized view every week by including the NEXT clause (with SYSDATE+7).

As shown in Figure 13-31, you can query the materialized view in the same way you query views with no physical properties (in other words, that don't store a copy of the data).

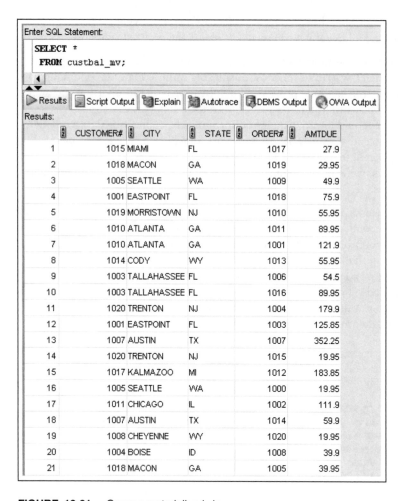

FIGURE 13-31 Query a materialized view

You remove a materialized view with a DROP command containing the keyword MATERIALIZED, as shown in Figure 13-32. Notice that the first DROP command fails because it doesn't include the MATERIALIZED keyword.

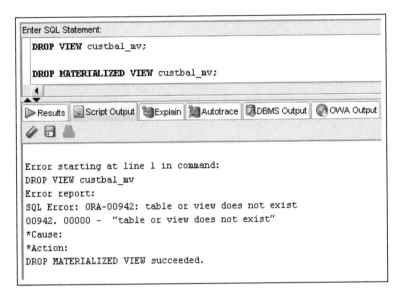

FIGURE 13-32 Drop a materialized view

501

NOTE

Tables in other databases, both Oracle and non-Oracle, can be queried by using database links. See the Oracle documentation for more information on creating and using database links.

Chapter Summary

- A view is a temporary or virtual table used to retrieve data stored in underlying database tables.
- The view query must be executed each time the view is used.
- A view can be used to simplify queries or restrict access to sensitive data.
- A view is created with the CREATE VIEW command.
- A view can't be modified. To change a view, it must be dropped and then re-created, or the CREATE OR REPLACE VIEW command must be used.
- Any DML operation can be performed on a simple query if it doesn't violate a constraint.
- A view containing expressions or functions or joining multiple tables is considered a complex view.
- A complex view can be used to update only one table. The table must be a key-preserved table.
- Data can't be added to a view column containing an expression.
- DML operations aren't permitted on non-key-preserved tables.
- DML operations aren't permitted on views that include group functions, a GROUP BY clause, the ROWNUM pseudocolumn, or the DISTINCT keyword.
- Oracle 11*g* assigns a row number to every row in a table to indicate its position in the table. The row number can be referenced by the keyword ROWNUM.
- A view can be dropped with the DROP VIEW command. The data isn't affected because it exists in the original tables.
- An inline view can be used only by the current statement and can include an ORDER BY clause.
- TOP-N analysis uses the ROWNUM of sorted data to determine a range of top values.
- Materialized views physically store view query results.

Chapter 13 Syntax Summary

The following table summarizes the syntax you have learned in this chapter. You can use the table as a study guide and reference.

Syntax Guide

Element	Command Syntax	Example
Command to create a new view	CREATE [FORCE|NOFORCE] VIEW *viewname* (*columnname*, . . .) AS *subquery* [WITH CHECK OPTION [CONSTRAINT *constraintname*]] [WITH READ ONLY];	CREATE VIEW inventory AS SELECT isbn, title, retail price FROM books WITH READ ONLY;

Element	Command Syntax	Example
Command to replace an existing view	CREATE OR REPLACE [FORCE|NOFORCE] VIEW *viewname* (*columnname*, ...) AS *subquery* [WITH CHECK OPTION [CONSTRAINT *constraintname*]] [WITH READ ONLY];	**CREATE OR REPLACE** **VIEW inventory** **AS SELECT isbn, title,** **retail price** **FROM books;**
Command to drop a view	DROP VIEW *viewname*;	**DROP VIEW inventory;**
Command to create an inline view	SELECT *columnname*, ... FROM (*subquery*) WHERE ROWNUM <= *N*;	**SELECT title, profit** **FROM (SELECT title,** **retail-cost profit** **FROM books** **ORDER BY retail-cost** **DESC)** **WHERE ROWNUM <= 5;**
Command to create a materialized view	CREATE MATERIALIZED VIEW *viewname* [REFRESH *option*] [START WITH *date*] AS *subquery*;	**CREATE MATERIALIZED VIEW** **custbal_mv** **REFRESH COMPLETE** **START WITH SYSDATE NEXT** **SYSDATE + 7** **AS SELECT customer#, city,** **state, order#,** **SUM(quantity*retail) Amtdue** **FROM customers JOIN orders** **USING (customer#)** **JOIN orderitems** **USING (order#)** **JOIN books USING (isbn)** **GROUP BY customer#, city,** **state, order#;**

503

Review Questions

1. How is a simple view different from a complex view?
2. Under what circumstances is a DML operation not allowed on a simple view?
3. When should the FORCE keyword be used in the CREATE VIEW command?
4. What's the purpose of the WITH CHECK OPTION constraint?

5. List the guidelines for DML operations on complex views.

6. How do you ensure that no user can change the data displayed by a view?

7. What's the difference between a key-preserved and a non-key-preserved table?

8. What command can be used to modify a view?

9. What's unique about materialized views compared with other views?

10. What happens to the data displayed by a view when the view is deleted?

Multiple Choice

To answer the following questions, refer to the tables in the JustLee Books database.

Questions 1–7 are based on successful execution of the following statement:

```
CREATE VIEW changeaddress
 AS SELECT customer#, lastname, firstname, order#,
  shipstreet, shipcity, shipstate, shipzip
 FROM customers JOIN orders USING (customer#)
 WHERE shipdate IS NULL
 WITH CHECK OPTION;
```

1. Which of the following statements is correct?
   a. No DML operations can be performed on the CHANGEADDRESS view.
   b. The CHANGEADDRESS view is a simple view.
   c. The CHANGEADDRESS view is a complex view.
   d. The CHANGEADDRESS view is an inline view.

2. Assuming there's only a primary key, and FOREIGN KEY constraints exist on the underlying tables, which of the following commands returns an error message?
   a. ```
      UPDATE changeaddress
        SET shipstreet = '958 ELM ROAD'
        WHERE customer# = 1020;
      ```
   b. ```
      INSERT INTO changeaddress
        VALUES (9999, 'LAST', 'FIRST', 9999,
         '123 HERE AVE', 'MYTOWN', 'AA', 99999);
      ```
   c. ```
      DELETE FROM changeaddress
        WHERE customer# = 1020;
      ```
   d. all of the above
   e. only a and b
   f. only a and c
   g. none of the above

3. Which of the following is the key-preserved table for the CHANGEADDRESS view?
   a. CUSTOMERS table
   b. ORDERS table
   c. Both tables together serve as a composite key-preserved table.
   d. none of the above
      ```

4. Which of the following columns serves as the primary key for the CHANGEADDRESS view?

    a. Customer#

    b. Lastname

    c. Firstname

    d. Order#

    e. Shipstreet

5. If a record is deleted from the CHANGEADDRESS view based on the Customer# column, the customer information is then deleted from which underlying table?

    a. CUSTOMERS

    b. ORDERS

    c. CUSTOMERS and ORDERS

    d. Neither—the DELETE command can't be used on the CHANGEADDRESS view.

6. Which of the following is correct?

    a. ROWNUM can't be used with the view because it isn't included in the results the subquery returns.

    b. The view is a simple view because it doesn't include a group function or a GROUP BY clause.

    c. The data in the view can't be displayed in descending order by customer number because an ORDER BY clause isn't allowed when working with views.

    d. all of the above

    e. none of the above

7. Assuming one of the orders has shipped, which of the following is true?

    a. The CHANGEADDRESS view can't be used to update an order's ship date because of the WITH CHECK OPTION constraint.

    b. The CHANGEADDRESS view can't be used to update an order's ship date because the Shipdate column isn't included in the view.

    c. The CHANGEADDRESS view can't be used to update an order's ship date because the ORDERS table is not the key-preserved table.

    d. The CHANGEADDRESS view can't be used to update an order's ship date because the UPDATE command can't be used on data in the view.

Questions 8–12 are based on successful execution of the following command:

```
CREATE VIEW changename
 AS SELECT customer#, lastname, firstname
 FROM customers
 WITH CHECK OPTION;
```

Assume that the only constraint on the CUSTOMERS table is a PRIMARY KEY constraint.

8. Which of the following is a correct statement?

    a. No DML operations can be performed on the CHANGENAME view.

    b. The CHANGENAME view is a simple view.

    c. The CHANGENAME view is a complex view.

    d. The CHANGENAME view is an inline view.

9. Which of the following columns serves as the primary key for the CHANGENAME view?

    a. Customer#

    b. Lastname

    c. Firstname

    d. The view doesn't have or need a primary key.

10. Which of the following DML operations could never be used on the CHANGENAME view?

    a. INSERT

    b. UPDATE

    c. DELETE

    d. All of the above are valid DML operations for the CHANGENAME view.

11. The INSERT command can't be used with the CHANGENAME view because:

    a. A key-preserved table isn't included in the view.

    b. The view was created with the WITH CHECK OPTION constraint.

    c. The inserted record couldn't be accessed by the view.

    d. None of the above—an INSERT command can be used on the table as long as the PRIMARY KEY constraint isn't violated.

12. If the CHANGENAME view needs to include the customer's zip code as a means of verifying the change (that is, to authenticate the user), which of the following is true?

    a. The CREATE OR REPLACE VIEW command can be used to re-create the view with the necessary column included in the new view.

    b. The ALTER VIEW ... ADD COLUMN command can be used to add the necessary column to the existing view.

    c. The CHANGENAME view can be dropped, and then the CREATE VIEW command can be used to re-create the view with the necessary column included in the new view.

    d. All of the above can be performed to include the customer's zip code in the view.

    e. Only a and b include the customer's zip code in the view.

    f. Only a and c include the customer's zip code in the view.

    g. None of the above includes the customer's zip code in the view.

13. Which of the following DML operations can't be performed on a view containing a group function?

    a. INSERT

    b. UPDATE

    c. DELETE

d. All of the above can be performed on a view containing a group function.

e. None of the above can be performed on a view containing a group function.

14. You can't perform any DML operations on which of the following?

a. views created with the WITH READ ONLY option

b. views that include the DISTINCT keyword

c. views that include a GROUP BY clause

d. All of the above allow DML operations.

e. None of the above allow DML operations.

15. A TOP-N analysis is performed by determining the rows with:

a. the highest ROWNUM values

b. a ROWNUM value greater than or equal to N

c. the lowest ROWNUM values

d. a ROWNUM value less than or equal to N

16. To assign names to the columns in a view, you can do which of the following?

a. Assign aliases in the subquery, and the aliases are used for the column names.

b. Use the ALTER VIEW command to change column names.

c. Assign names for up to three columns in the CREATE VIEW clause before the subquery is listed in the AS clause.

d. None of the above—columns can't be assigned names for a view; they must keep their original names.

17. Which of the following is correct?

a. The ORDER BY clause can't be used in the subquery of a CREATE VIEW command.

b. The ORDER BY clause can't be used in an inline view.

c. The DISTINCT keyword can't be used in an inline view.

d. The WITH READ ONLY option must be used with an inline view.

18. If you try to add a row to a complex view that includes a GROUP BY clause, you get which of the following error messages?

a. virtual column not allowed here

b. data manipulation operation not legal on this view

c. cannot map to a column in a non-key-preserved table

d. None of the above—no error message is returned.

19. A simple view can contain which of the following?

a. data from one or more tables

b. an expression

c. a GROUP BY clause for data retrieved from one table

d. five columns from one table

e. all of the above

f. none of the above

507

20. A complex view can contain which of the following?

    a.   data from one or more tables

    b.   an expression

    c.   a GROUP BY clause for data retrieved from one table

    d.   five columns from one table

    e.   all of the above

    f.   none of the above

Hands-On Assignments

To perform the following activities, refer to the tables in the JustLee Books database.

1. Create a view that lists the name and phone number of the contact person at each publisher. Don't include the publisher's ID in the view. Name the view CONTACT.

2. Change the CONTACT view so that no users can accidentally perform DML operations on the view.

3. Create a view called HOMEWORK13 that includes the columns named Col1 and Col2 from the FIRSTATTEMPT table. Make sure the view is created even if the FIRSTATTEMPT table doesn't exist.

4. Attempt to view the structure of the HOMEWORK13 view.

5. Create a view that lists the ISBN and title for each book in inventory along with the name and phone number of the person to contact if the book needs to be reordered. Name the view REORDERINFO.

6. Try to change the name of a contact person in the REORDERINFO view to your name. Was an error message displayed when performing this step? If so, what was the cause of the error message?

7. Select one of the books in the REORDERINFO view and try to change its ISBN. Was an error message displayed when performing this step? If so, what was the cause of the error message?

8. Delete the record in the REORDERINFO view containing your name. (If you weren't able to perform #6 successfully, delete one of the contacts already listed in the table.) Was an error message displayed when performing this step? If so, what was the cause of the error message?

9. Issue a rollback command to undo any changes made with the preceding DML operations.

10. Delete the REORDERINFO view.

Advanced Challenge

To perform the following activity, refer to the tables in the JustLee Books database.

    The Marketing Department of JustLee Books is about to begin an aggressive marketing campaign to generate sales to repeat customers. The strategy is to look at existing customers' previous purchases, and based on the categories from which these customers have made purchases, JustLee Books will send promotional information about other books in the same category that are highly profitable books for the company.

The Marketing Department has requested that you identify the five most frequently purchased books and the percentage of profit each book generates. The percentage of profit can be calculated by using the formula `((retail-cost)/cost*100)`. The employees in the Marketing Department will use the potential profitability of the marketing campaign to determine how much money to budget for the campaign.

Provide an SQL statement along with the output to respond to the Marketing Department request.

Case Study: *City Jail*

Note: This assignment assumes you have run the CityJail_8.sql script from Chapter 8, which makes all database objects available for completing this case study.

The City Jail Technologies Department is constructing an application to allow users in the Crime Analysis Unit to query data more easily. This system requires creating a number of views, described in the following list. Provide the SQL statement to perform each task and test your views with a query.

1. Create a statement that always returns the names of the three criminals with the highest number of crimes committed.
2. Create a view that includes details for all crimes, including criminal ID, criminal name, criminal parole status, crime ID, date of crime charge, crime status, charge ID, crime code, charge status, pay due date, and amount due. This view shouldn't allow performing any DML operations. Each time the view is used in the application, the data should be queried from the database. (For example, each use of the view should reflect the most current data in the database.)
3. Create a view that includes all data for officers, including the total number of crimes in which they participated in filing charges. To speed up the officer queries, store this view data and schedule the data to be updated every two weeks.

TABLES FOR THE JUSTLEE BOOKS DATABASE

The tables created by running the JLDB_Build.sql script include CUSTOMERS, BOOKS, ORDERS, ORDERITEMS, AUTHOR, BOOKAUTHOR, PUBLISHER, and PROMOTION. This appendix shows the initial structure and contents for each table.

NOTE

Keep in mind that you modify these tables' structures and data as you progress through the textbook.

CUSTOMERS Table

```
Name                            Null       Type
------------------------------- --------   -----------
CUSTOMER#                                  NUMBER(4)
LASTNAME                        NOT NULL   VARCHAR2(10)
FIRSTNAME                       NOT NULL   VARCHAR2(10)
ADDRESS                                    VARCHAR2(20)
CITY                                       VARCHAR2(12)
STATE                                      VARCHAR2(2)
ZIP                                        VARCHAR2(5)
REFERRED                                   NUMBER(4)
REGION                                     CHAR(2)
EMAIL                                      VARCHAR2(30)
```

FIGURE A-1   Structure of the CUSTOMERS table

CUSTOMER#	LASTNAME	FIRSTNAME	ADDRESS	CITY	STATE	ZIP	REFERRED	REGION	EMAIL
1001	MORALES	BONITA	P.O. BOX 651	EASTPOINT	FL	32328	(null)	SE	bm225@sat.net
1002	THOMPSON	RYAN	P.O. BOX 9835	SANTA MONICA	CA	90404	(null)	W	(null)
1003	SMITH	LEILA	P.O. BOX 66	TALLAHASSEE	FL	32306	(null)	SE	(null)
1004	PIERSON	THOMAS	69821 SOUTH AVENUE	BOISE	ID	83707	(null)	NW	tpier55@sat.net
1005	GIRARD	CINDY	P.O. BOX 851	SEATTLE	WA	98115	(null)	NW	cing101@zep.net
1006	CRUZ	MESHIA	82 DIRT ROAD	ALBANY	NY	12211	(null)	NE	cruztop@axe.com
1007	GIANA	TAMMY	9153 MAIN STREET	AUSTIN	TX	78710	1003	SW	treetop@zep.net
1008	JONES	KENNETH	P.O. BOX 137	CHEYENNE	WY	82003	(null)	N	kenask@sat.net
1009	PEREZ	JORGE	P.O. BOX 8564	BURBANK	CA	91510	1003	W	jperez@canet.com
1010	LUCAS	JAKE	114 EAST SAVANNAH	ATLANTA	GA	30314	(null)	SE	(null)
1011	MCGOVERN	REESE	P.O. BOX 18	CHICAGO	IL	60606	(null)	N	reesemc@sat.net
1012	MCKENZIE	WILLIAM	P.O. BOX 971	BOSTON	MA	02110	(null)	NE	will2244@axe.net
1013	NGUYEN	NICHOLAS	357 WHITE EAGLE AVE.	CLERMONT	FL	34711	1006	SE	nguy33@sat.net
1014	LEE	JASMINE	P.O. BOX 2947	CODY	WY	82414	(null)	N	jaslee@sat.net
1015	SCHELL	STEVE	P.O. BOX 677	MIAMI	FL	33111	(null)	SE	sschell3@sat.net
1016	DAUM	MICHELL	9851231 LONG ROAD	BURBANK	CA	91508	1010	W	(null)
1017	NELSON	BECCA	P.O. BOX 563	KALMAZOO	MI	49006	(null)	N	becca88@digs.com
1018	MONTIASA	GREG	1008 GRAND AVENUE	MACON	GA	31206	(null)	SE	greg336@sat.net
1019	SMITH	JENNIFER	P.O. BOX 1151	MORRISTOWN	NJ	07962	1003	NE	(null)
1020	FALAH	KENNETH	P.O. BOX 335	TRENTON	NJ	08607	(null)	NE	Kfalah@sat.net

FIGURE A-2   Data in the CUSTOMERS table

BOOKS Table

Name	Null	Type
ISBN	NOT NULL	VARCHAR2(10)
TITLE		VARCHAR2(30)
PUBDATE		DATE
PUBID		NUMBER(2)
COST		NUMBER(5,2)
RETAIL		NUMBER(5,2)
DISCOUNT		NUMBER(4,2)
CATEGORY		VARCHAR2(12)

FIGURE A-3   Structure of the BOOKS table

ISBN	TITLE	PUBDATE	PUBID	COST	RETAIL	DISCOUNT	CATEGORY
1059831198	BODYBUILD IN 10 MINUTES A DAY	21-JAN-05	4	18.75	30.95	(null)	FITNESS
0401140733	REVENGE OF MICKEY	14-DEC-05	1	14.2	22	(null)	FAMILY LIFE
4981341710	BUILDING A CAR WITH TOOTHPICKS	18-MAR-06	2	37.8	59.95	3	CHILDREN
8843172113	DATABASE IMPLEMENTATION	04-JUN-03	3	31.4	55.95	(null)	COMPUTER
3437212490	COOKING WITH MUSHROOMS	28-FEB-04	4	12.5	19.95	(null)	COOKING
3957136468	HOLY GRAIL OF ORACLE	31-DEC-05	3	47.25	75.95	3.8	COMPUTER
1915762492	HANDCRANKED COMPUTERS	21-JAN-05	3	21.8	25	(null)	COMPUTER
9959789321	E-BUSINESS THE EASY WAY	01-MAR-06	2	37.9	54.5	(null)	COMPUTER
2491748320	PAINLESS CHILD-REARING	17-JUL-04	5	48	89.95	4.5	FAMILY LIFE
0299282519	THE WOK WAY TO COOK	11-SEP-04	4	19	28.75	(null)	COOKING
8117949391	BIG BEAR AND LITTLE DOVE	08-NOV-05	5	5.32	8.95	(null)	CHILDREN
0132149871	HOW TO GET FASTER PIZZA	11-NOV-06	4	17.85	29.95	1.5	SELF HELP
9247381001	HOW TO MANAGE THE MANAGER	09-MAY-03	1	15.4	31.95	(null)	BUSINESS
2147428890	SHORTEST POEMS	01-MAY-05	5	21.85	39.95	(null)	LITERATURE

FIGURE A-4   Data in the BOOKS table

ORDERS Table

```
Name                          Null       Type
----------------------------  --------   ------------
ORDER#                        NOT NULL   NUMBER(4)
CUSTOMER#                                NUMBER(4)
ORDERDATE                     NOT NULL   DATE
SHIPDATE                                 DATE
SHIPSTREET                               VARCHAR2(18)
SHIPCITY                                 VARCHAR2(15)
SHIPSTATE                                VARCHAR2(2)
SHIPZIP                                  VARCHAR2(5)
SHIPCOST                                 NUMBER(4,2)
```

FIGURE A-5   Structure of the ORDERS table

ORDER#	CUSTOMER#	ORDERDATE	SHIPDATE	SHIPSTREET	SHIPCITY	SHIPSTATE	SHIPZIP	SHIPCOST
1000	1005	31-MAR-09	02-APR-09	1201 ORANGE AVE	SEATTLE	WA	98114	2
1001	1010	31-MAR-09	01-APR-09	114 EAST SAVANNAH	ATLANTA	GA	30314	3
1002	1011	31-MAR-09	01-APR-09	58 TILA CIRCLE	CHICAGO	IL	60605	3
1003	1001	01-APR-09	01-APR-09	958 MAGNOLIA LANE	EASTPOINT	FL	32328	4
1004	1020	01-APR-09	05-APR-09	561 ROUNDABOUT WAY	TRENTON	NJ	08601	(null)
1005	1018	01-APR-09	02-APR-09	1008 GRAND AVENUE	MACON	GA	31206	2
1006	1003	01-APR-09	02-APR-09	558A CAPITOL HWY.	TALLAHASSEE	FL	32307	2
1007	1007	02-APR-09	04-APR-09	9153 MAIN STREET	AUSTIN	TX	78710	7
1008	1004	02-APR-09	03-APR-09	69821 SOUTH AVENUE	BOISE	ID	83707	3
1009	1005	03-APR-09	05-APR-09	9 LIGHTENING RD.	SEATTLE	WA	98110	(null)
1010	1019	03-APR-09	04-APR-09	384 WRONG WAY HOME	MORRISTOWN	NJ	07960	2
1011	1010	03-APR-09	05-APR-09	102 WEST LAFAYETTE	ATLANTA	GA	30311	2
1012	1017	03-APR-09	(null)	1295 WINDY AVENUE	KALMAZOO	MI	49002	6
1013	1014	03-APR-09	04-APR-09	7618 MOUNTAIN RD.	CODY	WY	82414	2
1014	1007	04-APR-09	05-APR-09	9153 MAIN STREET	AUSTIN	TX	78710	3
1015	1020	04-APR-09	(null)	557 GLITTER ST.	TRENTON	NJ	08606	2
1016	1003	04-APR-09	(null)	9901 SEMINOLE WAY	TALLAHASSEE	FL	32307	2
1017	1015	04-APR-09	05-APR-09	887 HOT ASPHALT ST	MIAMI	FL	33112	3
1018	1001	05-APR-09	(null)	95812 HIGHWAY 98	EASTPOINT	FL	32328	(null)
1019	1018	05-APR-09	(null)	1008 GRAND AVENUE	MACON	GA	31206	2
1020	1008	05-APR-09	(null)	195 JAMISON LANE	CHEYENNE	WY	82003	2

FIGURE A-6   Data in the ORDERS table

ORDERITEMS Table

```
Name                              Null       Type
------------------------------    --------   ---------------
ORDER#                                       NUMBER(4)
ITEM#                                        NUMBER(2)
ISBN                                         VARCHAR2(10)
QUANTITY                          NOT NULL   NUMBER(3)
PAIDEACH                          NOT NULL   NUMBER(5,2)
```

FIGURE A-7   Structure of the ORDERITEMS table

ORDER#	ITEM#	ISBN	QUANTITY	PAIDEACH
1000	1	3437212490	1	19.95
1001	1	9247381001	1	31.95
1001	2	2491748320	1	85.45
1002	1	8843172113	2	55.95
1003	1	8843172113	1	55.95
1003	2	1059831198	1	30.95
1003	3	3437212490	1	19.95
1004	1	2491748320	2	85.45
1005	1	2147428890	1	39.95
1006	1	9959789321	1	54.5
1007	1	3957136468	3	72.15
1007	2	9959789321	1	54.5
1007	3	8117949391	1	8.95
1007	4	8843172113	1	55.95
1008	1	3437212490	2	19.95
1009	1	3437212490	1	19.95
1009	2	0401140733	1	22
1010	1	8843172113	1	55.95
1011	1	2491748320	1	85.45
1012	1	8117949391	1	8.95
1012	2	1915762492	2	25
1012	3	2491748320	1	85.45
1012	4	0401140733	1	22
1013	1	8843172113	1	55.95
1014	1	0401140733	2	22
1015	1	3437212490	1	19.95
1016	1	2491748320	1	85.45
1017	1	8117949391	2	8.95
1018	1	3437212490	1	19.95
1018	2	8843172113	1	55.95
1019	1	0401140733	1	22
1020	1	3437212490	1	19.95

FIGURE A-8 Data in the ORDERITEMS table

AUTHOR Table

```
Name                            Null       Type
------------------------------- --------   --------------
AUTHORID                        NOT NULL   VARCHAR2(4)
LNAME                                      VARCHAR2(10)
FNAME                                      VARCHAR2(10)
```

FIGURE A-9 Structure of the AUTHOR table

AUTHORID	LNAME	FNAME
S100	SMITH	SAM
J100	JONES	JANICE
A100	AUSTIN	JAMES
M100	MARTINEZ	SHEILA
K100	KZOCHSKY	TAMARA
P100	PORTER	LISA
A105	ADAMS	JUAN
B100	BAKER	JACK
P105	PETERSON	TINA
W100	WHITE	WILLIAM
W105	WHITE	LISA
R100	ROBINSON	ROBERT
F100	FIELDS	OSCAR
W110	WILKINSON	ANTHONY

FIGURE A-10   Data in the AUTHOR table

BOOKAUTHOR Table

```
Name                              Null      Type
------------------------------- -------- --------------

ISBN                            NOT NULL VARCHAR2(10)
AUTHORID                        NOT NULL VARCHAR2(4)
```

FIGURE A-11   Structure of the BOOKAUTHOR table

ISBN	AU...
0132149871	S100
0299282519	S100
0401140733	J100
1059831198	P100
1059831198	S100
1915762492	W100
1915762492	W105
2147428890	W105
2491748320	B100
2491748320	F100
2491748320	R100
3437212490	B100
3957136468	A100
4981341710	K100
8117949391	R100
8843172113	A100
8843172113	A105
8843172113	P105
9247381001	W100
9959789321	J100

FIGURE A-12   Data in the BOOKAUTHOR table

PUBLISHER Table

```
Name                               Null       Type
--------------------------------   --------   ---------------
PUBID                              NOT NULL   NUMBER(2)
NAME                                          VARCHAR2(23)
CONTACT                                       VARCHAR2(15)
PHONE                                         VARCHAR2(12)
```

FIGURE A-13 Structure of the PUBLISHER table

PUBID	NAME	CONTACT	PHONE
1	PRINTING IS US	TOMMIE SEYMOUR	000-714-8321
2	PUBLISH OUR WAY	JANE TOMLIN	010-410-0010
3	AMERICAN PUBLISHING	DAVID DAVIDSON	800-555-1211
4	READING MATERIALS INC.	RENEE SMITH	800-555-9743
5	REED-N-RITE	SEBASTIAN JONES	800-555-8284

FIGURE A-14 Data in the PUBLISHER table

PROMOTION Table

```
Name                               Null       Type
--------------------------------   --------   ---------------
GIFT                                          VARCHAR2(15)
MINRETAIL                                     NUMBER(5,2)
MAXRETAIL                                     NUMBER(5,2)
```

FIGURE A-15 Structure of the PROMOTION table

GIFT	MINRETAIL	MAXRETAIL
BOOKMARKER	0	12
BOOK LABELS	12.01	25
BOOK COVER	25.01	56
FREE SHIPPING	56.01	999.99

FIGURE A-16 Data in the PROMOTION table

APPENDIX **B**

SQL*PLUS AND SQL DEVELOPER OVERVIEW

INTRODUCTION

SQL*Plus and SQL Developer are two software tools for connecting to the Oracle database, executing SQL commands, and viewing the results. SQL*Plus is a basic command-line tool; SQL Developer has a graphical user interface. Both tools are installed by default with an Oracle 11g server installation and can also be installed on other (client) machines to connect to the Oracle server. In previous Oracle versions, Oracle 9i and Oracle 10g, a browser-based SQL*Plus tool called iSQL*Plus was available. iSQL*Plus is no longer supported by Oracle and isn't available with Oracle 11g. This appendix gives you an overview of using the SQL*Plus and SQL Developer software tools.

TIP

Keep in mind that the SQL language isn't case sensitive, no matter which client tool you're using.

SQL*PLUS

Because SQL*Plus is a command-line tool, it can seem a bit awkward for those accustomed to GUIs. However, this tool is almost always available for an Oracle installation, so it's worth becoming familiar with it. In the program listing on your operating system, you should find an SQL*Plus selection under the Oracle entries. When you start the software, you're prompted for login information, as shown in Figure B-1.

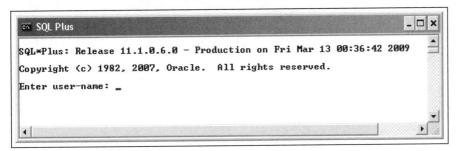

FIGURE B-1   The SQL*Plus login window

You need two pieces of information for this login: a username and a password. First, you're prompted to enter a username. If the account name is SCOTT, for example, simply type it and press Enter, if you're on the Oracle server machine. If you're on a client machine that connects to the Oracle server, you also need to enter a host string to identify the Oracle server to be used. The host string (also called service name) is set up when SQL*Plus is installed. To include the host string at login, use the format *username@host_string*, as shown in Figure B-2. After pressing Enter, you're prompted for a password. Type your password and press Enter again.

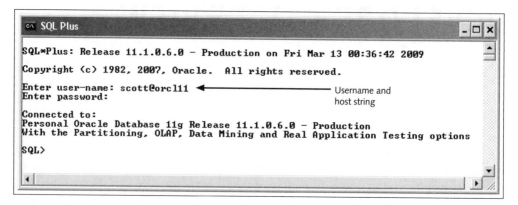

FIGURE B-2 The SQL*Plus command window

After you log in successfully, a "connected" message is displayed, followed by the SQL> prompt, where you enter the SQL query. As you type each line and press Enter, the line number is displayed. When you enter a semicolon and press Enter, the query is executed. Figure B-3 is an example of an executed query.

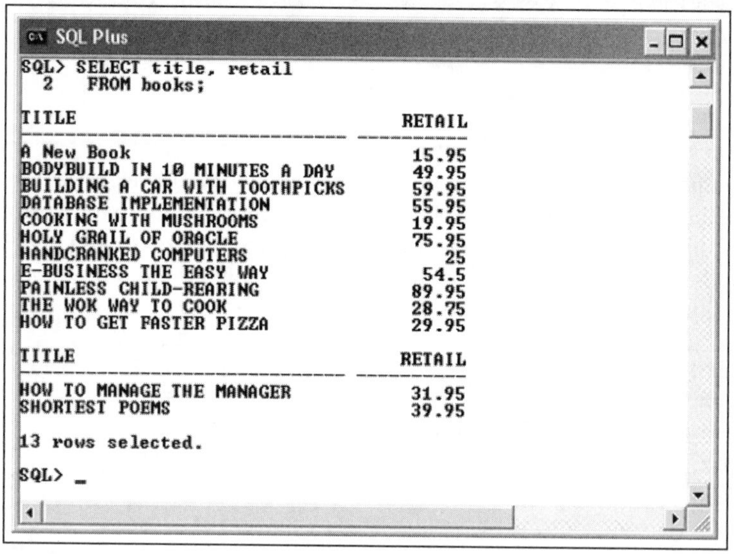

FIGURE B-3 Executing a query in SQL*Plus

Unfortunately, if you make an error, you can't just press Backspace or Delete to delete previous lines. There are a few methods for modifying statements. One is entering statements in a text editor; on most Windows systems, the default text editor is Notepad. If you want to use another program (or if no editor is defined), type DEFINE_EDITOR = NOTEPAD (or the name of the editor you want) at the SQL> prompt, and then press Enter.

NOTE

If you attempt to change the default editor and get a permissions error message, you might not be allowed to make changes to this system setting. If this happens, notify your instructor.

To enter commands in the text editor, follow these steps:

1. To access the editor, type **edit** at the SQL> prompt and press **Enter**. Because you didn't include a filename, the buffer's contents (the most recent SQL statement you executed) are displayed in the editor, which is using the default filename **afiedt.buf**. The buffer display enables you to use any editor features—such as cut and paste, insert, and delete—to make changes to your SQL statement. (*Note*: If you intend to save the file containing your statement, supply a filename when you enter the EDIT command, instead of leaving the default afiedt.buf filename.)

2. After you close the editor, the SQL statement that was in the editor when you exited is displayed in the SQL*Plus window.

3. To execute the corrected statement, type **/** (a slash) at the SQL> prompt and press **Enter.**

Some users take this method a step further and enter all their statements in a text file, and then copy and paste from the text file to the SQL*Plus window.

If you prefer not to use an editor, you can use the SQL*Plus editing commands. To view what's currently in the SQL*Plus buffer, do the following:

1. Type the letter **L**, the word **LIST**, or a semicolon (;) at the SQL> prompt and press **Enter.**

2. To display the last line stored in the buffer, type **LIST LAST** and press **Enter.**

3. If you need to delete a line from the buffer, type **DEL** followed by the line number and press **Enter.**

4. To add lines to the stored SQL statement, type **INPUT** (or the letter **I**) and press **Enter.** You can then add the new lines of text. To add text to the end of the current line in the buffer, type **APPEND** (or the letter **A**) and the text to be added, and then press **Enter.**

You can experiment with the editing commands by entering the SQL statement shown in Figure B-4.

```
SELECT title, cost
FROM books;
```

FIGURE B-4   A simple SELECT statement

Suppose that after pressing Enter, you realize that you wanted only the books from the publisher with the Pubid 4. To revise the query you just entered, do the following:

1. Type **LIST** to redisplay the SQL statement you just entered. Notice the asterisk (*) next to line 2, which indicates it's the current line.
2. Because you want to add another line to the SQL statement, type **INPUT** and press **Enter**. You can then enter additional lines immediately after the current line. Line 2 was the current line when you entered INPUT, so a 3 is displayed after you press Enter.
3. Type **WHERE pubid=4;** and press **Enter**.

After looking over the results, you realize you forgot to include the retail price for each book. To include the Retail column in the display, you need to alter the first line of the SQL statement you entered. Here's how to make this change:

1. Type **LIST** and press **Enter** to review the current statement in the buffer. Notice that the asterisk now indicates that line 3 is the current line. If you append the additional column name at this time, it's appended to line 3.
2. To set line 1 as the current line, type **1** (the numeral) at the SQL> prompt and press **Enter**. Line 1 is displayed with an asterisk, indicating that it's now the current line.
3. At the SQL> prompt, type **A , retail** and press **Enter** to add a comma and the word "retail" to the end of line 1. After pressing Enter, line 1 is displayed again with the additional text.
4. Type **L** (or **LIST**) and press **Enter** to see the revised SQL statement.
5. To execute the revised SQL statement, type **/** (or **RUN**) and press **Enter**.

After examining the new output, you realize that you need the ISBN of each book, not its title. You can use the CHANGE command to perform a simple search-and-replace operation to change Title to ISBN in the first line of the SQL statement. The syntax of the CHANGE command is C\ *old*\ *new*\. (The last backslash is optional.) Make this change by following these steps:

1. To correct the first line, you need to make line 1 the current line. Type **1** at the SQL> prompt and press **Enter**.
2. After the first line is displayed, type **C\title\ISBN** and press **Enter**.
3. The revision for line 1 is displayed. If it's correct, run the statement and examine the new output.

To close the SQL*Plus session, type EXIT; at the SQL prompt and press Enter.

NOTE

For more information on using SQL*Plus, review the SQL*Plus User's Guide and Reference in the Oracle documentation library at the OTN Web site.

SQL DEVELOPER

In the program listing on your operating system, you should find an SQL Developer selection under the Oracle entries. When you start the software, you see the initial window shown in Figure B-5.

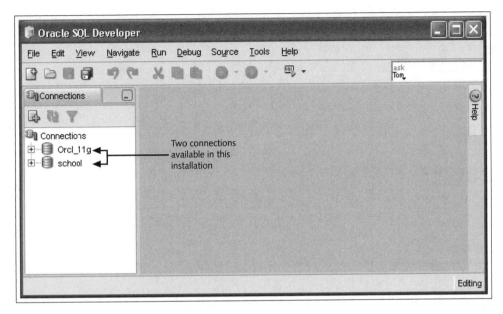

FIGURE B-5 The initial SQL Developer window

On the left is a list of available connections. These connections must be configured to identify the Oracle database to be used. To connect, double-click the connection you want, and the login window shown in Figure B-6 is displayed.

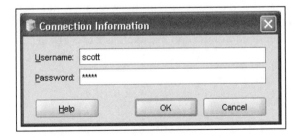

FIGURE B-6 The SQL Developer login window

TIP

If no connections are available, right-click the Connections node on the left and click New Connection to define the Oracle database connection information.

After you're connected, the object navigator pane on the left lists several database object types, as shown in Figure B-7. On the right is the work area where you enter SQL statements in the top pane and view the results in the bottom pane.

Enter SQL statements

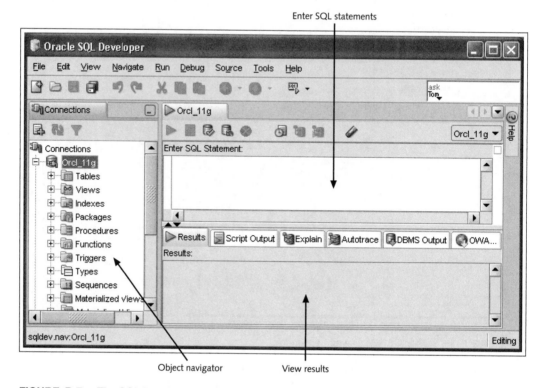

Object navigator   View results

FIGURE B-7   The SQL Developer interface

Enter the SQL statement shown in Figure B-8 in the top pane of the work area, and then click the Execute Statement button to process the statement and view the results. If you right-click the results pane, you see a number of additional options, such as Export Data.

Execute Statement button

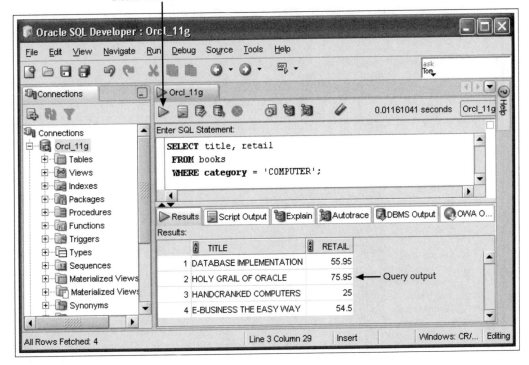

FIGURE B-8   Executing a query in SQL Developer

In this textbook, SELECT statements are processed with the Execute Statement button and all other statements are processed with the Run Script button. Figure B-9 shows using the Run Script option, which allows you to process multiple statements at once and helps you see execution messages more easily for statements, such as DML actions.

Run Script button

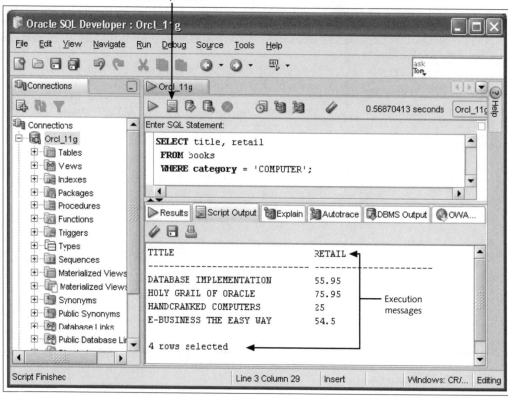

FIGURE B-9   Using the Run Script option

NOTE

For more information on using SQL Developer, review the SQL Developer User's Guide in the Oracle documentation library at the OTN Web site.

APPENDIX

ORACLE RESOURCES

The following list of resources has been provided to assist students receiving Oracle 11g database training or considering careers that require using Oracle products.

Oracle Academic Initiative (OAI)

The Oracle Academic Initiative (OAI) is a result of the Oracle Corporation's effort to provide curricula and other resources to the higher-education community. Students enrolled at an institution participating in the OAI have access to benefits such as discount vouchers for certification exams, discounts on certification preparation software, and a free subscription to *Oracle Magazine*. Visit the Oracle Web site at *http://oai.oracle.com* for more information.

Oracle Certification Program (OCP)

The Oracle Certification Program (OCP) provides certification paths for both database administrators and application developers. Certification is based on successful completion of a series of exams. Oracle offers several levels of certification, starting with the Oracle Certified Associate. You can find current information about the OCP at *www.oracle.com/education/certification*.

Oracle Technology Network (OTN)

The Oracle Technology Network (OTN; *http://otn.oracle.com*) provides several services to registered members. For example, members can download trial versions of Oracle software products and access discussion groups for help with technical issues. In addition, the site has an extensive documentation area, including reference manuals for SQL, PL/SQL, installation procedures, and so on. Most documentation is available in both HTML and PDF format and can be downloaded. Membership is free.

International Oracle Users Group (IOUG)

The International Oracle Users Group (IOUG) is composed of more than 100 local and regional user groups who meet regularly to share information about Oracle products. Members of IOUG can access the repository of knowledge accumulated from people who work with Oracle products on a daily basis. In addition, members can receive publications, discounts, and special offers from a variety of vendors and have access to discussion forums. Contact information for user groups, conferences, and more is available on the IOUG Web site at *www.ioug.org*.

D

SQL*LOADER

INTRODUCTION

SQL*Loader is a utility packaged with Oracle to load data from external files into tables. It gives you more flexibility for reading data in different formats and filtering and manipulating data during load operations. This appendix includes two examples of file loads to get you acquainted with the tool. You can explore more options in the SQL*Loader documentation at the Oracle Technology Network Web site.

READ A FIXED FILE FORMAT

SQL*Loader typically operates with two components: a control file with instructions for the data load and a data file. Both files are plaintext or ASCII files. The control file must have the extension .ctl, but there's no requirement for the data file's extension. Figure D-1 shows the contents of these two files and the command to call SQL*Loader and execute the data load.

The data is formatted in fixed placement, so instructions in the control file identify the specific positions to read for data that feeds into each column. For example, the value in character spaces 1 to 3 on each line in the data file is loaded (inserted into) the Prodno column.

Test.dat
Data file
101 ProdA 25.20
102 ProdB 32.40
103 ProdC 27.65
Test.ctl
Oracle table name
Indicates layout of data to read
Control file
LOAD D ATA
INFILE 'D:\test.dat'
INTO TABLE testprod
(prodno POSITION(01:03) INTEGER EXTERNAL,
pname POSITION(05:09) CHAR,
p_amt POSITION(13:17) DECIMAL EXTERNAL)
Execute the load
sqlldr USERID=scott/regitx CONTROL=D:\test.ctl LOG=D:\test.log

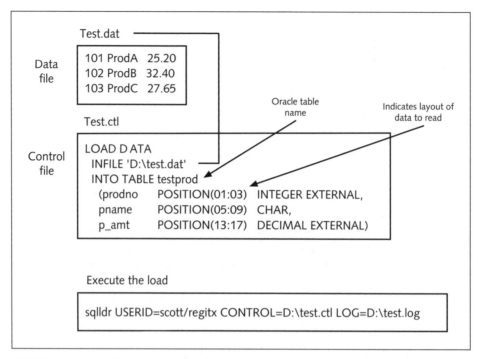

FIGURE D-1 SQL*Loader example—fixed file format

SQL*Loader is a command-line utility that uses a number of arguments. The command shown in Figure D-2 (the first line) contains arguments for specifying the user login, identifying the control file, and identifying the log file.

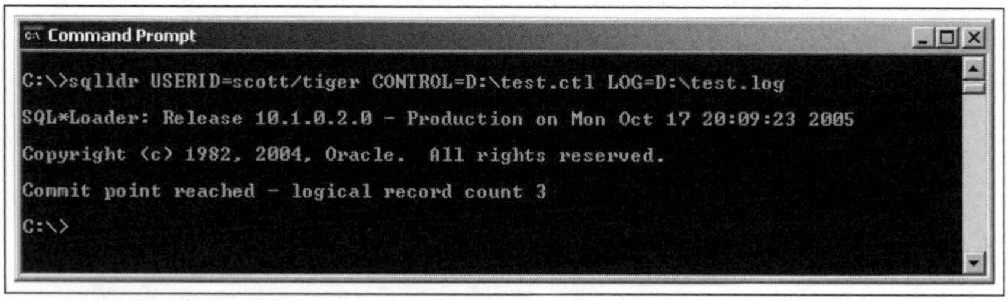

Command Prompt
C:\>sqlldr USERID=scott/tiger CONTROL=D:\test.ctl LOG=D:\test.log
SQL*Loader: Release 10.1.0.2.0 - Production on Mon Oct 17 20:09:23 2005
Copyright (c) 1982, 2004, Oracle. All rights reserved.
Commit point reached - logical record count 3
C:\>

FIGURE D-2 SQL*Loader command execution

The log file contains information on results of the load execution, as shown in Figure D-3. The list at the bottom indicates rows rejected based on errors, filtering conditions, or NULL values.

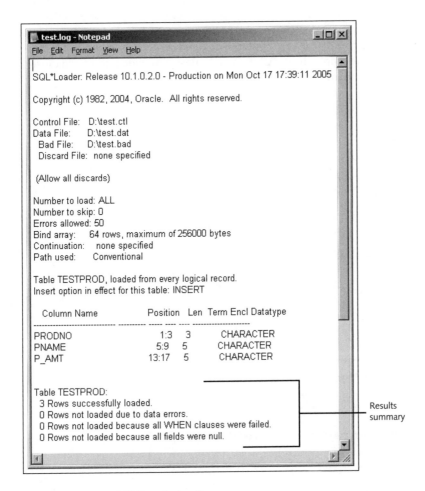

FIGURE D-3   An SQL*Loader log file

READ A DELIMITED FILE

Another common data format is a delimited file, in which data is separated by a specified character. The example in Figure D-4 loads a comma-delimited file. The field termination character is identified in the control file, which also specifies the order of columns into which the data should be read. In addition, if a table already contains data and you need the loader action to add the data to existing rows, you must include the APPEND option in the control file, as shown in this example. If the APPEND option isn't included, the table must be empty for SQL*Loader to operate correctly.

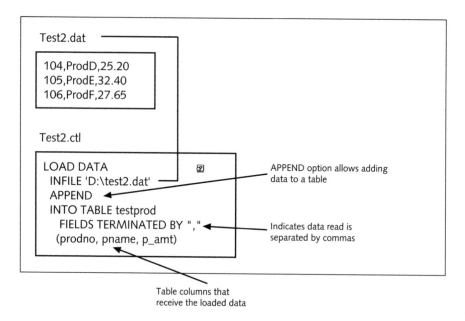

FIGURE D-4   SQL*Loader example—comma-delimited file format

SQL TUNING TOPICS

INTRODUCTION

Tuning is an ongoing effort to make your applications process efficiently so that resource use is minimized and response speed is increased. Typically, you should consider three general areas for optimizing application performance: hardware/networking, database configuration, and application source code. SQL is often a major component of application code tuning efforts because database interaction can represent a major portion of processing time. This appendix introduces SQL tuning topics in the context of an Oracle environment; however, these concepts also apply to tuning SQL in most databases. Keep in mind that this appendix is simply a brief introduction to SQL tuning to give you an overview of the basic concepts and a basis for further studies on SQL tuning.

TUNING CONCEPTS AND ISSUES

To begin tuning code, you must become familiar with methods that help identify issues affecting coding efficiency. You need to be able to identify which statements cause lengthy execution times and understand how the database server processes statements to determine what improvements could be applied. After attempting a modification, you need to be able to determine whether it succeeded in improving performance.

In this section, you explore some methods of identifying resource-intense SQL statements. Next, you examine how statement processing and options are managed in Oracle, and then investigate database features, such as the explain plan, to find processing information that's helpful in reviewing performance.

Identifying Problem Areas in Coding

To begin tuning efforts, developers need to identify the areas in source code that are likely candidates for tuning. The following basic methods are available for this task:

- The application-testing phase should simulate actual operations and include end users. Feedback from end users helps pinpoint problem areas, especially slow response times. Set up a procedure that makes it easy for testers to document where in the application they encountered trouble and what specific actions caused problems. This information can lead you to the coding that needs review.

- Oracle 11g includes some new automatic tuning capabilities. One is the SQL Tuning Advisor, which helps you identify problem SQL statements and makes recommendations for tasks such as collecting object statistics or creating an index. The SQL Tuning Advisor is available in the Oracle Enterprise Manager console.

NOTE

Oracle Enterprise Manager is an administrative GUI tool available with the Oracle installation files.

- The V_$SQLAREA view provides execution details, such as disk and memory reads, for all statements processed since the database startup. Use the DESCRIBE command on the view to list all available columns of data. Here are examples of statistics this view provides:
  - *Number of executions*—Indicates how many times a statement has been processed, which helps you identify frequently used statements
  - *Number of disk reads*—Reflects the total number of physical reads for a given statement
  - *Disk reads divided by number of executions*—Determines the number of reads per execution of a given statement

When you're deciding which statements need tuning, focus on the number of reads per statement execution. A high number of reads indicates that you might be able to improve performance by modifying the statement. Another important statistic in the V_$SQLAREA view is the number of buffer gets, which is the number of memory reads for the statement. You can query the Buffer_Gets column to see the number per statement execution. A high number of buffer reads can indicate that an index is needed or a join could be improved.

NOTE

DBA accounts (such as SYSTEM) have access to the V_$SQLAREA view; however, general users aren't typically granted access to this view by default. Usually, the DBA creates a public synonym named V$SQLAREA and then grants users access via the view or synonym.

You can also enable the SQL TRACE feature for a session, and statistics on SQL statement execution during that session can be stored for review. The statistics are saved in an operating system file that must be converted to a readable format with the TKPROF executable file. The converted file can then be viewed with simple word-processing software, such as WordPad.

After you've identified problem SQL statements, you need to determine possible tuning actions and explore questions such as "Could an index improve the performance of this statement?" Before examining some specific tuning examples, however, you must know how Oracle processes an SQL statement.

Processing and the Optimizer

Before tackling performance tuning, you need to understand the SQL processing architecture so that you can determine how statements can be modified to execute more efficiently. SQL processing contains the components shown in Figure E-1: parser, optimizer, row source generator, and execution engine.

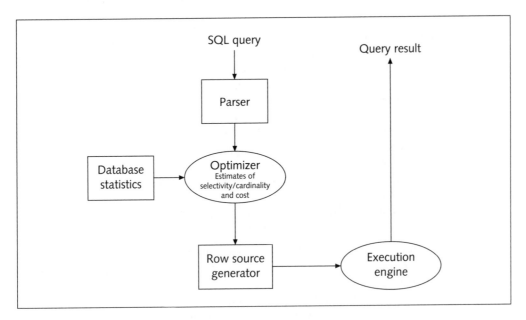

FIGURE E-1   SQL processing components

The parser checks for correct statement syntax and ensures that all referenced objects exist. The optimizer determines the most efficient way to process the statement and creates an execution plan for the statement to follow. For example, the optimizer decides whether an index is used and in what order tables are joined—in other words, factors that determine how efficiently your statements are processed. The row source generator sends the execution plan and the row source for each step in the plan to the execution engine. A row source returns a set of rows for the applicable step. The execution engine processes each row source and carries out the execution plan to produce the final results.

In versions before Oracle 10g, the Oracle database server contains two statement optimizers—rule-based and cost-based—that follow different methods to determine how a statement is processed. The rule-based optimizer, the older method, uses a list of rules to determine processing. For example, the rule-based engine typically uses an index if one is available, even if it might not be beneficial.

Oracle 11g now includes only the cost-based optimizer (CBO), a newer method that uses database object statistics, such as distribution of data, to determine how best to process statements. If only certain rows are being retrieved from a small table, for example, the CBO might decide not to use an index, as it wouldn't improve the query's performance. To determine the execution plan, the CBO estimates the following items based on database statistics:

- *Selectivity*—The proportion of rows from the row set to be used
- *Cardinality*—The distribution of data for all rows in the row set
- *Cost*—The units of resources used, including disk I/O, CPU use, and memory use

For the CBO to make the best decisions, it needs current database object statistics. By default, Oracle 11g gathers statistics on all database objects that have a daily scheduled job. The schedule for updating statistics can be modified as needed, depending on the database's volatility (frequency of changes). In addition, the analysis of objects used to gather statistics might need to be done manually by issuing an ANALYZE command or using the DBMS_STATS built-in package when new objects are added to the database. In versions before Oracle 11g, automatic statistic gathering isn't set up by default, and initiating database object analysis manually is essential to take advantage of the CBO.

If the CBO can't find any statistics for an object, dynamic sampling takes effect and performs random sampling during statement execution. This sampling can slow statement performance dramatically, however.

The CBO is one of many database settings, and the parameter is named OPTIMIZER_MODE. The following OPTIMIZER_MODE settings are available in Oracle 11g:

- `ALL_ROWS`: Optimizes with a goal of achieving the best throughput (use of the fewest resources). Use this setting, which is the default, to complete the entire statement.
- `FIRST_ROWS_n`: Optimizes with a goal of best response time to return the first *n* number of rows; *n* can equal 1, 10, 100, or 1000.
- `FIRST_ROWS`: Uses a mix of cost and heuristics to find the best plan for fast delivery of the first few rows.

The CBO uses a goal of best throughput by default. "Best throughput" means it chooses the execution path that uses the fewest resources to process all rows in the statement. However, the CBO can run with a goal of optimizing the response time. With this goal, it generates an execution path that uses the fewest resources to process the first row the statement accesses.

Applications involving large or batch requests, such as Oracle Reports, usually optimize for best throughput. Interactive or operational applications, such as those created with Oracle Forms, should be optimized for best response time because users are waiting to view feedback. If many rows are returned, users are still interested in getting the first rows of feedback fast so that they can begin analyzing the results.

So how do you determine how an SQL statement is being processed? The next section introduces some basic tools for examining statement processing.

The Explain Plan

One of the most important items to review for statement performance is the execution or explain plan for a statement. Recall that the optimizer develops an execution plan that outlines the specific steps to process a statement. Oracle provides several methods to review this plan. This section introduces the most common methods: the Tuning Advisor, the EXPLAIN PLAN FOR command, and the AUTOTRACE tool.

NOTE

The terms "execution plan" and "explain plan" are used interchangeably in this textbook.

You can access the Tuning Advisor in the Oracle Enterprise Manager console. After selecting a statement to review, you have the option to view the explain plan, as shown in Figure E-2.

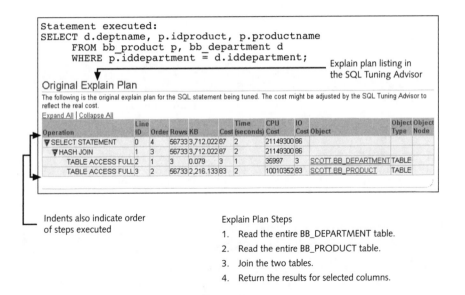

FIGURE E-2 The SQL Tuning Advisor displays the explain plan

Take a closer look at the processing steps outlined by the explain plan. First, the access path for each table is determined. The access path in this example is a full table scan. Another common access path is an index. If the query contained a WHERE clause limiting the rows returned from the BB_PRODUCT table and an index was available for the filtering column, the table might have been accessed via the index. (You see an example of this access path later in this section.)

Second, the order in which tables are accessed in a multiple-table query is determined. In this example, the BB_DEPARTMENT table is read first. Starting a join operation with the smallest table or the table with the most selective filter is the most efficient method.

Third, the join method for a multiple-table query is determined. Three common join methods are nested join loops, sort merges, and hash joins. (These join methods aren't joins you include in an SQL statement; they're methods the Oracle processor uses internally to handle join operations.) In Figure E-2, a hash join is indicated, which is appropriate when a large table is being joined to a small table via an equijoin. A hash table based on the small table is placed in memory and used to perform the join with the larger table more quickly. Nested join loops are typically used when joining small subsets of data, and the join condition is based on a column that can be accessed via an index. In this join method, one table is considered the outer table and the other is considered the inner table, which is accessed for each row in the outer table. A sort merge sorts both tables by the join key and then merges the tables. This join method can be used if the data is already sorted or if the join type isn't an equijoin.

With the Oracle AUTOTRACE tool, you can display both the execution plan and execution statistics. To start the AUTOTRACE tool, issue the SET AUTOTRACE ON command in SQL*Plus. Table E-1 lists the options you can use when starting this tool.

TABLE E-1 AUTOTRACE Tool Options

SQL*Plus Command	Description
SET AUTOTRACE ON	Displays explain plan, statistics, and result set
SET AUTOTRACE ON EXPLAIN	Displays explain plan and result set
SET AUTOTRACE ON STATISTICS	Displays statistics and result set
SET AUTOTRACE TRACEONLY	Displays explain plan and statistics

NOTE

To enable users to use the AUTOTRACE tool, two steps are required. First, a role named PLUSTRACE must be created and granted to users. The plustrce.sql script in the Oracle database directory *Your_Oracle_Home*\sqlplus\admin is used to create this role. Second, a table named PLAN_TABLE storing the explain plan information must be created in users' schemas. The utlxplan.sql script in the Oracle database directory *Your_Oracle_Home*\rdbms\admin creates this table.

Figure E-3 shows the same SQL statement used earlier with output from the AUTOTRACE tool.

```
SET AUTOTRACE TRACEONLY;
SELECT d.deptname, p.idproduct, p.productname
   FROM bb_product p, bb_department d
   WHERE p.iddepartment = d.iddepartment;
```

(Execute) (Load Script) (Save Script) (Cancel)

59991 rows selected.

Execution Plan

0		SELECT STATEMENT Optimizer=ALL_ROWS (Cost=88 Card=56733 Byte s=3801111)
1	0	HASH JOIN (Cost=88 Card=56733 Bytes=3801111)
2	1	TABLE ACCESS (FULL) OF 'BB_DEPAR TMENT' (TABLE) (Cost=3 Card=3 Bytes=81)
3	1	TABLE ACCESS (FULL) OF 'BB_PRODU CT' (TABLE) (Cost=84 Card=56733 Bytes=2269320)

Statistics

0	recursive calls
0	db block gets
4335	consistent gets
0	physical reads
0	redo size
1670592	bytes sent via SQL*Net to client
44501	bytes received via SQL*Net from client
4001	SQL*Net roundtrips to/from client
0	sorts (memory)
0	sorts (disk)

FIGURE E-3   Explain plan and statistics produced by AUTOTRACE

Table E-2 describes the statistics displayed by the AUTOTRACE tool. The explain plan this tool produces matches the explain plan displayed in the Tuning Advisor. Notice the SQL*Net statistics, which are helpful in identifying network traffic. In addition, the consistent reads are the same as the buffer gets covered earlier, and the physical reads are the same as disk reads.

TABLE E-2   Statistic Definitions

Statistic	Description
Recursive calls	Number of recursive calls generated at both the user and system level. Oracle maintains tables used for internal processing. When Oracle needs to change these tables, it generates an SQL statement internally, which generates a recursive call.
Db block gets	Number of times a CURRENT block was requested.
Consistent gets	Number of times a consistent read was requested for a block.

Statistic	Description
Physical reads	Total number of data blocks read from disk. This number equals the value of "physical reads direct" plus all reads into buffer cache.
Redo size	Total amount of redo operations generated in bytes.
Bytes sent via SQL*Net to client	Total number of bytes sent to the client from the foreground-to-client processes.
Bytes received via SQL*Net from client	Total number of bytes received from the client over Oracle Net.
SQL*Net roundtrips to/from client	Total number of Oracle Net messages sent to and received from the client.
Sorts (memory)	Number of sort operations that were performed completely in memory and didn't require any disk writes.
Sorts (disk)	Number of sort operations that required at least one disk write.
Rows processed	Number of rows processed during the operation.

The EXPLAIN PLAN FOR command produces an explain plan along with a list of conditions used to perform the query and any warnings. Figure E-4 shows the output of this command. Notice the warning at the bottom about the use of dynamic sampling. The tables used in the query were just created, so no statistics exist for them. Therefore, sampling had to be performed.

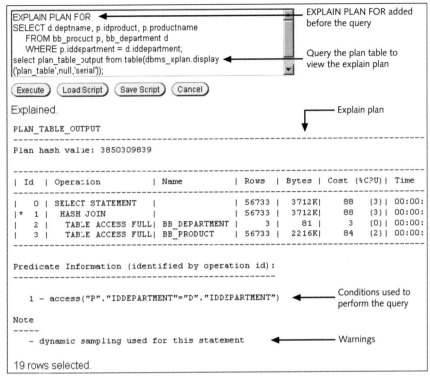

FIGURE E-4   Output of the EXPLAIN PLAN FOR command

Next, modify the query and examine the impact on the explain plan. Figure E-5 shows the explain plan with an additional condition in the WHERE clause. Also, tables are analyzed to provide statistics for the optimizer. Notice that because the query now processes fewer rows, the optimizer selects table indexes to access data and merge data rows with a sort merge. In addition, the dynamic sampling warning is no longer displayed.

The explain plan cost estimates can be used to determine whether a statement change has resulted in more efficient performance. However, another useful measure is the execution time. The next section introduces the timing feature that enables displaying the statement execution time, which gives you a simple check for seeing whether response time has improved.

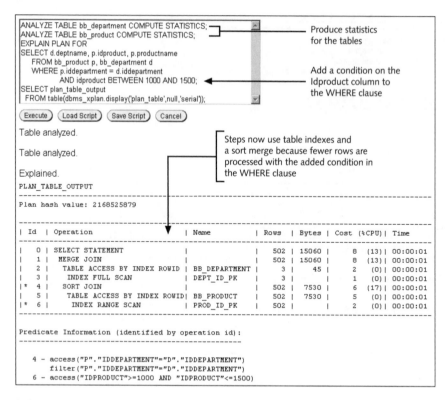

FIGURE E-5 Examining changes in the explain plan

Timing Feature

Before jumping into examples of SQL statement tuning, it's worth looking at another basic tuning tool you can use in SQL*Plus. The timing feature allows developers to measure a statement's execution time. Issue the SET TIMING ON command in SQL*Plus to display time elapsed in hours, minutes, and seconds for each statement. Figure E-6 shows three executions of the same query with the timing feature on.

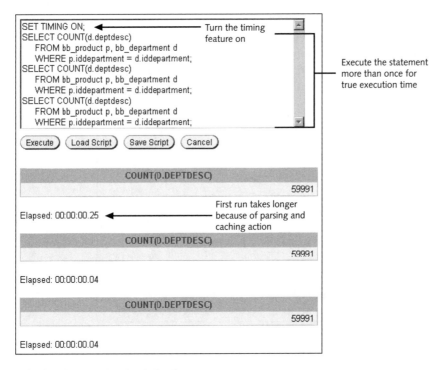

```
SET TIMING ON;                      Turn the timing
SELECT COUNT(d.deptdesc)            feature on
   FROM bb_product p, bb_department d
   WHERE p.iddepartment = d.iddepartment;
SELECT COUNT(d.deptdesc)
   FROM bb_product p, bb_department d
   WHERE p.iddepartment = d.iddepartment;
SELECT COUNT(d.deptdesc)
   FROM bb_product p, bb_department d
   WHERE p.iddepartment = d.iddepartment;
```

Execute the statement
more than once for
true execution time

543

(Execute) (Load Script) (Save Script) (Cancel)

COUNT(D.DEPTDESC)
59991

Elapsed: 00:00:00.25 ◄———— First run takes longer
because of parsing and
caching action

COUNT(D.DEPTDESC)
59991

Elapsed: 00:00:00.04

COUNT(D.DEPTDESC)
59991

Elapsed: 00:00:00.04

FIGURE E-6   Using the timing feature

In this query, the first execution takes .25 seconds, and the second and third queries take .04 seconds each. (Note that the execution time depends on the computer system being used.) Why the difference in execution times? The first run of a query is cached in the Oracle server's memory, so successive runs of the same query can skip parsing, creating the execution plan, loading into the SQL area, and storing the results in memory. Before modifying statements in tuning efforts, make sure you use the timing from a second run of the statement as the base time to improve. If you use the execution time from the first run as your baseline, you might think you have improved performance when your modification actually had no effect on the execution time.

SELECTED SQL TUNING GUIDELINES AND EXAMPLES

In this section, you focus on the explain plan and learn how to identify whether statement modifications affect the plan and potentially affect query performance. Trying modifications on a test database that resembles the production database in design and size is important. The database tables used in the following examples are rather small, and most statements are fairly simple so that you can understand the explain plan and the specific effects of statement modifications.

Avoiding Unnecessary Column Selection

Including columns that aren't actually needed in a select list can have a detrimental effect on performance. Take a look at the explain plan for a query on the BB_PRODUCT table. The AUTOTRACE and SET TIMING ON commands have been issued before the statements to display the explain plan and execution time. Figure E-7 shows a query on the BB_PRODUCT table that includes three columns.

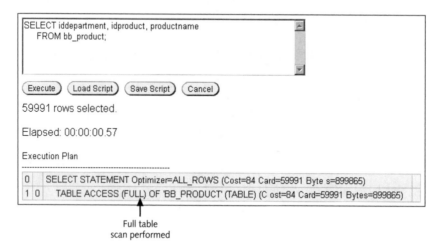

Full table
scan performed

FIGURE E-7   Explain plan and execution time for a query

Notice that a full table scan is performed on the BB_PRODUCT table, and the query execution time is .57 seconds. However, what if only the Idproduct column is actually needed for the query? Could this change affect performance? Figure E-8 shows another execution of this statement, which includes only the Idproduct column.

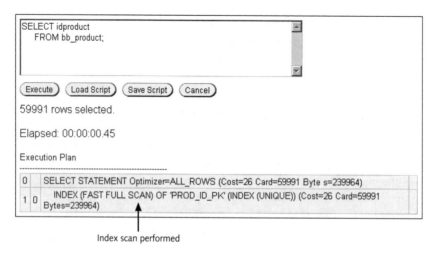

Index scan performed

FIGURE E-8   Explain plan and execution time for the modified query

Now the only step that needs to be processed is an index scan. A full table scan isn't necessary because the index can return the Idproduct value. Keep in mind that an index contains the value of the indexed columns and the ROWID for the associated row. The ROWID, a physical address for a table row, is the fastest method for retrieving rows. The execution time was reduced to .45 seconds. This reduction might not seem like a big difference, but you can imagine that with larger tables and more complex queries, the increase in performance can be significant.

Index Suppression

Because indexes are one of the central topics in performance tuning, you should learn to recognize when SQL statements suppress the optimizer from using an index. Index suppression can occur when a WHERE clause uses a function on a column or compares different datatype values. For example, Figure E-9 shows a query that includes a condition in the WHERE clause. No functions are used in the WHERE condition, and an index scan is used to perform the query.

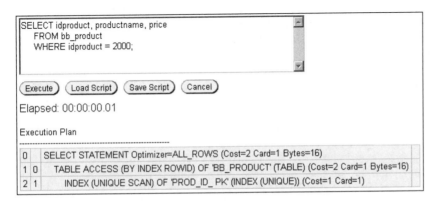

FIGURE E-9   With no function in the WHERE clause, an index scan is used

The next statement modifies the WHERE clause to add a function so that the query returns only products with an ID starting with 200. Figure E-10 shows the query with a modified WHERE clause, using the SUBSTR function on the Idproduct column. Notice that an index scan is no longer used, and the execution time has increased.

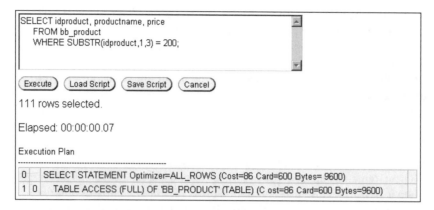

FIGURE E-10   With a function in the WHERE clause, an index scan is suppressed

How can you perform this query without using the SUBSTR function? In this example, the BETWEEN operator could be used to perform the same operation. Figure E-11 shows the query with the BETWEEN operator in the WHERE clause. Notice that an index scan is used, and execution time is improved.

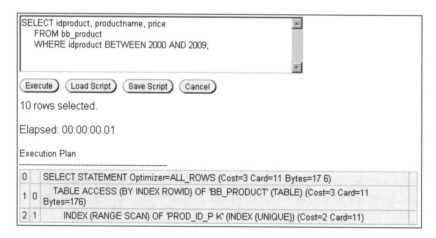

FIGURE E-11   BETWEEN operator in the WHERE clause, so an index scan is used

Comparing different datatypes can also suppress indexes in statement execution. If a character column is compared to a numeric value, the index is suppressed. In this case, the column value is converted internally to a numeric value, so the optimizer considers this activity the same as using a function on the column value. In Figure E-12, the Idprod2 column is a copy of the Idproduct column, but it's a character column, so the index is suppressed.

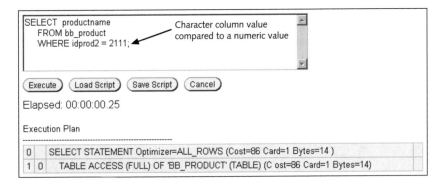

FIGURE E-12 An implicit conversion suppresses index use

Concatenated Indexes

In performance tuning, you should also consider the use of concatenated indexes (indexes involving more than one column). The optimizer uses these indexes only when the first column indexed is included in the WHERE clause's condition. For example, a concatenated index named BB_SHOPNAME_IDX that indexes the Lastname and Firstname columns of the BB_SHOPPER table is used only if the query uses the Lastname column and if the index is created with the Lastname column first.

Take a look at a few queries to see how the index is used or suppressed. First, the concatenated index is created with this statement:

```
CREATE INDEX bb_shopname_idx
  ON bb_shopper (lastname, firstname);
```

Figure E-13 shows a query on the BB_SHOPPER table, which uses the Lastname column in the SELECT clause. Notice that an index scan is performed because the first column of the index (Lastname) is used in the WHERE clause's condition. (The index would also be used if both the Lastname and Firstname columns were included.)

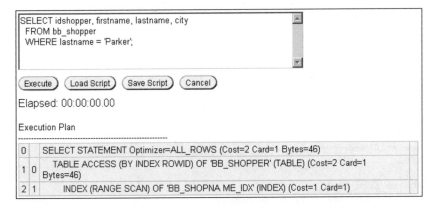

FIGURE E-13 A concatenated index is used because its first column is included in the WHERE clause

Now modify the WHERE clause so that only the Firstname column is used, as shown in Figure E-14. As you can see, the index is no longer used because the first column of the concatenated index, Lastname, isn't included in the query condition.

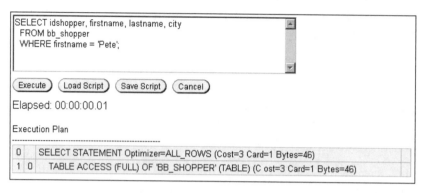

```
SELECT idshopper, firstname, lastname, city
  FROM bb_shopper
  WHERE firstname = 'Pete';
```

(Execute) (Load Script) (Save Script) (Cancel)

Elapsed: 00:00:00.01

Execution Plan

| 0 | | SELECT STATEMENT Optimizer=ALL_ROWS (Cost=3 Card=1 Bytes=46) |
| 1 | 0 | TABLE ACCESS (FULL) OF 'BB_SHOPPER' (TABLE) (C ost=3 Card=1 Bytes=46) |

FIGURE E-14 A concatenated index isn't used

Subqueries

Another area worth exploring is the type of subqueries used. Correlated subqueries are usually considered more efficient; however, this isn't always the case. If the row-filtering condition is in the subquery, an IN operator is typically most efficient. If the row-filtering condition is in the parent query, the EXISTS operator (correlated subquery) is typically more efficient. When a subquery isn't correlated, the inner query executes first and returns results that are treated like an IN list. Then the outer query is executed and compared to the inner query's results. In a correlated subquery, the outer query executes first, and then the inner query executes for each record returned from the outer query. Also, using the EXISTS operator terminates the inner query when a match is found, whereas the IN operator continues until all rows in the inner query have been processed. Figure E-15 shows whether using an uncorrelated or a correlated subquery is best.

```
SELECT productname, price
    FROM bb_product
        WHERE productname IN          Row-filtering condition in the subquery;
            (SELECT productname       use the IN operator
                FROM bb_prod_list
                WHERE price = 10);
```

```
SELECT productname, price
    FROM bb_product p
        WHERE price = 10          Row-filtering condition in the parent query;
            AND EXISTS            use the EXISTS operator
              (SELECT  'x
                  FROM bb_prod_list
                  WHERE productname = p.productname);
```

FIGURE E-15 Selecting the type of subquery

Optimizer Hints

Based on your knowledge of the data and a review of the execution plan, you might decide that the execution plan should be altered to improve performance. With SELECT, INSERT, UPDATE, and DELETE statements, you can alter the execution plan by using hints in the statement. Hints are included as comments preceded by a plus sign.

Numerous hints are available and are grouped in the following categories:

- The optimization approach for an SQL statement
- The goal of the cost-based optimizer for an SQL statement
- The access path for a table accessed by the statement
- The join order for a JOIN statement
- A join operation in a JOIN statement

One hint you can see in the access path category is FULL, which forces a full table scan, regardless of whether an index exists that could be used. The following statement shows the FULL hint included in a SELECT statement:

```
SELECT /*+FULL(b)*/ idshopper, firstname, lastname
  FROM bb_shopper b
  WHERE lastname = 'Parker';
```

As mentioned, hints are embedded by using comments. The opening comment symbol, /*, must be followed by a plus sign (+), and the table name or its alias must be included. Hints must be placed immediately after the keyword SELECT, INSERT, UPDATE, or DELETE. A hint ends with the closing comment symbol, */.

If a hint is added incorrectly, it doesn't usually raise an error; it's simply ignored. Therefore, to determine whether changing the execution plan resulted in improving query performance *and* to confirm that the hint is being used, make sure you review the execution plan after adding a hint.

This section gives you just a basic introduction to hints; the main idea is to recognize that you can control the optimizer by using hints. Use hints to modify execution plans and compare the efficiency of different execution plans. This method is an excellent way to become more familiar with the optimizer and determine which processing modifications improve efficiency.

TIP

You can use multiple hints in a single statement by listing all hints, separating each one with a space, in a single comment area.

NOTE

To explore these topics in more detail, read the Oracle Database Performance Tuning Guide at the OTN Web site for more in-depth coverage of tuning methods.

SQL IN VARIOUS DATABASES

INTRODUCTION

ANSI standards provide a common ground that all database vendors build on to include SQL capabilities in their products. Having common SQL standards makes application code portable and allows using an application with different databases with few or no changes required. This capability can be critical when developing an application to market to organizations that might have different databases. In addition, your department or company might switch database vendors because of cost or capabilities that are needed.

On the other hand, developing 100% portable SQL code is challenging. ANSI standards don't cover every SQL feature, and many vendors add SQL extensions to enhance capabilities and differentiate their products from others. Therefore, developers need to be aware of the nuances of SQL in various database products. Some of the most important differences are in SQL functions. The following charts are a sampling of specific SQL differences in Oracle, MySQL, and Microsoft SQL Server.

NOTE

The database versions used in the following SQL comparisons are Oracle 11*g*, MySQL 5, and SQL Server 2005.

Suppressing Duplicates

Oracle	MySQL	Microsoft SQL Server
DISTINCT or UNIQUE	DISTINCT	DISTINCT
----------------	---------------	----------------
SELECT UNIQUE category FROM books;	SELECT DISTINCT category FROM books;	SELECT DISTINCT category FROM books;

Locating a Value in a String

Oracle	MySQL	Microsoft SQL Server
INSTR	**LOCATE**	**CHARINDEX**
INSTR(*value*, *expression*)	LOCATE(*expression*, *value*, *start*)	CHARINDEX(*expression*, *value*, *start*)
-----------------	-----------------	-----------------
SELECT INSTR(title, 'SQL') FROM books;	SELECT LOCATE('SQL', title,1) FROM books;	SELECT CHARINDEX('SQL', title,1) FROM books;

N O T E

Each database provides a different function for locating a value in a string. In addition, the argument order varies, and Oracle doesn't provide a starting position argument.

Displaying the Current Date

Oracle	MySQL	Microsoft SQL Server
SYSDATE	**NOW()**	**GETDATE()**
---------------	-------------	-----------------
SELECT SYSDATE FROM DUAL;	SELECT NOW() FROM DUAL;	SELECT GETDATE();

N O T E

SQL Server doesn't use the DUAL table and allows executing a SELECT statement without a FROM clause.

Specifying a Default Date Format

Oracle	MySQL	Microsoft SQL Server
26-SEP-09	**2009-09-26**	**09-26-2009**
-----------------	-------------------	-----------------
SELECT * FROM orders WHERE orderdate = '26-SEP-09';	SELECT * FROM orders WHERE orderdate = '2009-09-26';	SELECT * FROM orders WHERE orderdate = '09-26-2009';

Replacing NULL Values in Text Data

Oracle	MySQL	Microsoft SQL Server
NVL2 NVL2(*value*, if NULL, if not NULL)	**IFNULL** IFNULL(*value*, if NULL)	**ISNULL** ISNULL(*value*, if NULL)

SELECT NVL2(referred,
   TO_CHAR(referred),
   'Not referred')
 FROM customers;
	SELECT IFNULL(referred,
	   'Not referred')
	  FROM customers;
		SELECT ISNULL(referred,
		   'Not referred')
		  FROM customers;

Adding Time to Dates

Oracle	MySQL	Microsoft SQL Server

ADD_MONTHS
ADD_MONTHS(*date*,
   # *months*)
	ADDDATE
	ADDDATE(*date*, *interval*)
		DATEADD
		DATEADD(*interval*,
		   # *months*, *date*)

SELECT ADD_MONTHS
   (SYSDATE, 12)
  FROM DUAL;
	SELECT ADDDATE(NOW(),
	   Interval 1 Year)
	  FROM DUAL;
		SELECT DATEADD(YYYY,
		   1, GETDATE());

NOTE

The sample statement adds one year to the current date.

Extracting Values from a String

Oracle	MySQL	Microsoft SQL Server

SUBSTR or **SUBSTRING**
SUBSTR(*value*, *start*,
   length)
	SUBSTR or **SUBSTRING**
	SUBSTR(*value*, *start*,
	   length)
		SUBSTRING
		SUBSTRING(*value*, *start*,
		   length)

SELECT SUBSTR(zip,
   1, 3)
  FROM customers;
	SELECT SUBSTR(zip,
	   1, 3)
	  FROM customers;
		SELECT SUBSTRING(zip,
		   1, 3)
		  FROM customers;

NOTE

SQL Server can use only SUBSTRING, not the abbreviated name SUBSTR.

Concatenating

Oracle	MySQL	Microsoft SQL Server
`||`	**CONCAT** `CONCAT(value, value, ...)`	`+`

`SELECT lastname ||`
 `', ' || firstname`
 `FROM customers;`
	`SELECT CONCAT(lastname,`
	 `', ', firstname)`
	 `FROM customers;`
		`SELECT lastname +`
		 `', ' + firstname`
		 `FROM customers;`

> **NOTE**
>
> Oracle also has a CONCAT function, but it's limited to two arguments. MySQL in ANSI mode treats the || symbol as concatenation rather than an OR operation.

Data Structures

In addition to functions, databases have a number of differences in data structures, such as datatypes and constraints. For example, Oracle has a datatype named NUMBER; however, MySQL and SQL Server don't use this name for their numeric datatypes. Another example is the Oracle datatype DATE, which is equivalent to DATETIME in MySQL and SQL Server. In terms of constraints, currently MySQL has no CHECK constraint capabilities, as in the other databases.

> **NOTE**
>
> CHECK constraint capabilities are planned for future versions of MySQL.

aggregate function *See* **group function.**

aggregate value A column included in a GROUP BY clause that exists outside a ROLLUP operation within the same query.

American National Standards Institute (ANSI) An industry-accepted committee that sets standards for SQL; another is the International Organization for Standardization (ISO).

application cluster environment A high-volume work environment, in which multiple users request data from a database simultaneously.

argument A value listed inside parentheses that specifies what a function should operate on. *See also* **function.**

authentication The process of validating the identity of computer users, typically based on a username and password.

authorization Granting object privileges to users based on their identities.

B-tree (balanced-tree) index The most common index used in Oracle; stores data in root node, branch, and leaf blocks. *See also* **index.**

bitmap index A two-dimensional array containing one column for each distinct value being indexed; useful with columns that have low cardinality. *See also* **cardinality** and **index.**

bridging entity An entity used to eliminate a many-to-many relationship by creating two one-to-many relationships.

buffer pool The shared cache memory area of the database server; improves data retrieval speed.

cardinality A term used to describe the level of distinct values in a column. If a column has many distinct values, it's said to have high cardinality.

Cartesian join A type of join that links table data so that each record in the first table is matched with each record in the second table; also called a Cartesian product or cross join.

Cartesian product *See* **Cartesian join.**

character The basic unit of data. It can be a letter, number, or special symbol.

character function A function used to alter the case of characters (called "case conversion functions") or to manipulate characters, such as substituting one character for another (called "character manipulation functions").

clause Each section of an SQL statement that begins with a keyword (SELECT clause, FROM clause, WHERE clause, and so on). *See also* **keyword.**

column In a physical database, fields are commonly referred to as columns. *See also* **field.**

column alias A name created in a query that's substituted for a column name in the query results; often used to more clearly indicate data displayed in the results.

column qualifier A prefix that indicates the table containing the column being referenced.

common column A column existing in two or more tables that contains equivalent data and is typically used to join tables.

comparison operator A search condition that indicates how data should relate to the search value (equal to, greater than, less than, and so forth). Common comparison operators include >, <, >=, and <=.

complex view A view that retrieves data from one or more tables; contains functions or grouped data.

composite column A collection of columns treated as a single unit in grouping operations.

composite (concatenated) index An index that includes multiple columns; can improve query performance. *See also* **index.**

composite primary key A combination of columns that uniquely identifies a record in a database table. *See also* **primary key.**

concatenated grouping An operation created by listing multiple grouping sets in a grouping operation.

concatenation Combining the contents of two or more columns or character strings. Two vertical bars, or pipes (||), instruct Oracle 11*g* to concatenate the columns in the query output.

condition A portion of an SQL statement that identifies what must exist or a requirement that must be met for a record to be included in query results.

constraint A rule used to ensure the accuracy and integrity of data. Constraints prevent adding data to tables that violates these rules. Constraints include PRIMARY KEY, FOREIGN KEY, UNIQUE, CHECK, and NOT NULL.

correlated subquery A subquery that references one or more columns in the outer query, and then the EXISTS operator tests whether the relationship or link is present before executing the query. It's processed once for each row in the outer query.

cross join *See* **Cartesian join.**

data anomaly An inconsistency in data stored in a database.

database A collection of interrelated files.

database management system (DBMS) A software product used to create and maintain the structure of a database and enable users to interact with a database to enter, manipulate, and retrieve the data it stores.

database object A defined, self-contained structure in Oracle 11*g*. Database objects include tables, sequences, indexes, and synonyms.

data definition language (DDL) Commands that create or modify database tables or other objects.

data dictionary A collection of objects Oracle 11*g* stores to maintain information about database objects. Stored information includes an object's name, type, structure, and owner and the identity of users who have access to the object.

data manipulation language (DML) Commands used to modify existing data. Changes to data made by DML commands aren't accessible to other users until the changes have been committed.

data mining Analyzing historical data and other information stored in an organization's database to support business functions, such as developing marketing campaigns.

data redundancy Having duplicate data in different places in a database, which wastes storage space and complicates updates and changes.

datatype Identifies the type of data Oracle 11*g* is expected to store in a column. Common datatypes include CHAR, NUMBER, and DATE.

default role A role that's enabled automatically when a user logs in to a database; usually consists of privileges a user needs frequently. *See also* **role.**

dimension Any category used in analyzing data, such as time, geography, and product line.

encryption Scrambling data to make it unreadable to anyone other than the sender and receiver.

entity Any person, place, or thing with characteristics or attributes to be included in a database. In an E-R model, an entity is usually represented as a square or rectangle.

entity-relationship (E-R) model A diagram that identifies the entities and data relationships in a database. The model is a logical representation of the physical database system to be built.

equality join A type of join that links table data in two or more tables having equivalent data stored in a common column; also called an equijoin or a simple join. *See also* **common column** and **inner join**.

equality operator A search condition that evaluates data for exact, or equal, values. The equality operator symbol is the equal sign (=).

equijoin *See* **equality join**.

exclusive lock When DDL operations are performed, Oracle 11*g* places this lock on a table so that no other users can alter the table or place a lock on it.

explain plan Another term for an execution plan, which identifies the steps Oracle takes to resolve a query and determine whether an index is used. *See also* **index**.

field A group of related characters that represents one attribute or characteristic of an entity. For example, a last name is one attribute of a customer.

file A group of records about the same type of entity. *See also* **record**.

first-normal form (1NF) The first step in the normalization process, in which repeating values are removed from database records.

foreign key When a common field exists in two tables being joined, it's called a primary key in one table and a foreign key in the second table. The foreign key appears on the "many" side of a one-to-many relationship.

format argument A series of elements representing exactly what data should look like in query results; must be enclosed in single quotation marks.

full table scan A scan in which each row of the table is read and a particular value is checked to determine whether it satisfies the condition. This scan is used if no index has been created on a table. *See also* **index**.

function A predefined block of code that accepts one or more arguments and returns one value as output.

function-based index An index created when a search is based on an expression or a function. *See also* **index**.

group function A function that processes groups of rows and returns only one result per group of rows processed; also called a multiple-row function or an aggregate function.

heap-organized table An unordered collection of data.

index A database object that stores a map of column values and ROWIDs (physical addresses of table rows) of matching table rows to improve data retrieval speed. An index can be created implicitly by Oracle 11*g* or explicitly by a user.

index organized table (IOT) A variation of a B-tree index that combines the index and table into a single structure with rows sorted in primary key order. *See also* **B-tree (balanced-tree) index**.

inline view A temporary view of underlying database tables that exists only while a command is being executed. It's not a permanent database object and can't be referenced again by a subsequent query.

inner join A join that displays data only if a corresponding record in each table is queried. Equality joins, non-equality joins, and self-joins are classified as inner joins.

International Organization for Standardization (ISO) An industry-accepted committee that sets standards for SQL. Another is the American National Standards Institute (ANSI).

join condition An instruction in a query for combining data from two tables.

Julian date Used in Oracle 11g, a numeric version of a date that represents the number of days that have passed between a specified date and January 1, 4712 B.C.

key-preserved table A table containing the primary key that a view is using to uniquely identify each record the view displays. *See also* **view.**

keyword A word used in an SQL query that has a predefined meaning in Oracle 11g. Common keywords include SELECT, FROM, and WHERE.

logical operator An operator used to combine search conditions. Logical operators include AND, OR, and NOT (which reverses the meaning of search conditions).

lookup table A common description for the table referenced in an foreign key relationship; typically used to identify descriptive data for a column value and ensure consistency of these descriptive values. *See also* **foreign key.**

materialized view A view that allows you to store data retrieved by the view query and reuse it without executing the view query again.

multiple-column subquery A nested subquery that returns more than one column of results to the outer query; can be listed in the outer query's FROM, WHERE, or HAVING clause.

multiple-row function *See* **group function.**

multiple-row subquery A nested subquery that returns more than one row of results to the outer query; used most commonly in WHERE and HAVING clauses and requires a multiple-row operator.

nesting Using one function as an argument inside another function; the inner function is resolved first and passed as input to the outer function.

non-equality join A type of join that links data in tables that don't have equivalent rows.

non-key-preserved table A table that doesn't uniquely identify the records in a view. *See also* **view.**

normal distribution A statistical concept of value dispersion, used in calculating standard deviation; it means that if you input many data values, they tend to cluster around an average value.

normalization A multistage process that designers use to develop raw data about an entity into a structured form that reduces data redundancy.

NULL value A value indicating that no data has been stored in a particular field; indicates the absence of data, *not* a blank space.

object privilege A type of privilege that allows users to perform DML operations on the data stored in database objects. Object privileges are assigned to specific database objects.

optimizer The Oracle feature that provides the logic a database system uses in determining the best path of execution, based on available information, and determines whether using an index is beneficial. *See also* **explain plan** and **index.**

optional keyword In an SQL query, a keyword that isn't required but is included to improve readability of SQL statements.

outer join A join that links data in tables that don't have equivalent rows; in other words, records existing in one table that don't have a matching record in the other table are included in the results.

outer join operator The plus symbol enclosed in parentheses (+), used in an outer join operation.

outer query The main query in an SQL statement; it incorporates the value passed from the subquery into its processing to determine the final output.

parent query *See* **outer query.**

partial dependency A problem that occurs when the fields in a record depend on only one portion of the primary key.

primary key A field that uniquely identifies a record in a database table.

primary sort The first column included in the ORDER BY clause.

private synonym An alias used by a user to reference objects he or she owns. *See also* **synonym.**

privileges Rights that allow users to perform certain types of actions on the database. Oracle 11*g* has system privileges and object privileges.

projection Choosing specific columns in a SELECT statement.

pseudocolumn Data that isn't physically stored in a database but is used to generate values (such as sequence values).

public synonym An alias used by others to access a user's database objects. *See also* **synonym.**

query A question posed to the database.

record A collection of fields describing the attributes of one database element. For example, name, address, and phone number fields make up a customer record.

recycle bin Starting in Oracle 10*g*, this storage area was added for recovering dropped tables (including the table structure and data).

referential integrity When a user refers to something existing in another table, the REFERENCES keyword is used to identify the table and column that must already contain the data being entered.

regular expression A type of operator that allows describing complex search patterns in textual data.

role A group, or collection, of privileges. In most organizations, roles correlate to users' job duties.

row In a physical database, records are commonly referred to as rows. *See also* **record.**

schema A collection of database objects owned by one user. By grouping objects according to owner, multiple objects that have the same object name can exist in the same database.

secondary sort A sort that specifies a second field to sort by if an exact match occurs between two or more rows in the primary sort.

second-normal form (2NF) The second step in the normalization process, in which partial dependencies are removed from database records by breaking a composite primary key into two parts, each representing a separate table. *See also* **partial dependency.**

selection A process that displays only records meeting certain conditions. *See also* **condition.**

self-join A type of join that links data in a table to other data in the same table.

sequence A database object that generates a series of integers, commonly used to create a unique primary key for a table or for an organization's internal controls (such as tracking accounting records).

set operator An operator used to combine the results of two (or more) SELECT statements. Valid set operators in Oracle 11*g* are UNION, UNION ALL, INTERSECT, and MINUS.

shared lock A table lock that allows other users to access portions of a table but not alter the table's structure.

simple join *See* **equality join.**

simple view A view that references only one table and doesn't include a group function, an expression, or a GROUP BY clause.

single value The output of a single-row subquery. *See also* **single-row subquery.**

single-row function A function that returns one row of results for each record processed.

single-row subquery A nested subquery that returns only one row of results consisting of only one column to the outer query. The output of a single-row subquery is a single value.

standard deviation A calculation used to determine how close each value in a group of numbers is to the mean, or average, of the group. *See also* **normal distribution.**

statistical group function A type of function used to perform basic statistical calculations, such as standard deviation and variance, for data analysis.

string literal Alphanumeric data enclosed in single quotation marks, which instructs Oracle 11g to interpret the data "literally" and not treat it as a keyword or command. String literals are displayed in the output exactly as they're entered.

Structured Query Language (SQL) The industry standard for interacting with a relational database. It's not considered a programming language; it's a data sublanguage with commands focused on creating database objects and manipulating data stored in an database.

subquery A complete query nested inside another query.

substitution variable Instructs Oracle 11g to prompt the user to enter a value in place of the substitution variable at the time a

command is executed; used to make SQL statements interactive and to simplify updating several records.

substring A portion of a string of data.

synonym An alternative name given to a database object with a complex name. Synonyms can be private or public.

syntax The basic structure or rules required for an SQL statement to execute.

system privilege A type of privilege that allows access to the Oracle 11g database and lets users perform DDL operations on database objects. An object privilege combined with the ANY keyword is also considered a system privilege.

Systems Development Life Cycle (SDLC) A series of steps for designing and developing a system.

table In a physical database, files are often referred to as tables. *See also* **file.**

table alias A temporary name for a table, established in the FROM clause. Table aliases are used to improve processing efficiency or reduce the number of keystrokes needed when specifying a table multiple times in an SQL statement.

third-normal form (3NF) The third step in the normalization process, in which transitive dependencies are removed from database records. *See also* **transitive dependency.**

TOP-N analysis A query that merges an inline view and a ROWNUM pseudocolumn to create a temporary list of records in a sorted order, and then the top "N," or number, of records are retrieved.

transaction A series of DML statements that should logically be performed together. In Oracle 11g, a transaction is simply a series of statements that have been issued but not committed. The duration of a transaction is defined by when a commit occurs implicitly or explicitly.

transaction control statement A command that saves modified data permanently or undoes uncommitted changes made in error.

transitive dependency A problem occurring when at least one value in a record isn't dependent on the primary key but on another field in the record.

uncorrelated subquery A subquery that follows this method of processing: The subquery is executed, its results are passed to the outer query, and then the outer query is executed.

unnormalized Refers to database records that contain repeating groups of data (multiple entries for a single column).

view A database object that stores a query statement. When a view is referenced, the query is executed, and the results are treated as a table of data. Views simplify complex queries for nontechnical users and can be used to restrict users' access to sensitive data.

virtual column A column that's not physically stored in a database but is derived based on other columns, such as performing a calculation.

wildcard character A symbol used to represent one or more alphanumeric characters in a pattern search. The wildcard characters in Oracle 11g are the percent sign (%) and the underscore symbol (_). The percent sign represents any number of characters; the underscore symbol represents only one character.

INDEX

Notes

Page numbers in bold type indicate definitions.
Page numbers followed by *(2)* indicate two separate
    discussions.
Page numbers followed by *n* indicate notes.
Page numbers followed by *t* indicate tables.
Page numbers followed by *tip* indicate tips.

Symbols

& (ampersand): substitution variable symbol, 153, 166*t*
*. *See* asterisk
@ (at sign): START keyword replacement character,
    28*tip*
\ (backslash): escape character, 261
^=. *See* <> (less than–greater than signs)
, (comma):
list delimiter, 35, 37*n*, 63
thousands position indicator, 249
$ (dollar sign): formatting character, 249
"" (double quotation marks): column alias delimiters, 37,
    38*n*, 50*t*
= (equal sign): equality operator, 245, 250*t*, 275*t*
= ANY operator, 439*t*, 442–443, 461*t*
!=. *See* <> (less than–greater than signs)
> (greater than sign): greater than operator, 113, 249,
    250*t*, 251, 252, 275*t*, 439
>= (greater than or equal to sign): greater than or equal
    to operator, 250*t*, 252–253, 275*t*
> ALL operator, 439*t*, 439–440, 461*t*
> ANY operator, 439*t*, 442, 461*t*
< (less than sign): less than operator, 113, 250*t*, 251,
    252, 275*t*, 439
<= (less than or equal to sign): less than or equal to oper-
    ator, 250*t*, 252, 275*t*
<> (less than–greater than signs): not equal to
    operator, 250*t*, 253–254, 275*t*
< ALL operator, 439*t*, 439, 440–441, 461*t*
< ANY operator, 439*t*, 441–442, 461*t*
(). *See* parentheses
%. *See* percent sign
. (period): decimal point, 249
+ (plus sign): outer join operator, 309–310, 312
; (semicolon): SQL statement terminator, 31
'. *See* single quotation marks

(space): in SELECT statements, 35
[] (square brackets): clause delimiters, 31, 245
_ (underscore). *See* underscore
|| (vertical bars): concatenation operator, 45

Numbers

1NF (first-normal form), 8–9
2NF (second-normal form), 9
3NF (third-normal form), 9

A

ABS function, 347–348, 374*t*
absolute value of numeric data: returning, 347–348
access order (for tables), 538
access privileges: for database objects, 63*n*
accessing sequence values, 183
ACCTBONUS table, 148
ACCTMANAGER table, 138*n*, 139, 148
    designing, 60, 61–63
ADD clause (ALTER TABLE command), 70
ADD_MONTHS function, 350–351, 374*t*
adding columns, 58*t*, 70–71, 88*t*
    virtual columns, 144–145
adding constraints, 103–116, 127–128*t*
    CHECK, 112–113, 114, 128*t*
    FOREIGN KEY, 106–107, 108–109, 109, 127*t*
    guidelines, 119
    NOT NULL, 114–115, 115, 116, 128*t*
    PRIMARY KEY, 103–104, 127*t*
    UNIQUE, 111, 127*t*. *See also* creating constraints
adding records. *See* inserting rows
adding tables, 6, 6–10, 10
advanced challenges:
    constraints, 134
    database concepts, 22
    database objects, 212
    DML and TC commands, 171–172
    group functions, 425
    join operations, 329–330
    privileges and roles, 241
    SELECT statements, 55
    selecting rows/records, 281
    single-row functions, 381

advanced challenges: (*continued*)
  sorting rows/records, 281
  subqueries, 468
  table creation and management, 93–94
  views, 508–509
aggregate functions. *See* group functions
aggregate values, **411**
aggregation. *See* grouping data; groups
aliases. *See* column aliases; table aliases
ALL keyword, 221, 386
ALL operators, 438, 439*t*, 439–441, 461*t*
ALL option (ALTER USER command), 229
ALL_ data dictionary view prefix, 207*t*
ALTER INDEX command … RENAME TO, 177*t*,
        200, 207*t*
ALTER privilege, 220
ALTER ROLE command, 215*t*, 230, 236*t*
ALTER SEQUENCE command, 176*t*, 185–186, 206*t*
ALTER TABLE command, 58*t*, 70
  command type, 12*t*
ALTER TABLE … ADD command:
  adding columns, 58*t*, 70–71, 88*t*
    virtual columns, 144–145
  adding constraints, 127–128*t*
    CHECK, 112–113, 114, 128*t*
    FOREIGN KEY, 106–107, 108–109, 109, 127*t*
    PRIMARY KEY, 103–104, 127*t*
    UNIQUE, 111, 127*t*
ALTER TABLE … DISABLE/ENABLE CONSTRAINT
        commands, 122–123
ALTER TABLE … DROP COLUMN command, 58*t*, 75–76
ALTER TABLE … DROP command, 88*t*, 108*n*, 123–125
ALTER TABLE … DROP UNUSED COLUMNS
        command, 59*t*, 76, 77, 78–79, 88*t*
ALTER TABLE … MODIFY command:
  adding a NOT NULL constraint, 114–115, 115, 116,
        128*t*
  modifying columns, 58*t*, 71–75, 88*t*
ALTER TABLE … SET UNUSED [COLUMN]
        command, 58*t*, 76–77, 88*t*
ALTER USER command, 214*t*, 215*t*, 224, 229, 235*t*, 236*t*
American National Standards Institute (ANSI), 13
ampersand (&): substitution variable symbol, 153, 166*t*
ANALYZE command, 536
AND operator, 244*t*, 262, 263, 276*t*
ANSI (American National Standards Institute), 13
ANSI-compliant joins: vs. traditional joins, 284
        See also JOIN method
ANY keyword, 218, 219
ANY operators, 438, 439*t*, 441–443, 461*t*
APPEND option (SQL*Loader), 532
application cluster environments, **180**
applications: optimization of, 536
arguments (to functions), **332**
arithmetic operation delimiters (()), 39
arithmetic operations:
  NULL values in, 41–42, 146*n*, 389–390
    substituting values for, 359–363

order of precedence, 39
performing, 26*t*, 39–40
syntax, 50*t*
artificial key(s), 8*n*
AS clause (CREATE TABLE command), 68
AS clause (CREATE VIEW command), 475
AS keyword, 37, 38*n*, 50*t*
ASC option (ORDER BY clause), 269*n*
ascending order, 267–268, 269*n*
assigning:
  aliases to columns, 26*t*, 36–39
  datatypes to columns, 61–63
  default values to columns, 63–64, 64–65
  privileges. *See* granting privileges
asterisk (*):
  COUNT function argument, 392
  include all columns symbol, 31–32
  line error indicator, 34*n*
at sign (@): START keyword replacement character,
        28*tip*
attributes of entities: representations of. *See* fields
authentication, **216**
  methods, 225
  usernames, 216–217. *See also* passwords
AUTHOR table, 15, 515–516
authorization, **216**. *See also* privileges; roles
Autocommit option (SQL Developer), 159*tip*
AUTOTRACE tool, 538
  enabling the use of, 538*n*
  execution plan output, 539
  set options, 538*t*
  statistics displayed by, 539, 539–540*t*
averaging column data values, 388–390
AVG function, 384*t*, 388–390, 418*t*

B

B-tree indexes (balanced-tree indexes), 190–196, 196*n*
  automatic creation of, 191, 194
  composite indexes, 195
  creating, 176*t*, 190, 194, 206*t*
  NULL values as not referenced in, 195. *See also*
        indexes
backslash (\): escape character, 261
best response time goal, 536
best throughput goal, 536
BETWEEN … AND operator, 113, 244*t*, 250*t*, 255–256,
        276*t*
  in non-equality joins, 303–305
bitmap indexes, 196–197, 196*n*
  creating, 176*t*, 196, 206*t*
  usefulness, 196–197
blank spaces vs. NULL values, 265
BOOKAUTHOR table, 15, 516
BOOKS table, 11, 14–15, 512–513
  normalizing, 7–10

branch blocks (data blocks), 191
bridging entities, **10**
buffer pool, **189**

C

cache memory area (buffer pool), **189**
CACHE option (CREATE SEQUENCE command), 180
calculating:
    column data value averages, 388–390
    column data value totals, 386–388
    dates *m* months ahead, 350–351
    gross pay, 360
    the number of days between dates, 348–349
    the number of months between dates, 349–350
    the standard deviation, 403–404
calculations-based indexes, 197–198
calling sequence values, 186–187
cardinality (of column data values), **189**, 536
    high selectivity and full table scan efficiency, 189
    low selectivity and bitmap index suitability, 196
Cartesian joins (cross joins), 284*t*, 285–289, 321*t*
    accidental generation of, 286
    JOIN method, 288–289, 321*t*
    traditional method, 286–288, 321*t*
CASCADE CONSTRAINTS option, 110
CASCADE option, 124–125
case conversion functions, 332*t*, 333–336, 372–373*t*
CASE expression, 369, 376*t*
case sensitivity:
    character data, 140, 246*n*
    column aliases, 37
    in SELECT statements, 34, 37
case study. *See* City Jail database
category lookup tables, 12
CBO (cost-based optimizer), 535
CHANGE command (SQL*Plus), 522
CHAR(*n*) datatype, 60*t*, 61, 62, 63
character data (nonnumeric data):
    case sensitivity, 140, 246*n*
    entering, 140(2), 246–247, 246*n*
    returning the largest value, 394
    returning the smallest value, 394. *See also* string
        literals
character datatypes. *See* CHAR(*n*) datatype;
        VARCHAR(*n*) datatype; VARCHAR2(*n*) datatype
character functions. *See* case conversion functions;
        character manipulation functions
character manipulation functions, 332*t*, 336–344,
        373–374*t*
character strings. *See* string literals
characters (literals), **2**
    determining the number of in character strings, 340
    wildcard characters as, 260–261. *See also* character
        data

CHECK constraint, 100*t*, 112–114, 119
    abbreviation/code for, 101*t*, 121
    adding, 112–113, 114, 128*t*
    creating, 127–128*t*
    syntax, 112, 127–128*t*
    testing, 113–114. *See also* constraints (on table data)
City Jail database, 13
    constraints, 134–135
    data entry and modification, 172–173
    database objects, 212
    group functions, 425
    initial design, 22
    join operations, 330
    privileges, 242
    selecting rows/records, 281–282
    single-row functions, 382
    sorting rows/records, 281
    subqueries, 469
    table creation and modification, 94–97, 172
    views, 509
clause delimiters ([]), 31, 245
clauses (SQL), **31**
    ADD clause (ALTER TABLE command), 70
    AS clause (CREATE TABLE command), 68
    AS clause (CREATE VIEW command), 475
    column aliases as allowed/not allowed in, 251*tip*
    DROP clauses (ALTER TABLE command), 75–76,
        76–77, 108*n*, 123–125
    FOR UPDATE clause (SELECT command), 164
    FROM. *See* FROM clause
    GROUP BY. *See* GROUP BY clause
    HAVING. *See* HAVING clause
    IDENTIFIED BY. *See* IDENTIFIED BY clause
    INCREMENT BY clause (CREATE SEQUENCE
        command), 178–179
    MAXVALUE clause (CREATE SEQUENCE
        command), 179, 185
    MINVALUE clause (CREATE SEQUENCE command), 179
    MODIFY clause (ALTER TABLE command), 71
    NEXT clause (CREATE MATERIALIZED VIEW
        command), 500
    ON. *See* ON clause
    ORDER BY. *See* ORDER BY clause
    REFRESH clause (CREATE MATERIALIZED VIEW
        command), 500
    SELECT. *See* SELECT clause
    SET clause (UPDATE command), 150
    SET UNUSED [COLUMN] clause (ALTER TABLE com-
        mand), 76
    START WITH. *See* START WITH clause
    for subqueries, 429
    TO clause (GRANT command), 219, 221
    USING clause. *See* JOIN … USING method
    VALUES clause (INSERT command), 140, 141–142, 183
    WHEN MATCHED THEN clause (MERGE
        command), 457–458, 459
    WHEN NOT MATCHED THEN clause (MERGE
        command), 457–458

clauses (SQL) (*continued*)
    WHERE. *See* WHERE clause
clearing column data values, 153
client software tools. *See* SQL Developer; SQL*Plus
coding problem areas, 533–535
column alias delimiters (" "), 37, 38*n*, 50*t*
column aliases:
    as allowed/not allowed in clauses, 251*tip*
    assigning, 26*t*, 36–39
    case sensitivity, 37
    in ORDER BY clauses, 269
    SQL*Plus displays, 389*n*
column ambiguity error messages, 292
column data values:
    alignment, 38
    assigning default values, 63–64, 64–65, 73–74
    averaging, 388–390
    cardinality. *See* cardinality (of column data values)
    changing, 150–152
        default values, 72–73
        using substitution variables, 152–155
    clearing, 153
    comparing for equality, 363–364
    comparing to listed values, 367–369
        in CASE expressions, 369
    default values: assigning, 63–64, 64–65, 73–74
        changing, 72–73
    grouping multiple-column values, 397–398
    highest. *See* highest values
    nulls. *See* NULL values
    searching for. *See* searching for column data values
    totaling, 386–388
column list delimiters (()), 63, 68
column names, 59
    assigning new names in views, 475
    listing, 33*tip*
        with constraints, 121–122
        in INSERT INTO statements, 141, 142, 143, 145–146, 149. (*See also under* table structure
    qualifying, 292, 293–294. *See also* column aliases
column qualifiers, **292**
    in equality joins, 292, 297, 299, 300
column-level constraints: creating, 102, 118–119, 120
columns (in tables), **3**, 11
    adding, 58*t*, 70–71, 88*t*
        virtual columns, 144–146
    arithmetic expression-based: DML operations on complex views with, 484, 486–488
    assigning aliases to, 26*t*, 36–39
    assigning datatypes to, 61–63
    assigning default values to, 63–64, 64–65, 73–74
    changing data values, 150–152
        default values, 72–73
        using substitution variables, 152–155
    common: identifying, 290
    composite columns, **413–414**
    concatenating, 26*t*, 44–48
    creating column-level constraints, 102, 118–119, 120

data value alignment in, 38
default values: assigning, 63–64, 64–65, 73–74
    changing, 72–73
defining, 63–65
deleting, 75–76
    unused columns, 76, 77, 78–79
indexes of. *See* indexes
listing. *See under* column names
listing DEFAULT settings and virtual column definitions, 67
marking for deletion (as unused), 76, 77
modifying, 71–75
multiple constraints on a single column, 120
names. *See* column names
narrowing, 72
primary. *See* primary key(s)
pseudocolumns, **183**–184, 185*n*. *See also* ROWNUM pseudocolumn
referring to, 33
    in ORDER BY clauses, 272–273, 272*tip*
renaming in views, 475
selecting. *See* selecting columns
virtual. *See* virtual columns
widening, 74–75. *See also* fields (in files)
comma (,):
    list delimiter, 35, 37*n*, 63
    thousands position indicator, 249
comma-delimited file format (SQL*Loader), 532
commands (SQL). *See* SQL commands; SQL*Plus commands
COMMIT command, 138*t*, 158, 158–159, 159–160, 161, 166*t*
    command type, 13*t*
    syntax, 166*t*
common columns: identifying, 290
common fields, **10**
comparing values for equality, 363–364
comparing values to listed values, 367–369
    in CASE expressions, 369
comparison operators, 113, 244*t*, 248–262, 275–276*t*
    ALL operators, 438, 439*t*, 439–441, 461*t*
    ANY operators, 438, 439*t*, 441–443, 461*t*
    BETWEEN … AND operator, 113, 244*t*, 250*t*, 255–256, 276*t*
    EXISTS operator, 451, 461*t*, 548
    IN. *See* IN operator
    IS NOT NULL operator, 251*t*, 262*n*, 276*t*
    IS NULL operator, 244*t*, 251*t*, 262*n*, 276*t*, 392*n*
        in subqueries, 450–451
    LIKE operator, 244*t*, 251*t*, 258–261, 276*t*
        ESCAPE option, 260–261
    mathematical operators, 113, 244*t*, 248–255, 275*t*
    multiple-row operators, 438–443, 439*t*, 461*t*
    NOT option, 250–251*t*, 257–258, 276*t*
    pattern operators, 357, 358*t*
    selecting rows/records with, 248–262, 265–266
        and logical operators, 257–258, 262–264, 276*t*
    single-row operators, 431, 437–438
    using with dates, 254–255

COMPLETE option (REFRESH clause), 500
complex views:
    creating, 484
    DML operations on: constraints and, 485, 488
        summary guidelines, 495
        views created with the DISTINCT keyword, 493
        views with a ROWNUM pseudocolumn, 494–495
        views with columns based on arithmetic
            expressions, 484, 486–488
        views with group functions or a GROUP BY
            clause, 491–493
        views with multiple tables, 489–491
    dropping, 495–496
composite columns, **413–414**
composite indexes, 195
composite primary key, **9**
    creating, 105–106
CONCAT function, 344, 374*t*
concatenated groupings, **414–415**
concatenated indexes, 195, 547–548
concatenating character strings, 344
concatenating columns/fields, 26*t*, 44–48
    syntax, 50*t*
concatenation operator (| |), 45
conditions, **245**
    join conditions, **284**
    querying databases using, 245–248. *See also*
        selecting rows/records. *See also* constraints
        (on table data)
CONNECT role, 228*t*
constraint names, 101, 119
constraint violations, 105, 152
    error messages, 105
constraints (on table data), 99–135, **100**, 100–101*n*
    abbreviations/codes for, 101*t*, 121
    adding, 103–116, 127–128*t*
        CHECK, 112–113, 114, 128*t*
        FOREIGN KEY, 106–107, 108–109, 109, 127*t*
        guidelines, 119
        NOT NULL, 114–115, 115, 116, 128*t*
        PRIMARY KEY, 103–104, 127*t*
        UNIQUE, 111, 127*t*. (*See also* creating, *below*)
    advanced challenge, 134
    creating, 101–103, 116–119, 126–128*t*
        without assigning names, 119
        at the column level, 102, 118–119, 120
        guidelines, 119
        multiple constraints on a single column, 120
        at the table level, 102–103, 117–118. (*See also*
            adding, *above*)
    deleting, 108*n*, 123–124
    disabling/enabling, 122–123
    and DML operations on complex views, 485, 488
    guidelines, 119
    hands-on assignments, 132–133
    identifying needed constraints, 116–117
    levels, 101–102, 117

    listing, 120–122
        with column names, 121–122
    multiple constraints on a single column, 120
    naming/not naming, 101, 119
    testing. *See* testing constraints
    types, 100*t*, 102
    values as not required for, 119. *See also* CHECK
        constraint; FOREIGN KEY constraint; NOT
        NULL constraint; PRIMARY KEY constraint;
        UNIQUE constraint; and *also* conditions
control files (SQL*Loader), 529
converting:
    character strings to lowercase letters, 333–334
    character strings to mixed case, 335–336
    character strings to uppercase letters, 334–335
    date displays to a particular format, 365–367
    date entries to the default format, 352–354
    dates or strings into numeric data, 370–371
copying rows from existing tables, 148–149
correlated subqueries, 428*t*, **451–453**, 460*t*, 548
cost (of resources), 536
cost-based optimizer (CBO), 535
COUNT function, 384*t*, 390–392, 418*t*
    asterisk argument, 392
counting records with non-NULL values, 390–392
CREATE BITMAP INDEX command, 176*t*, 196, 206*t*
CREATE INDEX command, 176*t*, 190, 206*t*
    ON clause, 190, 197
    UNIQUE keyword, 194
CREATE MATERIALIZED VIEW command, 474*t*,
    499–500, 503*t*
CREATE OR REPLACE VIEW command, 474*t*, 475, 503*t*
CREATE ROLE command, 215*t*, 226, 227, 228, 236*t*
CREATE SEQUENCE command, 176*t*, 178–180, 206*t*
CREATE SYNONYM (CREATE PUBLIC SYNONYM) com-
    mand, 177*t*, 202, 207*t*
CREATE TABLE command, 58*t*, 63–65, 88*t*, 126–128*t*
    command type, 12*t*
    and ORGANIZATION INDEX, 177*t*, 198, 206*t*
    table type created, 189
CREATE TABLE … AS command, 58*t*, 68–69, 88*t*
CREATE USER command, 214*t*, 216–217, 217*n*, 235*t*
CREATE VIEW command, 474*t*, 476, 496, 502*t*
    syntax elements, 474–475
creating constraints, 101–103, 116–119, 126–128*t*
    without assigning names, 119
    at the column level, 102, 118–119, 120
    guidelines, 119
    multiple constraints on a single column, 120
    at the table level, 102–103, 117–118. *See also* adding
        constraints
creating tables, 63–65
    advanced challenge, 94
    commands, 12*t*, 58*t*
    creating constraints when. *See* creating constraints
    from existing tables (with subqueries), 67–69
    hands-on assignments, 93
    syntax, 88*t*. *See also* designing tables

creating views, 474t, 474–475, 502t
   complex views, 484
   inline views, 446, 496–499
   materialized views, 499–501
   simple views, 476–480
CROSS JOIN keywords, 288–289, 321t
cross joins. See Cartesian joins
cross-tabular aggregations, 409–411
cube (data cube), 405–406
CUBE extension (GROUP BY clause), 385t, 409–411, 419t
   with the GROUPING function, 410
current date: entering, 142–143, 355–356
CURRENT_DATE function, 355–356
CURRVAL pseudocolumn, **183**–184, 184, 185n
CURRVAL values, 184–185
   calling, 186–187
CUSTOMERS table, 14, 511–512
CYCLE option (CREATE SEQUENCE command), 179

D

data:
   cardinality. See cardinality (of column data values)
   constraints on. See constraints (on table data)
   dimensions, **405**
      multi-dimensional analysis, 405–416
   eliminating duplicate listings, 26t, 42–44
   eliminating test data, 82tip
   encryption, **225**
   entering. See entering data
   exporting to external files, 150n
   grouping. See grouping data
   importing from external files, 150n;
      with SQL*Loader, 529–532
   listing table data, 31–32, 141, 144, 148. See also searching for column data values
   retrieval speed, 189
   retrieving from databases. See querying databases
   sorting. See sorting rows/records
   standard deviation function, 403–404
   VARIANCE function, 385t, 404–405, 418t. See also character data; column data values; dates; NULL values; numeric data
data analysis functions, 403–405
data anomalies. See data inconsistencies
data block storage, **191**
data control language commands. See DCL (data control language) commands
data cube, 405–406
data definition language commands. See DDL (data definition language) commands
data dictionary, 3, **28**
data dictionary views (objects), 28, 29n
   prefixes, 207t
   privilege views, 230, 231t
   querying. See querying data dictionary views
   V_$SQLAREA view, 534, 534n. See also specific DBA_ …, ROLE_ …, SESSION_ …, and USER_ … views
data entry. See entering data
data inconsistencies (data anomalies), 6–7
   preventing during updating, 163–164
data integrity, 3
   ensuring. See constraints
data manipulation language commands. See DML (data manipulation language) commands
data mining, 14
data redundancy, 6–7
data security, 215–216. See also passwords; privileges; roles
data structures: differences in various database products, 554. See also constraints; datatypes; tables
data values. See character data; column data values; dates; NULL values; numeric data
data verification, 100–101n
database design, 4–12
   designing tables, 59–63
   entity-relationship model. See E-R model
   identifying needed constraints, 116–117
   normalizing tables, 6–10
   single-table approach, 6–7
database management systems (DBMSs), **2**
   functionality, 3
database normalization, 6–10
database object statistics, 536
database objects, **58**, 175–212, **176**
   access privileges, 63n
   advanced challenge, 212
   dynamic sampling of, 536
      warning about, 540
   hands-on assignments, 211–212
   referencing in others' schemas, 203–204, 220n, 222tip
   statistics, 536
   synonyms for, **176**, 201–204. See also indexes; sequences; tables; views
databases, **2**
   basic concepts, 1–23
      advanced challenge, 22
      hands-on assignments, 21
   basic terminology, 2–3
   designing. See database design
   marketing value, 14
   normalizing, 6–10
   querying. See querying databases
   relating tables in, 10–12
   requirements, 4
   SQL differences in various products, 551–554
datatypes, 60–61t, **61**
   additional available types, 61tip
   assigning to columns, 61–63

listing for all columns, 29–30
time zone datatype, 356*n*
DATE datatype, 61*t*, 62
date functions, 332*t*, 348–356, 374–375*t*
dates:
    calculating dates *m* months ahead, 350–351
    calculating the number of days between, 348–349
    calculating the number of months between, 349–350
    converting displays to a particular format, 365–367
    converting entries to the default format, 352–354
    converting into numeric data, 370–371
    default format, 348, 348*n*
    determining the date of the last day of the month,
        351–352
    determining the date of the next specific day of the
        week, 351
    entering, 248
        the current date, 142–143, 355–356
    format arguments: TO_CHAR function, 365–367*t*
        TO_DATE function, 352–353*t*
    Julian date, 348–349
    as numeric data, 348
    returning the current date, 355–356
    rounding, 354
    truncating, 355
    using comparison operators with, 254–255
DBA accounts, 534*n*
DBA role, 228*t*
DBA_ data dictionary view prefix, 207*t*
DBA_ROLE_PRIVS view, 231*t*
DBA_ROLES view, 231*t*
DBA_TAB_PRIVS view, 231*t*
DBMS-STATS built-in package, 536
DBMSs. *See* database management systems
DCL (data control language) commands, 13*t*. *See also*
    GRANT command; REVOKE command
DDL (data definition language) commands, 12*t*, **58**,
    58–59*t*, 63–65
    executing, 64*n*
    privileges for issuing. *See* system privileges.
        See also ALTER TABLE *commands;* CREATE
        TABLE *commands;* DROP TABLE command;
        FLASHBACK TABLE … TO BEFORE DROP
        command; PURGE TABLE command;
        RENAME … TO command; TRUNCATE
        TABLE command
deadlocks, 163
decimal point (.), 249
DECODE function, 367–369, 376*t*, 410–411
default installation tables, 29*n*, 65*n*, 117
DEFAULT ROLE option (ALTER USER command), 229
default roles: setting, 229
default values for columns:
    assigning, 63–64, 64–65, 73–74
    changing, 72–73
DELETE command, 138*t*, 156–157, 166*t*
    command type, 12*t*
DELETE privilege, 220

deleting:
    columns, 75–76
        unused columns, 76, 77, 78–79
    constraints, 108*n*, 123–124
    indexes, 177*t*, 201, 207*t*
    materialized views, 501
    parent tables, 109–110
    primary keys, 124–125
    roles, 215*t*, 234, 236*t*
    rows (records), 156–157
        from parent tables, 108–109
    sequences, 176*t*, 188, 206*t*
    synonyms, 177*t*, 203, 204, 207*t*
    tables, 83–84
        commands for, 12*t*, 59*t*
        parent tables, 109–110
        permanently, 85–86
        syntax, 89*t*
    users, 214*t*, 234, 235*t*
    views, 474*t*, 495–496, 503*t*
delimited file format (SQL*Loader), 532
DEPT default installation table, 117
DESC option (ORDER BY clause), 268
descending order, 268
    in indexes, 195
DESCRIBE command (SQL*Plus), 26*t*, 29–30, 30*tip*,
    33*tip*, 50*t*, 115*n*, 534
    syntax, 26*t*, 50*t*
    verifying table creation and modification, 66–67,
        68–69, 77–78, 83–84
designing constraints. *See* adding constraints;
        creating constraints
designing databases. *See* database design
designing tables, 59–63
determining:
    the date of the last day of the month, 351–352
    the date of the next specific day of the week, 351
    entities, 4, 6, 6–10, 10
    the number of characters in character strings, 340
    primary key values, 62*tip*. *See also* listing; returning
dimensions (of data), **405**
    multi-dimensional analysis, 405–416
disabling/enabling constraints, 122–123
disk I/O: reducing, 189, 191
displaying. *See* listing
DISTINCT keyword, 42–44, 50*t*, 386, 391
    DML operations on views created with, 493
division operations: returning the remainders of,
    346–347
DML (data manipulation language) commands (actions),
    12*t*, **138**, 138*t*
    advanced challenge, 171–172
    hands-on assignments, 170–171
    indexes and DML command performance, 192, 194,
        197, 198
    with MERGE statements, 428*t*, 456–459, 461*t*
    privileges for issuing. *See* object privileges
    reversing/undoing, 158, 483*n*

DML (data manipulation language) commands (actions)
(*continued*)
    back to a point, 159–162
  saving, 158
    up to a point, 159–162
  as transactions, 157. *See also* DELETE command;
    INSERT command; UPDATE command
DML operations on views:
  complex views: constraints and, 485, 488
    summary guidelines, 495
    views created with the DISTINCT keyword, 493
    views with a ROWNUM pseudocolumn, 494–495
    views with columns based on arithmetic
      expressions, 484, 486–488
    views with group functions or a GROUP BY
      clause, 491–493
    views with multiple tables, 489–491
  simple views, 480–483
DML subqueries, 428*t*, 455
dollar sign ($): formatting character, 249
double quotation marks (""): column alias delimiters, 37,
  38*n*, 50*t*
DROP clauses (ALTER TABLE command), 75–76, 76–77,
  108*n*, 123–125
DROP INDEX command, 177*t*, 201, 207*t*
DROP MATERIALIZED VIEW command, 501
DROP ROLE command, 215*t*, 234, 236*t*
DROP SEQUENCE command, 176*t*, 188, 206*t*
DROP SYNONYM (DROP PUBLIC SYNONYM)
  command, 177*t*, 204, 207*t*
DROP TABLE command, 59*t*, 83–84
  command type, 12*t*
  FOREIGN KEY constraint and, 109–110
DROP TABLE ... PURGE command, 85, 86, 89*t*
DROP USER command, 214*t*, 234, 235*t*
DROP VIEW command, 474*t*, 495–496, 501, 503*t*
dropping. *See* deleting
DUAL table, 186–187, 371
dynamic sampling of database objects, 536
  warning about, 540

E

E-R model (entity-relationship model), **4**
  JustLee Books model, 5–6, 290
  notations, 5*n*
  relationships defined in, 6
  structure, 5–6
eliminating:
  duplicate grouping results, 415–416
  duplicate listings, 26*t*, 42–44
  many-to-many relationships, 10
  test data, 82*tip*
enabling roles after login, 215*t*, 230, 236*t*
encryption, **225**

entering data, 65, 140–143
  character data (nonnumeric data), 140(2), 246–247, 246*n*
  dates, 248
    the current date, 142–143, 355–356
  NULL values, 141–142, 143*tip*
  numeric data, 141, 246–247. *See also* inserting rows
    (adding records)
entering statements, 31
  in SQL Developer, 524
  in SQL*Plus: with the editing commands, 521–522
    in the text editor, 521
entities, **4**
  attribute representations. *See* fields
  bridging entities, **10**
  identifying (determining), 4, 6, 6–10, 10
  relationships between, 6. *See also* table relationships
  representations of. *See* tables
entity relationships, 6. *See also* table relationships
entity-relationship model. *See* E-R model
equality: comparing values for, 363–364
equality joins (equijoins), 284*t*, 289–302, 321*t*
  JOIN methods, 296–302, 320*t*, 321*t*
  traditional method, 291–296, 308–309, 321*t*
equality operator (equal to operator) (=), 245, 250*t*, 275*t*
ERD. *See* E-R model
error messages:
  column ambiguity messages, 292
  constraint violation messages, 105
  "insufficient privileges", 222, 223
  line error indicator (*), 34*n*
  "object does not exist", 30, 138
  "ORA-00922", 65*tip*
  "ORA-00942", 80
  "ORA-00955", 65*n*
  "table or view does not exist", 80
escape character (\), 261
ESCAPE option (LIKE operator), 260–261
EXCEPT option (ALTER USER command), 229
exclusive locks, **163**
Execute Statement button (SQL Developer), 29, 64*n*,
  70*tip*, 525
executing scripts, 28*tip*, 155*tip*
executing statements:
  in SQL Developer, 29, 64*n*, 70*tip*, 524–526
  in SQL*Plus, 520
execution engine, 535
execution plan. *See* explain plan
execution times: measuring, 542–543
EXISTS operator, 451, 461*t*
  efficiency, 548
explain plan (execution plan) (for queries), **192**,
  192–193, 535
  AUTOTRACE tool output, 539
  determination of, 536
  modifying queries and assessing, 541
  review methods, 537–542
  storage of, 538*n*
EXPLAIN PLAN FOR command, 540

explicit COMMIT command, 158
exponential powers: raising numeric data to, 39*n*, 348
exporting data to external files, 150*n*
expressions, regular. *See* regular expressions
expressions-based indexes, 197–198
extracting substrings from character strings, 336–338

F

FAST option (REFRESH clause), 500
fields (in files), **3**, 11
    common fields, **10**
    concatenating, 26*t*, 44–48
    data value alignment in, 38
    primary. *See* primary key(s). *See also* columns (in tables)
files, **3**
    collections of. *See* databases
    control files (SQL*Loader), 529
    exporting data to external files, 150*n*
    importing data from external files, 150*n*
        with SQL*Loader, 529–532
    log file (SQL*Loader), 529–530
    TKPROF executable file, 534. *See also* fields; records
filtering records. *See* selecting rows/records
finding. *See* searching for column data values; *and also* determining; listing; returning
first-normal form (1NF), 8–9
fixed file format (SQL*Loader), 529–530
FLASHBACK TABLE … TO BEFORE DROP command, 59*t*, 85, 89*t*
FOR UPDATE clause (SELECT command), 164
FORCE keyword, 475
foreign key(s), **10**
FOREIGN KEY constraint, 100*t*, 106–110, 109*n*, 119
    abbreviation/code for, 101*t*, 121
    adding, 106–107, 108–109, 109, 127*t*
    creating, 127*t*
    deleting, 108*n*
    and the DROP TABLE command, 109–110
    in join operations, 290, 311*n*
    syntax, 106, 108, 127*t*
    testing, 107. *See also* constraints (on table data)
format arguments, **352**
    TO_CHAR function, 365–367*t*
    TO_DATE function, 352–353*t*
formatting characters: dollar sign ($), 249
FROM clause (SELECT statement), 31, 49*t*
    multiple-column subqueries in, 445–447
FULL hint, 549
FULL keyword, 311
full outer joins, 310, 312, 314*n*, 320*t*
full table scans, **189***(2)*
function-based indexes, 195, 197–198
    creating, 176*t*, 197, 198, 206*t*
    usefulness, 197–198

functions (SQL), **332**
    differences in various database products, 551–554
    nesting, **339, 402**. *See also* group functions (multiple-row functions); single-row functions

G

GRANT command, 214*t*, 215*t*, 236*t*
    command type, 13*t*. *See also* granting …
granting privileges to roles, 214*t*, 227*(2)*, 236*t*
granting privileges to users, 214*t*, 236*t*
    object privileges, 220–223
    system privileges, 219–220
granting roles to users, 215*t*, 227–228*(2)*, 236*t*
greater than operator (>), 113, 249, 250*t*, 251, 252, 275*t*, 439
greater than or equal to operator (>=), 250*t*, 252–253, 275*t*
gross pay: calculating, 360
GROUP BY clause, 386, 393–394, **395**–397, 399, 418*t*
    DML operations on views with, 491–493
    extensions, 385*t*, 405–416, 406*n*, 419*t*
    unsuitable use of, 396–397
group functions (multiple-row functions), **332, 384**, 384–385*t*, 386–394, 417–418*t*
    advanced challenge, 425
    DML operations on views with, 491–493
    GROUP_ID function, 415–416
    GROUPING function, 410
    hands-on assignments, 424–425
    HAVING clause, 386, **398**–401, 418*t*
    statistical functions, 403–405. *See also* GROUP BY clause
GROUP_ID function, 415–416
grouped data: DML operations on views with, 491–493
grouping data, 395–397
    across multiple dimensions, 405–416
        cross-tabular aggregations, 409–411
        subtotals, 405, 407–416
    concatenated groupings, **414**–415
    eliminating duplicate results, 415–416
    restricting groups, 398–401
GROUPING function: CUBE extension with, 410
GROUPING SETS expression (GROUP BY clause), 385*t*, 406, 407–409, 419*t*
groups (aggregated output): restricting, 398–401

H

hands-on assignments:
    constraints, 132–133
    database concepts, 21

hands-on assignments: (*continued*)
database objects, 211–212
DML and TC commands, 170–171
group functions, 424–425
join operations, 329
SELECT statements, 54–55
selecting rows/records, 281
single-row functions, 381
sorting rows/records, 281
subqueries, 467–468
table creation and management, 93
user accounts, privileges, and roles, 241
views, 508
hash joins, 538
HAVING clause (SELECT statement), 386, **398–401,** 418*t*
multiple-row subqueries in, 443–445
single-row subqueries in, 434–435
heap-organized table, **189**
highest values:
listing the n highest values, 496–499
returning the largest value, 393–394

I

ID codes, 10, 60
IDENTIFIED BY clause (ALTER ROLE command), 230
IDENTIFIED BY clause (ALTER USER command), 224
IDENTIFIED BY clause (CREATE USER command), 217
identifying:
common columns, 290
entities, 4, 6, 6–10, 10
resource-intensive SQL statements, 533–535
subtotal rows, 410
implicit COMMIT command, 158
importing data from external files, 150*n*
with SQL*Loader, 529–532
IN operator, 113, 244*t*, 251*t*, 256–257, 276*t*
as a series of OR operators, 263*tip*
efficiency, 548
in multiple-row subqueries, 438, 443
include all columns symbol (*), 31–32
INCREMENT BY clause (CREATE SEQUENCE command), 178–179
index organized tables (IOTs), 177*t*, 198, 206*t*
INDEX privilege, 220
index suppression, 545–547
indexes, **176, 188–201**
automatic creation of, 191, 194
B-tree. *See* B-tree indexes
bitmap indexes, 196–197
composite indexes, 195
concatenated indexes, 547–548
confirming creation of, 191*tip*
creating, 176–177*t*, 190, 194, 196, 197, 198
data block *storage*, 191
deleting, 177*t*, 201, 207*t*
descending order in, 195
and DML command performance, 192, 194, 197, 198
function-based indexes, 195, 197–198
inappropriate use of, 192–193, 194
IOTs (index-organized tables), 177*t*, 198, 206*t*
NULL value references, 195, 197–198
and query performance, 197–198
renaming, 177*t*, 200, 207*t*
and sorting rows, 194*n*, 195, 198
suppression of, 545–547
testing, 194
unique indexes, 194
usefulness, 188–189, 191, 193–194, 194*n*
bitmapped indexes, 196–197
function-based indexes, 197–198
IOTs, 198
verifying, 199–200
weighing the benefits, 193–194
industry standards committees for SQL, 13
INITCAP function, 335–336, 373*t*
inline views (temporary tables), 445, 447, 496
creating, 446, 474*t*, 496–499, 503*t*
inner joins, 308–309. *See also* equality joins
INNER keyword, 308
INSERT All command, 148*n*
INSERT command, 138*t*, 166*t*
command type, 12*t*
syntax, 139–140, 166*t*
VALUES clause, 140, 141–142, 183. *See also* INSERT INTO statements
INSERT INTO statements, 139–149, 166*t*
copying rows from existing tables, 148–149
inserting single rows, 139–144
handling virtual columns, 144, 145–146
including single quotation marks in strings, 146–147
listing column names, 141, 142, 143, 145–146, 149
using sequence values, 183–185
testing constraints with: CHECK, 113–114
FOREIGN KEY, 107
PRIMARY KEY, 104–105
UNIQUE, 111–112
INSERT privilege, 220
inserting rows (adding records), 139–150
CHECK constraint on, 112–113
multiple rows, 148*n*
single rows, 139–144
handling virtual columns, 144, 145–146
including single quotation marks in strings, 146–147
listing column names, 141, 142, 143, 145–146, 149
using sequence values, 183–185
inserting single quotation marks in strings, 146–147

INSTR function, 338–339, 373*t*
"insufficient privileges" error message, 222, 223
integrity: referential, 106–107. *See also* data integrity
intelligent key(s), 8*n*
International Oracle Users Group (IOUG), 527
International Organization for Standardization
      (ISO), 13
INTERSECT set operation (operator), **312**, 313*t*, 317
IOTs (index organized tables), 177*t*, 198, 206*t*
IOUG (International Oracle Users Group), 527
IS NOT NULL operator, 251*t*, 262*n*, 276*t*
IS NULL operator, 244*t*, 251*t*, 262*n*, 276*t*, 392*n*
   in subqueries, 450–451
ISO (International Organization for Standardization), 13
iSQL*Plus, 519

J

JLDB_Build_5.sql script, 138, 138*n*
join conditions, **284**
JOIN keyword, 284
JOIN methods (for joins):
   ANSI-compliant joins vs. traditional joins, 284
   Cartesian joins, 288–289, 321*t*
   equality joins (equijoins), 296–302, 320*t*, 321*t*
   non-equality joins, 304–305, 321*t*
   outer joins, 310–312, 320*t*, 322*t*
   self-joins, 307–308, 322*t*
join methods (for multiple-table queries), 538
join operations (joins), 283–330
   advanced challenge, 329–330
   Cartesian joins, 284*t*, 285–289, 321*t*
   equality joins (equijoins), 284*t*, 289–302, 308–309,
      321*t*
   FOREIGN KEY constraint in, 290, 311*n*
   hands-on assignments, 329
   multiple operations in WHERE clauses, 295–296
   non-equality joins, 284*t*, 302–305, 321*t*
   number needed, 295*tip*
   outer joins, 285*t*, 308–312, 314*n*, 320*t*, 322*t*
   retaining nonmatching rows in, 312*n*
   self-joins, 285*t*, 305–308, 322*t*
   sorting rows in, 194*n*, 294
   table relationships and, 295
   types, 284–285*t*
   verifying, 303*tip*
JOIN ... ON method:
   equality joins, 296, 299–301, 320*t*, 321*t*
   non-equality joins, 304–305, 321*t*
   self-joins, 307–308, 322*t*
JOIN ... USING method:
   equality joins, 296, 298–299, 301, 320*t*, 321*t*
   outer joins, 311–312, 322*t*
   self-joins, 307
joining tables. *See* join operations (joins)
Julian date, 348–349

JustLee Books database, 13–16
   basic assumptions, 13–14
   BOOKS table, 7–10
   creating, 27–30
   E-R model, 5–6, 290
   marketing value, 14
   sequence example, 181–182
   table structures, 10–11
   tables in, 14–16, 511–517

K

key-preserved tables, **490**
keys: foreign key(s), **10**. *See also* primary key(s)
keywords (SQL), **31**
   ALL, 221
   ANY, 218, 219
   AS, 37, 38*n*, 50*t*
   DISTINCT, 42–44, 50*t*
   REFERENCES, 106
   as reserved words, 59
   UNIQUE, 44, 50*t*
keywords (SQL*Plus): START keyword replacement char-
     acter (@), 28*tip*

L

labeling subtotal rows, 410–411
largest value: returning, 393–394. *See also* highest
     values
LAST_DAY function (date), 351–352, 375*t*
leaf blocks (data blocks), 191, 198
LEFT keyword, 311
left outer joins, 310, 311, 320*t*
LENGTH function, 340, 373*t*
less than operator (<), 113, 250*t*, 251, 252, 275*t*, 439
less than or equal to operator (<=), 250*t*, 252, 275*t*
LIKE operator, 244*t*, 251*t*, 258–261, 276*t*
   ESCAPE option, 260–261
line error indicator (*), 34*n*
list delimiter (,), 35, 37*n*, 63
listing:
   column names, 33*tip*
     with constraints, 121–122
     in INSERT INTO statements, 141, 142, 143,
       145–146, 149. (*See also* table structure,
       below)
   constraints, 120–122
     with column names, 121–122
   the n highest values, 496–499
     returning the largest value, 393–394
   table data, 31–32, 141, 144, 148. *See also* searching
     for column data values

listing: (*continued*)
   table structure, 26t, 29–30, 33tip, 50t, 115n
     verifying table creation and modification, 66–67, 68–69, 77–78, 83–84
   tables, 28–29, 50t, 65–66
   virtual column definitions, 67
literals. *See* characters; string literals
LOCK TABLE command, 138t, 162–163, 166t
locking tables, 162–164
log file (SQL*Loader), 529–530
logical operators, 244t, 257–258, 262–264, 276t
lookup tables, **12**
LOWER function, 333–334, 372t
lowercase letters: converting character strings to, 333–334
LPAD function, 340–341, 373t
LTRIM function, 342, 373t

M

many-to-many relationships, 6
   eliminating, 10
materialized views, 473, **499**
   advantages and disadvantages, 499
   creating, 474t, 499–501, 503t
   deleting, 501
   querying, 500
mathematical comparison operators, 113, 244t, 248–255, 275t
MAX function, 384t, 393–394, 418t
MAXVALUE clause (CREATE SEQUENCE command), 179, 185
MERGE command, 152n
   clauses, 457–458, 459
MERGE statements: DML actions with, 428t, 456–459, 461t
merges, sort, 538
metadata: table information, 29n
Microsoft Excel pivot tables, 406–407
MIN function, 384t, 394, 418t
MINUS set operation (operator), **312**, 313t, 318
MINVALUE clause (CREATE SEQUENCE command), 179
mixed case: converting character strings to, 335–336
MOD function, 346–347, 374t
MODIFY clause (ALTER TABLE command), 71
modifying columns, 71–75
   syntax, 88t
modifying rows, 150–151
   multiple rows, 151–152
   preventing data inconsistencies during, 163–164
   using substitution variables, 152–155
modifying tables, 70–82
   advanced challenge, 93
   commands, 12t, 58–59t
   hands-on assignments, 93
   syntax, 88t

MONTHS_BETWEEN function, 349–350, 355, 374t
multi-dimensional analysis, 405–416
   cross-tabular aggregations, 409–411
   subtotals, 405, 407–416
multiple columns: selecting, 26t, 34–36
multiple constraints on a single column, 120
multiple joins in WHERE clauses, 295–296
multiple roles: granting to users, 228
multiple rows/records:
   inserting/adding, 148n
   modifying/updating, 151–152
multiple tables: DML operations on complex views with, 489–491
multiple-column indexes, 195
multiple-column set operations, 315–316
   guidelines for, 317
multiple-column subqueries, 428t, 445–448
   in a FROM clause, 445–447
   in a WHERE clause, 447–448
multiple-column values: grouping, 397–398
multiple-row comparison operators, 438–443, 439t, 461t
multiple-row functions. *See* group functions
multiple-row subqueries, 428t, 437–445
   in a HAVING clause, 443–445
   ALL operators in, 438, 439t, 439–441, 461t
   ANY operators in, 438, 439t, 441–443, 461t
   IN operator in, 438, 443
multiple-table queries: join methods, 538

N

names:
   column table names, 59
   constraint names, 101, 119
   sequence names, 178
   table names, 59–60
   tables with the same name, 65n. *See also* column aliases; column names; table aliases
naming constraints, 101
   not naming constraints vs., 119
narrowing columns, 72
NATURAL JOIN method, 296, 297–298, 320t
natural key(s), 8n, 177n
NATURAL keyword, 297
nested loop joins, 538
nested subqueries, 453–454
nesting functions, **339, 402**
NEXT clause (CREATE MATERIALIZED VIEW command), 500
NEXT_DAY function (date), 351, 375t
NEXTVAL pseudocolumn, **183**–184, 184
NEXTVAL values, 185
   calling, 186–187

NLV function, 197–198
NOCACHE option (CREATE SEQUENCE command), 180
NOCYCLE option (CREATE SEQUENCE command), 179–180
NOFORCE keyword, 475
NOMAXVALUE option (CREATE SEQUENCE command), 179
NOMINVALUE option (CREATE SEQUENCE command), 179
non-equality joins, 284t, 302–305, 321t
    JOIN method, 304–305, 321t
    traditional method, 303–304, 321t
non-key-preserved tables, **490**
    DML operations on views with, 490–491
non-NULL values: counting records with, 390–392
NONE option (ALTER USER command), 229
nonnumeric data. *See* character data
NOORDER option (CREATE SEQUENCE command), 180
normal distribution, **404**
normalizing databases/tables, 6–10
not equal to operator (<>), 250t, 253–254, 275t
NOT NULL constraint, 100t, 109, 114–116, 117
    abbreviation/code for, 101t, 121
    adding, 114–115, 115, 116, 128t
    assignment to primary key error, 119tip
    creating, 128t
    and DML operations on complex views, 488
    syntax, 115, 128t
    testing, 115–116. *See also* constraints (on table data)
NOT operator, 244t, 257–258
NULL values, 32n, 40–42, 102n
    in arithmetic operations, 41–42, 146n, 389–390
        substituting values for, 359–363
    vs. blank spaces, 265
    entering, 141–142, 143tip
    index references to, 195, 197–198
    searching for, 265–266, 392tip
    in subqueries, 448–451. *See also* non-NULL values
NULLIF function, 363–364, 375t
NULLS FIRST option (ORDER BY clause), 269–270
NULLS LAST option (ORDER BY clause), 269–270
NUMBER(p, s) datatype, 60t, 62
number of buffer reads statistic, 534
number of reads per statement transaction statistic, 534
numeric data:
    converting dates or strings into, 370–371
    dates as, 348
    entering, 141, 246–247
    listing the n highest values, 496–499
    precision and scale, **60**
    raising to exponential powers, 39n, 348
    returning the absolute value of, 347–348
    returning the largest value, 393–394
    returning the smallest value, 394
    rounding, 344–345
    truncating, 345–346

numeric datatype. *See* NUMBER(p, s) datatype
numeric fields: data value alignment in, 38
numeric functions, 332t, 344–348, 374t
NVL function, 359–362, 375t
    in subqueries, 449–450
NVL2 function, 362–363, 364, 375t

O

OAI (Oracle Academic Initiative), 527
"object does not exist" error message, 30, 138
object privileges, **218**, 220
    granting, 214t, 220–223, 236t
    revoking, 214t, 233, 236t
OCP (Oracle Certification Program), 527
ON clause (CREATE INDEX command), 190, 197
ON clause (GRANT command), 221
ON clause (with JOIN). *See* JOIN ... ON method
ON DELETE CASCADE option, 108–109
one-to-many relationships, 6
one-to-one relationships, 6
operators:
    comparison. *See* comparison operators
    concatenation operator (||), 45
    include all columns symbol (*), 31–32
    logical operators, 244t, 257–258, 262–264, 276t
    order of, 39, 263
        overriding, 39, 263–264
    outer join operator (+), 309–310, 312
    pattern operators, 357, 358t. *See also* set operations (operators)
optimizer (for queries), **192**, 535
    old and new methods, 535–536
optimizer hints, 549
OPTIMIZER_MODE settings, 536
OR operator, 244t, 263(2), 276t
"ORA-00922" error message, 65tip
"ORA-00942" error message, 80
"ORA-00955" error message, 65n
Oracle 11g database system, 16
Oracle Academic Initiative (OAI), 527
Oracle Certification Program (OCP), 527
Oracle Enterprise Manager, 534n
Oracle resources, 527
Oracle Technology Network (OTN), 527
ORDER BY clause (SELECT statement), 244t, 251tip, 267–273, 275t, 496
    column references in, 272–273, 272tip
    default order (ASC option), 267–268, 269n
    DESC option, 268
    NULLS FIRST/LAST options, 269–270. *See also* sorting rows/records
order of operations (operators), 39, 263
    overriding, 39, 263–264

order of values in sort operations:
    ascending order, 267–268, 269*n*
    descending order, 268
ORDER option (CREATE SEQUENCE command), 180
ORDERITEMS table, 15, 514–515
ORDERS table, 15, 513–514
ORGANIZATION INDEX keywords: CREATE TABLE and, 198
other functions, 332*t*, 359–371, 375–376*t*
OTN (Oracle Technology Network), 527
outer join operator (+), 309–310, 312
outer joins, 285*t*, 308–312, 314*n*, 320*t*, 322*t*
    JOIN method, 310–312, 320*t*, 322*t*
    traditional method, 309–310, 322*t*
OUTER keyword, 311
outer queries, **429**

P

padding character strings on the left/right, 340–342
parent queries, **429**
parent tables:
    deleting, 109–110
    deleting records from, 108–109
parentheses (()):
    column list delimiters, 63, 68
    operation delimiters, 39, 263–264
    subquery delimiters, 429, 454*tip*
    total aggregation argument, 407
parser, 535
partial dependency, 9
partial ROLLUPs, 411, 412
PASSWORD command (SQL*Plus), 225
password expiration prompt bug in SQL Developer, 220*n*
PASSWORD EXPIRE option (CREATE USER command), 217, 220
passwords:
    adding to roles, 215*t*, 230, 236*t*
    caution about, 225
    creating, 217
    management methods, 225
    resetting, 214*t*, 224–225, 235*t*
pattern matching:
    with the LIKE operator, 244*t*, 251*t*, 258–261, 276*t*
    with regular expressions, 357–359
pattern operators, 357, 358*t*
percent sign (%):
    as a string literal, 259*tip*
    as a wildcard character, 258–260, 276*t*, 478
performance:
    explain plan review methods, 537–542
    goals, 536
    identifying resource-intensive SQL statements, 533–535
    indexes and DML command performance, 192, 194, 197, 198
    indexes and query performance, 197–198
    SQL processing architecture, 535–536
    SQL processing steps, 537–538
    statistics. *See* performance statistics
    timing queries, 542–543
    tuning guidelines, 543–549
performance statistics, 534
    from the AUTOTRACE tool, 539, 539–540*t*
    execution times, 542–543
period (.): decimal point, 249
phonetic representations: searching for, 369–370
pivot tables (Microsoft Excel), 406–407
PLAN_TABLE table, 538*n*
plus sign (+): outer join operator, 309–310, 312
PLUSTRACE role, 538*n*
Portable Operating System Interface for UNIX (POSIX) standards, 357
POSIX (Portable Operating System Interface for UNIX) standards, 357
POWER function, 348, 374*t*
precedence. *See* order of operations (operators)
precision (of numbers), **60**
predefined roles, 228–229
primary key(s), **7–8**, 10
    blank entries in, 67
    deleting, 124–125
    determining appropriate values for, 62*tip*
    NOT NULL assignment error, 119*tip*
    operations involving, 198
    sequence values as, 177
    types, 8*n*, 177*n*
PRIMARY KEY constraint, 100*t*, 103–106, 106*tip*
    abbreviation/code for, 101*t*, 121
    adding, 103–104, 127*t*
    creating, 126–127*t*
    syntax, 103, 126–127*t*
    testing, 104–105. *See also* constraints (on table data)
private synonyms, 202–203, 203–204
privilege views, 230, 231*t*
privileges, 63*n*, **218**
    advanced challenge, 241
    checking/confirming, 231–232
    collections of. *See* roles
    data dictionary views, 230, 231*t*
    for deleting synonyms, 203
    granting to roles, 214*t*, 227(2), 236*t*
    granting to users, 214*t*, 219–220, 220–223, 236*t*
    hands-on assignments, 241
    revoking, 214*t*, 232–233, 236*t*

types, 218. *See also* object privileges; system privileges
profit criterion-based indexes, 197
projection, **33**
PROMOTION table, 15, 517
pseudocolumns, **183**–184, 185*n*. *See also* ROWNUM pseudocolumn
PUBLIC keyword, 202–203, 203, 204, 221
public synonyms, 202–203
   deleting, 204
PUBLISHER table, 15, 517
PURGE TABLE command, 59*t*, 86, 89*t*

Q

qualifying column names, 292, 293–294
queries (SELECT statements), 12*t*, 26*t*, **30**–31
   arithmetic operations in, 26*t*, 39–40
   case sensitivity in, 34, 37
   executing: in SQL Developer, 29, 64*n*, 70*tip*, 525–526 in SQL*Plus, 520
   explain plan (statement execution plan), **192**, 192–193
   indexes and performance of, 197–198
   modifying, 541
   optimizer, **192**, 535–536
   outer/parent queries, **429**
   syntax, 26*t*, 31, 35, 49–50*t*
   timing, 542–543. *See also* querying databases; SELECT command/statements; subqueries
querying data dictionary views:
   USER_CONS_COLUMNS object, 121–122
   USER_IND_COLUMNS object, 200
   USER_INDEXES object, 199–200
   USER_OBJECTS object, 181–182
   USER_SEQUENCES object, 182, 186, 188
   USER_TAB_COLUMNS object, 67
   USER_TABLES object, 28–29, 65–66
   USER_UNUSED_COL_TABS object, 77, 78
querying databases, 25–55
   advanced challenge, 55
   assigning aliases to columns, 26*t*, 36–39
   concatenating columns/fields, 26*t*, 44–48
   eliminating duplicate listings, 26*t*, 42–44
   filtering records. *See* selecting rows/records
   hands-on assignments, 54–55
   selecting all columns from a table, 26*t*, 31–33
   selecting multiple columns from a table, 26*t*, 34–36
   selecting one column from a table, 26*t*, 33–34
   using arithmetic operations, 26*t*, 39–40
   using conditions, 245–248. *See also* selecting rows/records. *See also* join operations; querying data dictionary views; set operations
quotation marks. *See* double quotation marks; single quotation marks

raising numeric data to exponential powers, 39*n*, 348
rebuilding tables, 30
records (in files), **3**, 11
   adding. *See* inserting rows
   copying from existing tables, 148–149
   counting records with non-NULL values, 390–392
   deleting, 156–157
      from parent tables, 108–109
   filtering. *See* selecting rows/records
   matching. *See* join operations
   retaining nonmatching records in join operations, 312*n*
   sorting. *See* sorting rows/records
   updating. *See* updating records. *See also* rows (in tables)
recovering tables, 84–85
   syntax, 89*t*
REFERENCES keyword, 106
REFERENCES privilege, 220
referential integrity, 106–107, 290
REFRESH clause (CREATE MATERIALIZED VIEW command), 500
REGEXP_LIKE, 260*n*, 357–358, 375*t*
REGEXP_SUBSTR, 359, 375*t*
regular expressions, 332*t*, 357–359, 375*t*
relating tables, 10–12
relational operators. *See* comparison operators
relationships between entities, 6. *See also* table relationships
remainders of division operations: returning, 346–347
RENAME … TO command, 59*t*, 79–81, 88*t*
renaming indexes, 177*t*, 200, 207*t*
renaming tables, 79–81, 88*t*
REPLACE function, 342–343, 373*t*
replacing character strings, 342–343
replacing single characters in character strings, 343
replacing views (re-creating views), 474*t*, 475, 477*n*, 480, 503*t*
reserved words: keywords as, 59
resetting passwords, 214*t*, 224–225, 235*t*
RESOURCE role, 228*t*
resource-intensive SQL statements: identifying, 533–535
resources, Oracle, 527
restoring tables. *See* recovering tables
restricting groups, 398–401
restricting rows. *See* selecting rows/records
retaining nonmatching rows in join operations, 312*n*
retrieving data from databases. *See* querying databases
returning:
   the absolute value of numeric data, 347–348
   the current date, 355–356
   the largest value, 393–394
   remainders of division operations, 346–347
   the smallest value, 394. *See also* determining

reversing DML commands, 158, 483*n*
  back to a point, 159–162
REVOKE command, 214*t*, 215*t*, 232–233, 236*t*
  command type, 13*t*
revoking privileges/roles, 214*t*, 215*t*, 232–233, 236*t*
RIGHT keyword, 311
right outer joins, 310, 311–312, 320*t*
ROLE_SYS_PRIVS view, 231*t*
ROLE_TAB_PRIVS view, 231*t*
roles, **225**–230
  adding passwords to, 215*t*, 230, 236*t*
  advanced challenge, 241
  combining into one, 228
  creating, 215*t*, 226, 227, 228, 236*t*
  default roles, 229
  deleting (dropping), 215*t*, 234, 236*t*
  enabling after login, 215*t*, 230, 236*t*
  granting privileges to, 214*t*, 227*(2)*, 236*t*
  granting to users, 215*t*, 227–228*(2)*, 236*t*
  hands-on assignments, 241
  PLUSTRACE role, 538*n*
  predefined roles, 228–229
  revoking, 215*t*, 232–233, 236*t*
  revoking privileges from, 233
ROLLBACK command, 138*t*, 158, 160–161, 161–162,
    166*t*
  command type, 13*t*
  syntax, 166*t*
ROLLBACK TO SAVEPOINT command, 159, 161
ROLLUP extension (GROUP BY clause), 385*t*, 411–416,
    419*t*
root node block (data block), 191
ROUND function (date), 354, 374*t*
ROUND function (number), 344–345, 374*t*
rounding dates, 354
rounding numeric data, 344–345
row source, 535
row source generator, 535
row-level locks, 162
ROWIDs, 188
ROWNUM pseudocolumn, **493**
  DML operations on views with, 494–495
  inline views and, 496–499
rows (in tables), **3**, 11
  copying from existing tables, 148–149
  deleting, 156–157
    from parent tables, 108–109
  groups. *See* groups (aggregated output)
  inserting. *See* inserting rows
  matching. *See* join operations
  modifying, 150–151
    multiple rows, 151–152
    using substitution variables, 152–155
  restricting. *See* selecting rows/records
  retaining nonmatching rows in join operations, 312*n*
  sorting. *See* sorting rows/records. *See also* records
    (in files); subtotal rows
RPAD function, 341–342, 373*t*

RTRIM function, 342, 373*t*
rule-based optimizer, 535
Run Script button (SQL Developer), 29, 64*n*, 70*tip*,
    525–526

S

SAVEPOINT command, 138*t*, 159–162
saving DML commands, 158
  up to a point, 159–162
scale (of numbers), **60**
schemas:
  including when creating tables, 63
  referencing objects in others', 203–204, 220*n*, 222*tip*
  SCOTT user schema, 65*n*, 117
SCOTT user schema, 65*n*, 117
scripts (in SQL), **28***n*, 155
  executing, 28*tip*, 155*tip*
  extension, 155
  JLDB_Build_5.sql script, 138, 138*n*
  update scripts, 155
SDLC (Systems Development Life Cycle), 4
searched CASE expression, 369, 376*t*
searching: for values. *See* searching for column data
    values. *See also* determining; listing; returning
searching for column data values, 245–248
  with comparison operators, 248–260, 265–266
    and logical operators, 257–258, 262–264, 276*t*
  largest value, 393–394
  matching a listed value, 256–257
  n highest values, 496–499
  NULL values, 265–266, 392*tip*
  phonetic representations, 369–370
  smallest value, 394
  within a specified range, 255–256
  substrings, 338–339
  with wildcard characters, 258–260. *See also* pattern
    matching; selecting rows/records
second-normal form (2NF), 9
secondary sorts, **270**–272
SELECT * statements, 31–32, 141, 144, 148
SELECT clause, 31, 49*t*
  column references in, 272*tip*
  single-row subqueries in, 435–437
SELECT command/statements, 31
  clauses: FOR UPDATE clause, 164. *See also* FROM
    clause; HAVING clause; ORDER BY clause;
    WHERE clause (SELECT statement)
  command type, 12*t*
  as queries. *See* queries (SELECT statements)
  syntax, 26*t*, 31, 35, 49*t*, 245, 267, 386
SELECT DISTINCT statement, 42–44, 50*t*
  ORDER BY clause in, 272*tip*
SELECT privilege, 220
SELECT UNIQUE statement, 42–44, 50*t*
  ORDER BY clause in, 272*tip*

SELECT ... AS statement, 37, 38*n*, 50*t*
SELECT ... FOR UPDATE command, 138*t*, 163–164, 166*t*
selecting columns:
    all columns, 26*t*, 31–33
    avoiding unnecessary selection, 544–545
    multiple columns, 26*t*, 34–36
    one column, 26*t*, 33–34
selecting rows/records (restricting/filtering), 244,
        245–248, 478
    advanced challenge, 281
    with comparison operators, 248–260, 248–262,
        265–266
        and logical operators, 257–258, 262–264, 276*t*
    in groups, 398–401
    hands-on assignments, 281. *See also* searching for col-
        umn data values
selection, **244**. *See also* selecting rows/records
selectivity (of rows), 536. *See also* cardinality (of
        column data values)
self-joins, 285*t*, 305–308, 322*t*
    JOIN method, 307–308, 322*t*
    traditional method, 306–307, 322*t*
semicolon (;): SQL statement terminator, 31
sequence values, 177
    accessing, 183
    calling, 186–187
    inserting rows using, 183–185
    as primary key(s), 177
sequences, **176**, 177–182
    altering (changing settings), 176*t*, 185–188, 206*t*
    creating (generating), 176*t*, 178–182, 206*t*
    deleting, 176*t*, 188, 206*t*
    gaps in, 180
    JustLee Books example, 181–182
    naming convention, 178
    uses, 177–178
    values. *See* sequence values
SESSION_PRIVS view, 231*t*
SESSION_ROLES view, 231*t*
SET AUTOTRACE ON command, 538
SET clause (UPDATE command), 150
set operations (operators), 284, 285*t*, **312**–318, 313*t*, 322*t*
    multiple-column operations, 315–316
        guidelines for, 317
    sorting in, 314–315
SET ROLE command, 215*t*, 230, 236*t*
SET TIMING ON command, 542–543
SET UNUSED [COLUMN] clause (ALTER TABLE
        command), 76
shared cache memory area (buffer pool), **189**
shared locks, **162**–163
simple joins. *See* equality joins
simple views:
    creating, 476–480
    DML operations on, 480–483
single quotation marks ('):
    entering NULL values with, 141–142
    inserting in strings, 146–147

string literal delimiters, 46, 47*n*, 50*t*, 140, 141,
        246–247
single-row comparison operators, 431, 437–438
single-row functions, 331–382, **332**
    advanced challenge, 381
    case conversion functions, 332*t*, 333–336, 372–373*t*
    character manipulation functions, 332*t*, 336–344,
        373–374*t*
    date functions, 332*t*, 348–356, 374–375*t*
    hands-on assignments, 381
    nesting, **339**
    numeric functions, 332*t*, 344–348, 374*t*
    other functions, 332*t*, 359–371, 375–376*t*
    regular expressions, 332*t*, 357–359, 375*t*
single-row subqueries, 428*t*, 429–437
    in a HAVING clause, 434–435
    in a SELECT clause, 435–437
    in a WHERE clause, 429–434, 437*tip*
    operators, 431
single-table approach to database design, 6–7
smallest value: returning, 394
sort merges, 538
sorting rows/records, 194*n*
    advanced challenge, 281
    in ascending order, 267–268, 269*n*
    in descending order, 268
    hands-on assignments, 281
    indexes and, 194*n*, 195, 198
    in join operations, 194*n*, 294
    secondary sorts, **270**–272
    in set operations, 314–315. *See also* ORDER BY clause
        (SELECT statement)
SOUNDEX function, 369–370, 376*t*
space ( ): in SELECT statements, 35
SQL (Structured Query Language), 12–13
    coding problem areas, 533–535
    differences in various database products,
        551–554
    industry standards committees for, 13
    processing architecture, 535–536. *See also* clauses
        (SQL); keywords (SQL); SQL commands; SQL
        statements; syntax (SQL)
SQL commands:
    for creating tables, 12*t*, 58*t*
    for modifying tables, 12*t*, 58–59*t*
    for querying databases. *See* queries (SELECT
        statements); SELECT command/statements
    types, 12–13*t*. *See also* DCL (data control language)
        commands; DDL (data definition language)
        commands; DML (data manipulation language)
        commands; TC (transaction control)
        commands. *See also* SQL statements; *and*
        specific SQL commands
SQL Developer, 16, 519, 523–526
    Autocommit option, 159*tip*
    connection and login, 523
    entering statements, 524
    executing statements, 29, 64*n*, 70*tip*, 524–526

SQL Developer (*continued*)
   interface, 27, 28, 32*n*, 523, 524
   password expiration prompt bug, 220*n*
.sql extension, 155
SQL statement terminator (;), 31
SQL statements, 12
   entering, 31
   executing: in SQL Developer, 29, 64*n*, 70*tip*, 524–526
      in SQL*Plus, 520
   execution plan. *See* explain plan
   identifying resource-intensive statements, 533–535
   measuring execution time, 542–543
   storing in scripts, 155*tip*
   syntax. *See* syntax (SQL). *See also* INSERT INTO
      statements; queries (SELECT statements);
      scripts; SQL commands
SQL TRACE feature, 534
SQL Tuning Advisor, 534, 537
SQL*Loader, 529–532
SQL*Net statistics, 539, 539–540*t*
SQL*Plus, 16, 519, 519–522
   column alias displays, 389*n*
   commands. *See* SQL*Plus commands
   entering statements: with the editing commands,
      521–522
      in the text editor, 521
   executing queries, 520
   interface, 27, 28, 32*n*
   login, 519–520
   scripts, 28*tip*, 155*tip*
   START keyword replacement character (@), 28*tip*.
      See also substitution variables
SQL*Plus commands:
   CHANGE command, 522
   PASSWORD command, 225. *See also* DESCRIBE
      command
square brackets ([]): clause delimiters, 31, 245
standard deviation: calculating, 403–404
START keyword replacement character (@) (SQL*Plus),
   28*tip*
START WITH clause (CREATE MATERIALIZED VIEW
   command), 500
START WITH clause (CREATE SEQUENCE command),
   179, 179*tip*, 185
statement execution plan. *See* explain plan
statements. *See* SQL statements
statistical group functions, 403–405
statistics: database object statistics, 536. *See also*
   performance statistics
STDDEV function, 384*t*, 403–404, 418*t*
storing SQL statements in scripts, 155*tip*
string literal delimiters ('), 46, 47*n*, 50*t*, 140, 141,
   246–247
string literals (character strings), 50*t*, 246–247
   concatenating, 344
   concatenating fields with, 46–48
   converting into numeric data, 370–371
   converting to lowercase letters, 333–334

   converting to mixed case, 335–336
   converting to uppercase letters, 334–335
   determining the number of characters in, 340
   extracting substrings from, 336–338
   inserting single quotation marks in, 146–147
   padding on the left/right, 340–342
   replacing, 342–343
   replacing single characters in, 343
   searching for substrings in, 338–339
   trimming on the left/right, 342
   wildcard characters as, 259*tip*, 260–261. *See also* char-
      acter data (nonnumeric data)
strings. *See* string literals
Structured Query Language. *See* SQL
subqueries, **67–68**, **428–455**
   advanced challenge, 468
   clauses for, 429
   copying rows from existing tables, 148–149
   correlated subqueries, 428*t*, **451–453**, 460*t*, 548
   creating tables with, 67–69
   DML subqueries, 428*t*, 455
   efficiency: by type, 548
   hands-on assignments, 467–468
   IS NULL operator in, 450–451
   multiple-column subqueries, 428*t*, 445–448
   multiple-row subqueries, 428*t*, 437–445
   nested subqueries, 453–454
   NULL values in, 448–451
   NVL function in, 449–450
   rules for, 429
   single-row subqueries, 428*t*, 429–437
   types, 428, 428*t*
   uncorrelated subqueries, 428*t*, **451**, 461*t*, 548
substituting values for NULL values in arithmetic operations, 359–363
substitution variable symbol (&), 153, 166*t*
substitution variables, **153**
   modifying rows/updating records using, 152–155
SUBSTR function, 336–338, 373*t*
substrings:
   extracting from character strings, 336–338
   searching for, 338–339
subtotal row indicator, 407
subtotal rows:
   identifying, 410
   labeling, 410–411
subtotals: generating, 405, 407–416
SUM function, 384*t*, 386–388, 417*t*
surrogate key(s), 8*n*, 177*n*
synonyms (for database objects), **176**, 201–204
   creating, 177*t*, 202, 207*t*
   deleting, 177*t*, 203, 204, 207*t*
   private vs. public, 202–203, 203–204
   uses, 210–212
syntax (SQL), 31
   ADD_MONTHS function, 351
   adding constraints, 102, 127–128*t*
      CHECK, 112, 128*t*

FOREIGN KEY, 106, 108, 127*t*
NOT NULL, 115, 128*t*
PRIMARY KEY, 103, 127*t*
UNIQUE, 111, 127*t*
ALTER INDEX command ... RENAME TO, 177*t*, 207*t*
ALTER ROLE command, 215*t*, 230, 236*t*
ALTER SEQUENCE command, 176*t*, 185–186, 206*t*
ALTER TABLE command, 70
ALTER TABLE ... ADD command: adding columns, 70, 88*t*
    adding constraints, 103, 106, 108, 111, 112, 127–128*t*
ALTER TABLE ... DISABLE/ENABLE CONSTRAINT commands, 122
ALTER TABLE ... DROP COLUMN command, 75, 88*t*
ALTER TABLE ... DROP UNUSED COLUMNS command, 77, 88*t*
ALTER TABLE ... MODIFY command, 71
    adding a NOT NULL constraint, 115, 128*t*
ALTER TABLE ... SET UNUSED [COLUMN] command, 76, 88*t*
ALTER USER command, 214*t*, 215*t*, 224, 229, 235*t*, 236*t*
arithmetic operations, 50*t*
AS keyword, 50*t*
AVG function, 384*t*, 388, 418*t*
Cartesian joins, 321*t*
CHECK constraint, 112, 127–128*t*
COMMIT command, 166*t*
CONCAT function, 344
concatenating columns/fields, 50*t*
correlated subqueries, 460*t*
COUNT function, 384*t*, 390, 418*t*
CREATE BITMAP INDEX command, 176*t*, 206*t*
CREATE INDEX command, 176*t*, 206*t*
CREATE MATERIALIZED VIEW command, 474*t*, 503*t*
CREATE OR REPLACE VIEW command, 474*t*, 475, 503*t*
CREATE ROLE command, 215*t*, 226, 236*t*
CREATE SEQUENCE command, 176*t*, 178, 206*t*
CREATE SYNONYM (CREATE PUBLIC SYNONYM) command, 177*t*, 202, 207*t*
CREATE SYNONYM command, 202
CREATE TABLE command, 63, 88*t*, 126–128*t*
    and ORGANIZATION INDEX, 177*t*, 206*t*
CREATE TABLE ... AS command, 68, 88*t*
CREATE USER command, 214*t*, 216, 235*t*
CREATE VIEW command, 474*t*, 474–475, 502*t*
creating constraints, 102, 126–128*t*
CUBE extension, 385*t*, 419*t*
DECODE function, 367
DELETE command, 156, 166*t*
disabling/enabling constraints, 122–123
DISTINCT keyword, 50*t*
DROP INDEX command, 177*t*, 200, 201, 207*t*
DROP ROLE command, 215*t*, 234, 236*t*
DROP SEQUENCE command, 176*t*, 188, 206*t*
DROP SYNONYM (DROP PUBLIC SYNONYM) command, 177*t*, 204, 207*t*

DROP TABLE command, 83
DROP TABLE ... PURGE command, 89*t*
DROP USER command, 214*t*, 234, 235*t*
DROP VIEW command, 474*t*, 495, 503*t*
equality joins (equijoins), 321*t*
FLASHBACK TABLE ... TO BEFORE DROP command, 89*t*
FOREIGN KEY constraint, 106, 108, 127*t*
GRANT command, 214*t*, 215*t*, 219, 221, 236*t*
GROUP BY clause, 395, 418*t*
GROUP BY extensions, 385*t*, 419*t*
group functions, 384–385*t*, 417–418*t*
GROUPING SETS extension, 385*t*, 419*t*
HAVING clause, 398, 418*t*
INITCAP function, 335
INSERT INTO statements, 139–140, 166*t*
INSTR function, 338
JOIN ... ON method, 320*t*
JOIN ... USING method, 320*t*
LENGTH function, 340
listing table structure, 50*t*
listing tables, 50*t*
LOCK TABLE command, 162, 166*t*
LOWER function, 333
LPAD function, 341
LTRIM function, 342
MAX function, 384*t*, 393, 418*t*
MERGE statements, 457, 461*t*
MIN function, 384*t*, 394, 418*t*
modifying columns, 88*t*
modifying tables, 88*t*
MONTHS_BETWEEN function, 349
NATURAL JOIN method, 320*t*
NEXT_DAY function (date), 351
non-equality joins, 321*t*
NOT NULL constraint, 115, 128*t*
NVL function, 359
NVL2 function, 362
outer joins, 320*t*, 322*t*
POWER function, 348
PRIMARY KEY constraint, 103, 126–127*t*
PURGE TABLE command, 89*t*
RENAME ... TO command, 79, 88*t*
REPLACE function, 342
REVOKE command, 214*t*, 215*t*, 232, 236*t*
ROLLBACK command, 166*t*
ROLLBACK TO SAVEPOINT command, 159
ROLLUP extension, 385*t*, 419*t*
ROUND function (date), 354
ROUND function (number), 344
RPAD function, 341–342
RTRIM function, 342
SELECT command/statements, 26*t*, 31, 35, 49–50*t*, 245, 267
SELECT ... FOR UPDATE command, 164, 166*t*
self-joins, 322*t*
set operations, 322*t*
SET ROLE command, 215*t*, 230, 236*t*
SOUNDEX function, 370

syntax SQL (*continued*)
   STDDEV function, 384*t*, 403, 418*t*
   SUBSTR function, 336
   SUM function, 384*t*, 386, 417*t*
   TO_CHAR function, 365
   TO_DATE function, 352
   TRANSLATE function, 343
   TRUNC function (number), 345–346
   TRUNCATE TABLE command, 82, 88*t*
   uncorrelated subqueries, 461*t*
   UNIQUE constraint, 111, 127*t*
   UNIQUE keyword, 50*t*
   UPDATE command, 150, 166*t*
   UPPER function, 334
   USER_TABLES object, 50*t*
   VARIANCE function, 385*t*, 404, 418*t*
   WHERE clause (SELECT statement), 245–248, 447
   WHERE clause joins, 320*t*
syntax (SQL*Plus):
   CHANGE command, 522
   DESCRIBE command, 26*t*, 50*t*
SYS_C*n* format, 101
SYSDATE function, 355–356
SYSDATE value:
   as a default, 64, 143
   entering, 142–143
system crashes: rollbacks after, 158
system implementation, 4
system privileges, **218**–219, 219*n*
   granting, 214*t*, 219–220, 236*t*
   revoking, 232, 236*t*
SYSTEM user (administrator account),
      182*n*, 215
systems analysis, 4
systems deployment, 4
systems design, 4
Systems Development Life Cycle (SDLC),
      4 models, 4*n*
systems integration and testing, 4
systems investigation, 4
systems maintenance and review, 4

T

table aliases, **294**
   in join operations, 293–294, 306
   qualifying column names with, 293–294
table data:
   entering. *See* entering data
   listing, 31–32, 141, 144, 148. *See also* searching for
      column data values
table information (metadata), 29*n*
table locks, 162–164
table names, 59–60. *See also* table aliases
"table or view does not exist" error message, 80
table relationships (entity relationships), 6

   establishing, 10–12
   and joins, 295
table structure:
   JustLee Books tables, 10–11
   listing, 26*t*, 29–30, 33*tip*, 50*t*, 115*n*
      verifying table creation and modification, 66–67,
         68–69, 77–78, 83–84
table-level constraints: creating, 102–103, 117–118
tables, 3, 6
   access order, 538
   access paths, 537
   adding, 6, 6–10, 10
   adding constraints to. *See* adding constraints
   copying rows from, 148–149
   creating 63–65
      advanced challenge, 94
      commands, 12*t*, 58*t*
      creating constraints when. *See* creating
         constraints; from existing tables (with
         subqueries), 67–69
      hands-on assignments, 93
      syntax, 88*t*. (*See also* designing, *below*)
   creating table-level constraints, 102–103, 117–118
   data block storage, 191
   deadlocks, 163
   default installation tables, 29*n*, 65*n*, 117
   deleting, 83–84
      commands, 12*t*, 59*t*
      parent tables, 109–110
      permanently, 85–86
      syntax, 89*t*
   designing, 59–63
   entering data in. *See* entering data
   full scans, 189(2)
   heap-organized table, **189**
   joining. *See* join operations
   in the JustLee Books database, 14–16, 511–517
   key-preserved tables, **490**
   listing, 28–29, 50*t*, 65–66
   listing data in, 31–32, 141, 144, 148. *See also* search-
      ing for column data values; selecting rows/
      records
   locking, 162–164
   lookup tables, **12**
   modifying, 70–82
      advanced challenge, 93
      commands, 12*t*, 58–59*t*
      hands-on assignments, 93
      syntax, 88*t*
   multiple tables: DML operations on complex views
      with, 489–491
   names, 59–60
   non-key-preserved tables, **490**
      DML operations on views with, 490–491
   normalizing, 6–10
   PLAN_TABLE table, 538*n*
   rebuilding, 30

recovering, 84–85
  commands, 59*t*
  syntax, 89*t*
relating, 10–12
renaming, 79–81, 88*t*
with the same name, 65*n*
selecting columns. *See* selecting columns
selecting rows. *See* selecting rows/records
structure. *See* table structure
temporary. *See* inline views
truncating, 81–82, 88*t*
updating one based upon another, 456–459
virtual multiple tables. *See* views. *See also* columns;
  rows
TC (transaction control) commands, 13*t*, 138, 138*t*,
  157–162
  hands-on assignments, 170–171. *See also* COMMIT
    command; LOCK TABLE command;
    ROLLBACK command; SAVEPOINT
    command; SELECT … FOR UPDATE
    command
temporary tables. *See* inline views
test data: eliminating, 82*tip*
testing constraints, 104
  CHECK, 113–114
  FOREIGN KEY, 107
  NOT NULL, 115–116
  PRIMARY KEY, 104–105
  UNIQUE, 111–112
testing indexes, 194
text fields: data value alignment in, 38
third-normal form (3NF), 9
time zone datatype, 356*n*
TIMESTAMP datatype, 356*n*
timing queries, 542–543
TKPROF executable file, 534
TO clause (GRANT command), 219, 221
TO_CHAR function, 365–367, 376*t*, 388, 402*n*
TO_DATE function, 352–354, 375*t*
  format arguments, 352–353*t*
TO_NUMBER function, 370–371, 376*t*
TOP-N analysis, 496–499
total aggregation argument (()), 407
totaling column data values, 386–388
TRACE feature, 534
traditional joins:
  ANSI-compliant joins vs., 284
  Cartesian joins, 286–288, 321*t*
  equality joins (equijoins), 291–296, 308–309, 321*t*
  non-equality joins, 303–304, 321*t*
  outer joins, 309–310, 322*t*
  self-joins, 306–307, 322*t*
transaction control commands. *See* TC (transaction control) commands
transactions, **157**, **158**
  DML commands as, 157
transitive dependency, 9
TRANSLATE function, 343, 373*t*

trimming character strings on the left/right, 342
TRUNC function (date), 355
TRUNC function (number), 345–346, 374*t*
TRUNCATE TABLE command, 59*t*, 81–82, 88*t*
truncating dates, 355
truncating numeric data, 345–346
truncating tables, 81–82, 88*t*
Tuning Advisor, 534, 537

U

uncorrelated subqueries, 428*t*, **451**, 461*t*, 548
underscore (_):
  as a string literal, 259*tip*
  as a wildcard character, 258–260, 276*t*
undoing DML commands, 158, 483*n*
  back to a point, 159–162
UNION ALL set operation (operator), **312**, 313*t*, 314
UNION set operation (operator), **312**, 313, 313*t*, 314*n*
  multiple-column operations, 316, 406, 407, 408
UNIQUE constraint, 100*t*, 110–112, 119
  abbreviation/code for, 101*t*, 121
  adding, 111, 127*t*
  creating, 127*t*
  syntax, 111, 127*t*
  testing, 111–112. *See also* constraints (on table data)
unique indexes, 194
UNIQUE keyword:
  with CREATE INDEX, 194
  with SELECT, 44, 50*t*
unnormalized data: storing difficulties, 7
UPDATE command, 138*t*, 150–152
  command type, 12*t*
  using substitution variables with, 152–155
UPDATE privilege, 220
update scripts, 155
updating records, 150–151
  multiple records, 151–152
  preventing data inconsistencies during, 163–164
  using substitution variables, 152–155
updating tables: one based upon another, 456–459
UPPER function, 334–335, 372*t*
uppercase letters: converting character strings to,
  334–335
UPSERT statements, 457*n*
user accounts: creating, 216–217
USER_ data dictionary view prefix, 207*t*
USER_CONS_COLUMNS view: querying, 121–122
USER_CONSTRAINTS view, 121
USER_IND_COLUMNS view: querying, 200
USER_INDEXES view: querying, 199–200
USER_OBJECTS view, 182*tip*
  querying, 181–182
USER_ROLE_PRIVS view, 231*t*
USER_SEQUENCES view: querying, 182, 186, 188
USER_SYS_PRIVS view, 231*t*

USER_TAB_COLUMNS view: querying, 67
USER_TAB_PRIVS view, 231*t*
USER_TABLES view, 28, 50*t*, 65
 querying, 28–29, 65–66
USER_UNUSED_COL_TABS view: querying, 77, 78
usernames: creating, 216–217
users (user accounts):
 creating, 214*t*, 216–217, 235*t*
 deleting (dropping), 214*t*, 234, 235*t*
 granting privileges to, 214*t*, 219–220, 220–223, 236*t*
 granting roles to, 215*t*, 227–228(2), 236*t*
 hands-on assignments, 241
 privileges. *See* privileges
 revoking privileges/roles from, 214*t*, 232–233, 236*t*
 verifying, 224*tip*
USING clause. *See* JOIN … USING method

V

V_$SQLAREA view, 534, 534*n*
V$ data dictionary view prefix, 207*t*
values. *See* column data values; sequence values
VALUES clause (INSERT command), 140, 141–142, 183
VARCHAR(*n*) datatype, 61*n*
VARCHAR2(*n*) datatype, 60*t*, 61*n*, 61–62
VARIANCE function, 385*t*, 404–405, 418*t*
verification. *See* listing; verifying; *and specific
  procedures needing verification*
verifying:
 indexes, 199–200
 join operations, 303*tip*
 user accounts, 224*tip*
vertical bars (||): concatenation operator, 45
viewing. *See* listing
views, 471–509, **472**
 advanced challenge, 508–509
 complex. *See* complex views
 creating, 474*t*, 474–475, 502*t*
  complex views, 484
  inline views, 446, 496–499
  materialized views, 499–501
  simple views, 476–480
 deleting (dropping), 474*t*, 495–496, 503*t*
 DML operations on. *See* DML operations on views
 hands-on assignments, 508
 inline. *See* inline views
 materialized. *See* materialized views
 processing of, 472
 referencing, 472, 477

renaming columns in, 475
replacing (re-creating), 474*t*, 475, 477*n*, 480, 503*t*
simple views: creating, 476–480
 DML operations on, 480–483
types, 473*t*
usefulness (purposes), 472, 473. *See also* data
 dictionary views
virtual columns, **62**
 adding, 144–145
 handling in INSERT INTO statements, 144, 145–146
 listing the definitions of, 67
virtual multiple tables. *See* views

W

warning about dynamic sampling of database objects,
 540
WHEN MATCHED THEN clause (MERGE command),
 457–458, 459
WHEN NOT MATCHED THEN clause (MERGE
 command), 457–458
WHERE clause (DELETE command), 156, 157
WHERE clause (SELECT statement), 244*t*, 245–248,
 251*tip*, 275*t*
 joins. *See* WHERE clause joins
 multiple-column subqueries in, 447–448
 restricting groups with, 398, 399, 400, 400–401
 single-row subqueries in, 429–434, 437*tip*
WHERE clause (UPDATE command), 150
WHERE clause joins, 320*t*
 ANSI-compliant joins vs., 284
 equality joins (equijoins), 291–296, 308–309, 321*t*
 multiple joins, 295–296
widening columns, 74–75
wildcard characters, 258–260, 276*t*
 as string literals, 259*tip*, 260–261
WITH ADMIN OPTION (GRANT command), 219, 233
WITH CHECK OPTION (CREATE VIEW command), 475
WITH GRANT OPTION (GRANT command), 221, 222*n*,
 233
WITH READ ONLY (CREATE VIEW command), 475,
 478, 480

Z

zeros, insignificant: as not displayed, 39